Applications of Neuro-Symbolic Artificial Intelligence

Boris Galitsky

Applications of Neuro-Symbolic Artificial Intelligence

Springer

Boris Galitsky
Knowledge-Trail
San Jose, CA, USA

ISBN 978-3-032-16523-7 ISBN 978-3-032-16524-4 (eBook)
https://doi.org/10.1007/978-3-032-16524-4

This Springer imprint is published by the registered company Springer Nature Switzerland AG
The registered company address is: Gewerbestrasse 11, 6330 Cham, Switzerland

Preface

Over the last few years, large language models (LLMs) have redefined the contours of artificial intelligence. Their uncanny ability to generate coherent text, summarize documents, converse fluently, and perform a wide range of cognitive tasks has launched them into every corner of digital life—from customer service and creative writing to scientific research and legal analysis. Yet, beneath this fluid linguistic surface, a deeper question remains unresolved: Can these models truly reason, adapt, and engage in humanlike understanding?

This book explores a new generation of research and experimentation that seeks to answer that question not just in the affirmative but in principled, adversarial, and integrated ways. Rather than relying solely on statistical generalization from massive datasets, we investigate how the symbolic, structured, and logical traditions of AI can be fused with the probabilistic, data-driven world of LLMs—and how that fusion can mitigate hallucination, improve reasoning, and unlock more humanlike interaction.

The ten chapters in this book present a diverse but tightly interrelated set of approaches. They offer a roadmap toward a neuro-symbolic future where LLMs can do more than mimic intelligence—they can engage, explain, invent, and co-create in profoundly new ways.

Bridging neural and symbolic AI through adversarial integration: Chapter "Adversarial Integration of LLM and Logic Program," lays the theoretical and algorithmic foundation for this synthesis. Traditional symbolic logic programming, rooted in rule-based inference and provability, is framed as a scaffolding to test and improve the generative responses of LLMs. Rather than treating logic as a post-processing filter or as a detached explanation layer, the authors advocate for an adversarial interaction: LLM outputs are continuously challenged, refined, or corrected by logic modules, turning symbolic reasoning into a dynamic opponent and collaborator. This integration is not simply about correctness—it is about fostering a process of argumentative refinement, much like a human expert's evolving thought under peer review.

Hallucination as a failure of abductive coherence: Building upon the above integration, chapter "Adversarial Abductive Dialogue Framework with Reinforcement for Tackling LLM Hallucination" introduces the adversarial abductive dialogue framework—a reinforcement-driven architecture for diagnosing and countering hallucination in LLMs. Hallucination, often cast as a stochastic defect, is reconceived here as a failure of abductive plausibility: the model lacks a coherent explanation for why a claim should be true given its premises.

By embedding the LLM in a dialogic setting where it must generate, defend, or revise hypotheses, and where an abductive logic engine rewards or penalizes responses based on internal coherence and informativeness, this framework transforms the hallucination problem into an epistemic game. The model must not just predict the next word—it must justify its belief.

Theory of Mind and humanlike intent modeling: The ability to attribute beliefs, intentions, and emotions to others—Theory of Mind (ToM)—is a cornerstone of human cognition. Yet it remains a weak spot for most LLMs, whose responses often falter in multi-agent scenarios or emotionally charged dialogue. Chapter "Improving ToM Capabilities of LLMs in Applied Domains" tackles this head-on, offering methods for enhancing the ToM capabilities of LLMs in applied domains like law, medicine, and education.

By integrating ToM modeling as an explicit latent variable within dialogue planning and using reinforcement learning guided by human preference data, the authors show that LLMs can be trained to simulate intentional states, anticipate interlocutor reactions, and adapt their communication accordingly. This is not mere empathy engineering—it is a step toward socially aware AI.

Rehumanizing AI: From functional output to intriguing engagement. While much of AI focuses on utility, chapter "Rehumanizer: Enhancing LLM Recommendations for Human-Like Engagement and Intriguing Interactions"—Rehumanizer—asks a deeper question: Can machines be interesting? Drawing on theories of curiosity, novelty, and user engagement, the Rehumanizer is an LLM wrapper that steers responses toward unexpected but relevant replies, simulating the kinds of serendipitous turns and emotionally intelligent pivots that characterize compelling human conversation.

This framework does not discard functional accuracy but rather augments it with heuristics for emotional valence, engagement trajectory, and discourse flow. The result is an AI that doesn't just answer but connects.

LLM as a copilot *for human agency and creativity*: The subsequent chapters explore applied frameworks where these foundational ideas are operationalized in practical systems. Chapter "A Copilot for Social Promotion" presents a copilot for social promotion, where LLMs assist users in crafting persuasive messages, social media content, and personal branding materials with tone-aware, goal-directed generation.

Chapter "Abduction-Based Personalized Recommender for Web 3.0" explores *abduction-based personalized recommendations for Web 3.0*, applying logic-based personalization to decentralized environments where user data is fragmented and implicit. By treating each user interaction as a partially observed explanation, the system generates not just probable content but intentionally grounded suggestions.

Designing narrative, invention, and simulation systems: Chapters "Creating Ad Sequences with Optimal Discourse Structure by LLM", "A Computational Framework for Analyzing and Supporting Invention Dialogues" and "A System for Distributed Chemical Engineering Simulations" expand the application sphere further. Chapter "Creating Ad Sequences with Optimal Discourse Structure by LLM" describes how to *generate advertising sequences with optimal discourse structures*, using LLMs trained on rhetorical schemas and persuasive storytelling arcs. Chapter "A Computational Framework for Analyzing and Supporting Invention Dialogues" proposes a computational framework for invention dialogues—conversational agents that help users brainstorm, iterate, and refine novel ideas across science, law, and art. This is a vision of the LLM as co-innovator. Chapter "A System for Distributed Chemical Engineering Simulations" turns to the scientific domain with a system for distributed chemical engineering simulations, where natural language interfaces mediate collaboration between human experts and numerical solvers. Here, LLMs do not replace scientific reasoning—they amplify and orchestrate it.

Reducing tail embedding risk through stepwise embedding approximation: Finally, chapter "Step Wise Approximation of CBOW Reduces Hallucinations in Tail Cases" addresses a key technical bottleneck in model generalization: embedding performance degradation in tail cases. The authors propose a stepwise approximation of CBOW embeddings that bridges the gap between static word vectors and fully contextualized embeddings—retaining interpretability while mitigating hallucination. Particularly in domains with rare entities or ambiguous terms, this hybrid representation improves grounding, coherence, and factuality.

A vision for the future: Neuro-symbolic, adversarial, human aligned. Taken together, the chapters in this book propose a shift in how we build and think about language models. Rather than treat them as monolithic oracles, we can design modular, dialogic, and adversarial systems that challenge the model to reason, explain, adapt, and co-create. The goal is not to eliminate error but to structure the space in which error becomes signal—a teaching moment for the system to grow more aligned, more accountable, and more human.

This is not a return to symbolic AI, nor a rejection of deep learning, but an adversarial fusion of both—one that takes seriously the human dimensions of language: its ambiguity, its creativity, and its purpose. As AI becomes ever more pervasive in our lives, these systems must not just understand our words—they must understand us.

This book is designed for researchers, practitioners, and advanced students in artificial intelligence, cognitive science, and human-computer interaction who are seeking to move beyond the limitations of current large language models. It speaks directly to those invested in the next frontier of AI—systems that not only generate fluent text but also reason logically, engage meaningfully, and align with human intentions. Readers familiar with the foundations of neural networks and symbolic AI will find a rich synthesis of both paradigms, grounded in concrete architectures and experimental frameworks. The technical depth of chapters on adversarial integration, abductive reasoning, and embedding approximation makes it particularly valuable for machine learning engineers and AI architects aiming to build more robust, interpretable, and trustworthy systems.

Equally, this book appeals to domain experts in law, medicine, education, creative industries, and scientific research who are exploring how LLMs can evolve from passive tools into active collaborators. By illustrating how neuro-symbolic integration enhances theory of mind, supports persuasive communication, and enables inventive co-creation, the text offers practical pathways for deploying AI in high-stakes, human-centered environments. Whether the reader is interested in reducing hallucinations in clinical decision support, designing emotionally intelligent dialogue agents, or building AI-augmented workflows for innovation, this volume provides both conceptual clarity and implementable insights—bridging the gap between cutting-edge research and real-world application.

San Jose, CA, USA Boris Galitsky

Acknowledgments

The author is grateful to Dmitri Ilvovsky, Tatyana Machalova, Saveli Goldberg, Sergey O. Kuznetsov, Dina Pisarevskaya, Daniel Usikov, Alexander Rybalov, and other collaborators for fruitful discussions on the topics of this book.

Competing Interests The author has no competing interests to declare that are relevant to the content of this manuscript.

Contents

Introduction

1 Reasoning and Learning

Deduction and induction are two fundamental modes of reasoning at the heart of both human cognition and artificial intelligence. Deductive reasoning involves drawing conclusions from known or formally specified knowledge. This places emphasis on the expressive power of formal languages used to represent such knowledge, as well as the design of robust proof systems that enable systematic inference. Inductive reasoning, by contrast, focuses on learning general patterns from observations—extracting knowledge from partial or incomplete data. Here, the central concerns are the semantics of generalization, the choice of a suitable representation language for the inferred knowledge, and the development of algorithms capable of computing meaningful and accurate generalizations from noisy or sparse inputs.

The integration of deduction and induction is of fundamental importance to both the theory and practice of intelligence, natural or artificial. A unified framework that supports both forms of reasoning is necessary not only for modeling human cognition but also for building more adaptive, general-purpose AI systems (Mooney 2000). From a conceptual perspective, this integration is relevant to fields like philosophy, cognitive science, and computational epistemology, where understanding how humans balance rules and experience is a long-standing challenge. From a practical standpoint, that is a focus of this book—the ability to harmonize deductive expertise with inductive learning is critical across scientific disciplines, where empirical data must be interpreted in light of structured theories or domain-specific knowledge. Whether in medical diagnostics, financial forecasting, or autonomous systems, systems that can combine logic with data are better equipped to navigate uncertainty, adapt to new information, and provide transparent justifications for their conclusions.

B. Galitsky, *Applications of Neuro-Symbolic Artificial Intelligence*,
https://doi.org/10.1007/978-3-032-16524-4_1

Historically, however, research in AI has often treated reasoning and learning as separate pursuits. Deductive logic and symbolic AI focused on rule-based manipulation of knowledge, while statistical and machine learning methods emphasized pattern recognition from large datasets. Yet this separation has proven increasingly limiting. Emerging fields such as statistical relational learning (Getoor and Taskar 2007; Raedt et al. 2016), high-level decision-making frameworks, and, most notably, neuro-symbolic systems have begun to bridge this divide. These approaches demonstrate that the traditional dichotomy between symbolic and statistical methods may be both artificial and unproductive. Logic-based systems offer the ability to model complex relationships, structure the world into interpretable entities and their interactions, and support compositional reasoning. Meanwhile, purely inductive models often assume that data points are independent and identically distributed—an assumption that breaks down when modeling relational or hierarchical structures common in real-world environments. Integrating the strengths of both paradigms leads to systems that are not only more accurate but also more interpretable, flexible, and aligned with humanlike cognition (Fig. 1).

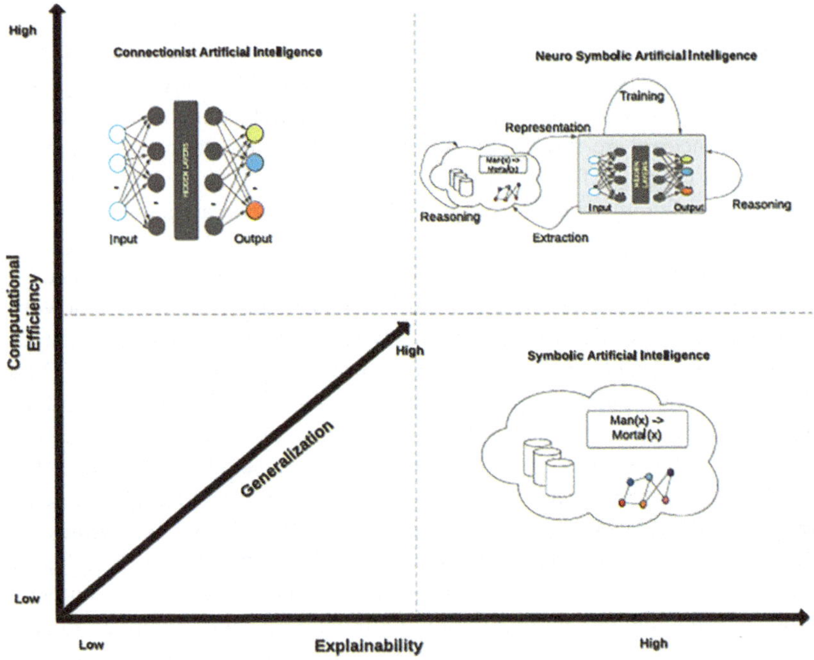

Fig. 1 Explainability, efficiency, and generalization of neuro-symbolic systems

2 Logical Interdependence of Chapters

We have the sequence chapters so that previous ones help understand the consecutive and also additional dependences shown in arrows in Fig. 2.

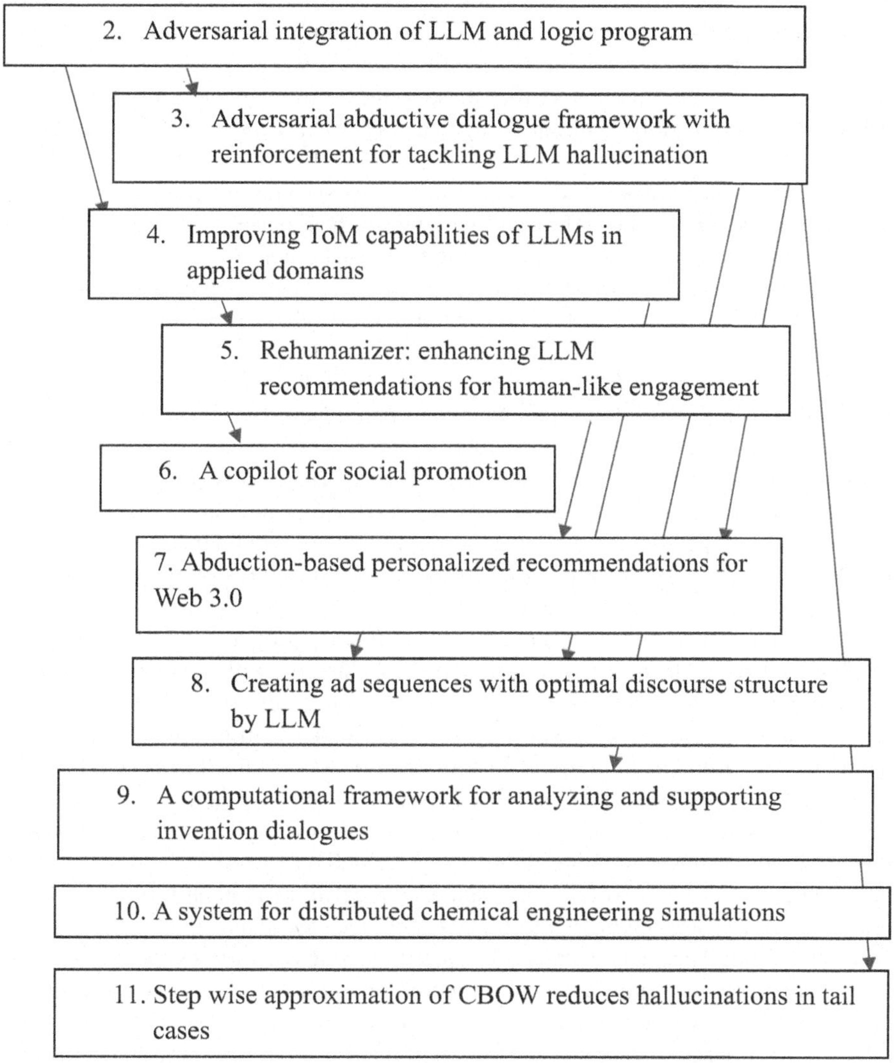

Fig. 2 Logical dependence of chapters in the book

3 Applications in Finance, Legal, Health, and Engineering

A successful integration of AI into financial systems requires more than technological advancement; it also demands systemic and institutional adaptation. In highly regulated environments, policymakers and regulators must be willing to evolve existing frameworks to enable the safe and effective deployment of AI tools. This may include updating compliance guidelines, risk assessment protocols, and audit standards to accommodate AI-driven decision-making. In under-resourced financial institutions, targeted investment in AI infrastructure, workforce training, and data governance is essential to ensure that these tools are implemented responsibly and deliver measurable value without compromising trust, transparency, or regulatory compliance.

Moreover, transparency and explainability of AI systems are crucial for building trust among healthcare providers and patients. In healthcare, where decisions can have life-or-death consequences, AI systems must be able to justify their recommendations clearly. This is particularly important in systems where resources are scarce, as AI decisions must be made with the highest level of accuracy and accountability.

The legal domain's foremost priority must remain the fair and equitable administration of justice. Any adoption of AI technologies should be driven by their ability to enhance legal reasoning, access to justice, and procedural integrity—not by treating the legal system as a convenient testbed for experimental AI tools.

The financial sector's core responsibility is to ensure market stability, fiduciary integrity, and the protection of client assets. AI innovation in finance must be guided by these principles, rather than using financial infrastructures or customer interactions as a proving ground for untested or speculative AI applications.

This book embarks on a comprehensive exploration of how LLMs and neuro-symbolic AI can be leveraged to address these challenges, with a particular focus on their applications in engineering, finance, and health. It introduces innovative techniques that combine the flexibility of neural networks with the interpretability of symbolic reasoning, leading to the development of AI systems that are both powerful and explainable. From assisting in legal research and case analysis to optimizing resource allocation within courts and law firms, the AI architectures discussed in this book are designed to enhance legal service delivery while upholding the trust and confidence of legal professionals and the integrity of judicial processes. These architectures are also applicable to finance, engineering, and health.

Deep learning, predictive analytics, natural language processing, anomaly detection, and personalized medicine have ushered in a new era of proactive healthcare, resulting in improved patient outcomes, fewer misdiagnoses, and more cost-effective treatments. Systematic reviews highlight the significant impact of machine learning across a wide range of medical fields. Looking ahead, the future of healthcare is promising, with advancements in data integration, interdisciplinary collaboration, explainable AI, real-time monitoring, global accessibility, ethical considerations, and continuous learning—all poised to reshape the healthcare landscape for the better. The ultimate goal is to develop AI systems that can not only assist medical professionals but do so in a manner that is trustworthy, transparent, and aligned with human values.

3.1 Limitations

Despite the promise demonstrated by neuro-symbolic approaches in various applications, including legal, several critical challenges persist (Hossain and Chen 2025) in advancing their use within the legal domain, as outlined below (Fig. 3):

1. **Knowledge representation**: Encoding legal knowledge—such as statutory hierarchies, case law precedents, and procedural rules—into neuro-symbolic models is challenging due to the nuanced, hierarchical, and context-sensitive nature of legal information required for accurate legal reasoning and decision support.
2. **Lack of standardized benchmarks and evaluation**: The absence of consistent benchmarks for evaluating neuro-symbolic systems in legal tasks (e.g., contract analysis, legal outcome prediction, or compliance checking) hinders the fair and

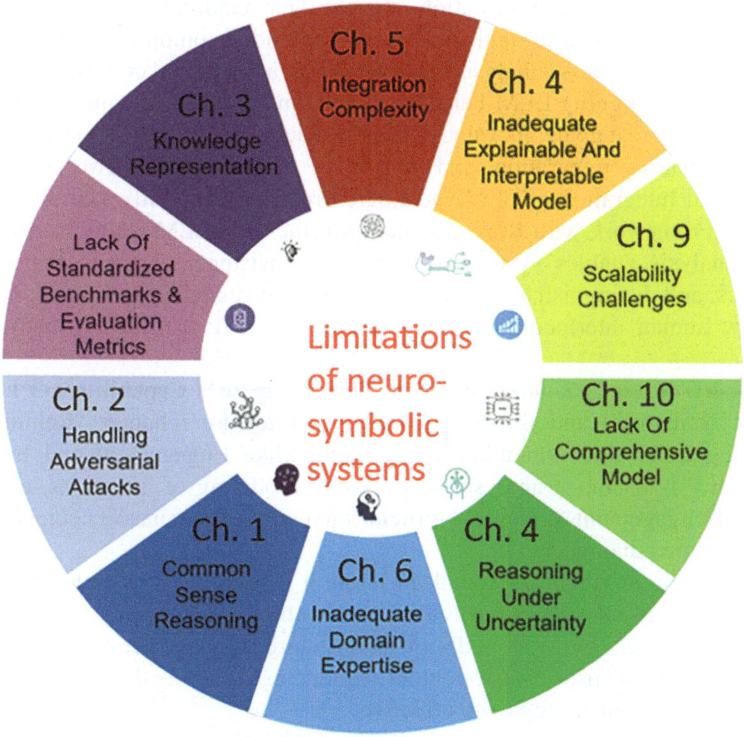

Fig. 3 Limitations and challenges of neuro-symbolic AI

rigorous assessment of model performance across jurisdictions, legal traditions, and domains of law.

3. **Handling adversarial attacks**: Legal neuro-symbolic models—particularly those deployed for tasks like evidence classification or regulatory compliance—must be robust against adversarial manipulation, such as subtle changes in language or citations that could alter interpretations or legal conclusions.
4. **Common sense and legal reasoning**: Integrating legal common sense reasoning—such as assumptions about intent, burden of proof, or the reasonable person standard—is essential for NeSy systems to function reliably in complex scenarios like tort litigation or criminal defense. Knowledge graphs (KGs) may offer a promising path toward building robust symbolic infrastructures to support such reasoning.

4 Discourse Analysis and Neuro-Symbolic AI

Discourse analysis plays a foundational role throughout this book, serving as both a methodological lens and a structural design principle for advancing the next generation of language model applications. While LLMs excel at surface-level fluency, much of their power—and limitation—lies in their handling of discourse: the patterns, intentions, structures, and coherence of extended communication.

In this book, discourse analysis is not treated as a peripheral concern but as a core mechanism for aligning LLM behavior with human reasoning, interaction norms, and communicative goals. Its value is reflected across multiple dimensions.

In adversarial integration and abductive reasoning frameworks (chapters "Adversarial Integration of LLM and Logic Program" and "Adversarial Abductive Dialogue Framework with Reinforcement for Tackling LLM Hallucination"), discourse analysis enables systems to detect inconsistencies, trace argumentation structures, and distinguish between plausible inference and hallucination. This mirrors how human interlocutors evaluate credibility and justification in multi-turn dialogue.

In theory of mind modeling (chapter "Improving ToM Capabilities of LLMs in Applied Domains") and the Rehumanizer architecture (chapter "Rehumanizer: Enhancing LLM Recommendations for Humanlike Engagement and Intriguing Interactions"), discourse analysis supports the modeling of intentions, emotional cues, and engagement strategies—crucial elements of persuasive, empathetic, or socially attuned dialogue.

In applications such as social promotion (chapter "A Copilot for Social Promotion") and ad generation (chapter "Creating Ad Sequences with Optimal Discourse Structure by LLM"), discourse structure informs narrative flow, rhetorical progression, and user-centric content sequencing, ensuring that LLM outputs are not just coherent but strategically effective.

In the invention dialogue framework (chapter "A Computational Framework for Analyzing and Supporting Invention Dialogues"), discourse analysis enables systems to scaffold creative ideation over time, track the evolution of hypotheses, and preserve coherence across exploratory conversational turns.

At a broader level, the book treats discourse as the interface between logic and language, cognition, and communication. Discourse analysis allows AI systems not only to understand what is being said but why, how, and to what end. It enables models to move beyond word prediction toward goal-aware, trust-building, and human-aligned language generation.

By weaving discourse-level reasoning into adversarial, abductive, and personalized AI architectures, this book demonstrates that discourse analysis is not a post-hoc interpretive tool—it is a generative force that can shape how language models think, reason, and co-create with humans.

Much of the work in applying computational linguistics to finance, legal, and health hinges on understanding the syntax and semantics of language. Syntax is crucial for tasks such as information extraction, drug matching, and processing electronic health records, where structural language patterns help identify and organize key medical data. Semantic analysis, on the other hand, plays a vital role in finance recommenders, medical decision support systems, and solving complex problems that require intelligence-driven insights (chapters "Adversarial Integration of LLM and Logic Program," "Adversarial Abductive Dialogue Framework with Reinforcement for Tackling LLM Hallucination" and "Step-wise Approximation of CBOW Reduces Hallucinations in Tail Cases"). While semantic analysis creates and manages a logical representation of text, discourse analysis elevates this to a higher level of abstraction (chapters "Improving ToM Capabilities of LLMs in Applied Domains," "Rehumanizer: Enhancing LLM Recommendations for Humanlike Engagement and Intriguing Interactions," "Creating Ad Sequences with Optimal Discourse Structure by LLM," and "A Computational Framework for Analyzing and Supporting Invention Dialogues").

Discourse analysis is particularly significant in healthcare because it mirrors how physicians organize their thoughts when diagnosing conditions and recommending treatments (chapter "Creating Ad Sequences with Optimal Discourse Structure by LLM"). These high-level representations of medical reasoning are indispensable for developing automated diagnosis systems and supporting patient care through AI-driven decision-making. Health discourse forms the foundation of this book, underpinning essential techniques like supported decision trees, multi-case matching, dialogue management, multi-modal representation, and content generation (Galitsky and Goldberg 2022).

By focusing on discourse analysis, we can better capture the nuances of legal and medical reasoning and integrate them into AI systems, enhancing their ability to handle complex tasks (chapters "A Copilot for Social Promotion" and "Abduction-Based Personalized Recommender for Web 3.0"). This approach allows us to build systems that not only process financial and legal information but also understand the broader context of needs of different users.

5 Commonalities in Neuro-Symbolic AI Across Domains

Even though finance, law, engineering, and healthcare are very different fields, they all share some common challenges where neuro-symbolic AI—which combines learning from data (neural networks) with structured logic (symbolic reasoning)—can help in powerful ways. We spell out what they have in common and how neuro-symbolic AI addresses those needs:

1. Decisions must be clear and trustworthy (chapter "Adversarial Abductive Dialogue Framework with Reinforcement for Tackling LLM Hallucination"). In all these fields, decisions can have serious consequences. People need to understand why a decision was made—whether it's a loan approval, a legal argument, a diagnosis, or a design choice. Neuro-symbolic AI helps by explaining its reasoning clearly using rules and logic, not just hidden calculations (Table 1).
2. Working with uncertainty. Real-world situations are often messy or incomplete (chapter "Creating Ad Sequences with Optimal Discourse Structure by LLM"). For example, a financial risk might be unknown, or a patient's symptoms might not match a common diagnosis. Neuro-symbolic AI can combine patterns from data with logical reasoning to make the best possible decision even when some information is missing.
3. Using expert knowledge. Each domain has a lot of important rules and guidelines—laws, safety codes, medical protocols, and financial regulations. Symbolic reasoning lets AI use this knowledge directly, rather than trying to "learn" it from scratch like a neural network would (chapter "Abduction-Based Personalized Recommender for Web 3.0").
4. Making personalized decisions. No two clients, patients, or cases are the same. AI systems must adapt their advice to the person or situation. Logic-based reasoning allows AI to adjust recommendations based on specific user goals, histories, or constraints (chapter "Rehumanizer: Enhancing LLM Recommendations for Humanlike Engagement and Intriguing Interactions").
5. Following the rules. These fields are heavily regulated. One cannot just make decisions that "usually work"—they must follow strict standards. Symbolic

Table 1 How neuro-symbolic approach addresses the need of complex decision making domains

Common need	Neuro-symbolic contribution
Explainability and trust	Rule-based justification, traceable logic
Uncertainty handling	Hybrid probabilistic-symbolic reasoning
Domain knowledge usage	Logic rules, ontologies, and formal models
Personalization	Constraint-based adaptation and abductive reasoning
Regulatory compliance	Rule enforcement and verification
Human-AI collaboration	Dialogic reasoning, argumentation frameworks
Long-tail generalization	Structured extrapolation using symbolic rules
Structured text and dialogue	Discourse modeling and procedural planning

components help AI stay within legal and ethical boundaries, by checking decisions against formal rules (chapter "Improving ToM Capabilities of LLMs in Applied Domains").

6. Working with people, not replacing them. Professionals like doctors, lawyers, engineers, and analysts need tools that support their work, not replace their expertise. Neuro-symbolic AI is built to collaborate with humans, providing suggestions, explanations, and options in a way experts can trust and understand (chapter "A Copilot for Social Promotion").
7. Handling rare or unusual cases. Sometimes, systems face rare diseases, legal edge cases, or unexpected financial events—things that are not in the training data. Symbolic reasoning and an approximation of embeddings help AI think through these rare cases using logic, even if it has never seen them before (chapter "Step-Wise Approximation of CBOW Reduces Hallucinations in Tail Cases").
8. Understanding complex communication. These domains often use structured documents and multi-step reasoning—like legal contracts, medical records, engineering plans, or financial reports. Neuro-symbolic AI can follow and generate structured communication, helping it make sense of complicated situations (chapter "Adversarial Integration of LLM and Logic Program").

In our previous book, *AI for Health Care Applications and Management* (Galitsky and Goldberg 2022), we delved into how AI algorithms can significantly enhance the efficiency of patient care through scheduling, prediction, automated support, personalization, and diagnostics. We introduced the concept of medical discourse analysis to provide a high-level representation of medical texts, enabling AI systems to better understand and process complex medical language. By focusing on these areas, we demonstrated how AI could streamline various healthcare processes and improve patient outcomes. In the consecutive book, *Health Applications of Neuro-Symbolic AI* (Galitsky 2025), we explored how neuro-symbolic AI helps in making diagnoses, answering complex health questions and managing and scheduling healthcare services in an efficient manner, relying on large language models.

In this current book, we build upon the foundations laid in the first volume, with a focus on the transformative power of LLMs. While LLMs have revolutionized natural language processing, they come with inherent limitations, especially when dealing with intricate medical reasoning and decision-making tasks. Recognizing these challenges, we explore how different forms of reasoning, as well as complementary non-reasoning techniques, can be integrated with LLMs to further enhance the effectiveness of AI in medical systems. By bridging the gap between LLMs and more sophisticated reasoning approaches, we aim to push the boundaries of what AI can achieve in engineering, finance, and health.

6 Neuro-Symbolic AI and Humanities

The twenty-first century has witnessed remarkable progress in the fields of computer science and engineering, particularly within Western universities across Europe and North America. These institutions continue to lead innovation in areas such as artificial intelligence, robotics, and neuro-symbolic reasoning—pioneering the integration of symbolic logic and neural architectures that define the cutting edge of intelligent systems. This book, *Applications of Neuro-Symbolic AI*, is a testament to that advancement, offering a multidisciplinary perspective on how logic-based formalisms can enrich data-driven models.

However, this pace of innovation and adaptation is not uniformly present across all academic disciplines. In contrast to their STEM counterparts, many humanities departments in the same institutions have struggled to evolve in parallel with the modern intellectual climate. While computer science departments are embracing cross-cultural learning, empirical rigor, and forward-looking technologies, segments of the humanities remain anchored in ideologies and frameworks that have lost traction in many parts of the world.

In particular, certain Western humanities faculties continue to emphasize twentieth-century paradigms—such as classical Marxism or postcolonial activism—that are increasingly disconnected from both global academic discourse and current societal needs. Meanwhile, many Asian universities and religious institutions have moved toward frameworks that prioritize ethical scholarship, community-centered development, and peaceful civilizational progress. The global academic landscape is shifting, and there is a growing call for intellectual renewal—one that realigns humanities education with values that foster innovation, critical dialogue, and social cohesion.

As this book explores the intersection of logic and learning in artificial intelligence, it also gestures toward a broader aspiration: that all academic fields, including the humanities, reflect the intellectual courage and openness found in the most dynamic scientific disciplines. Just as neuro-symbolic AI harmonizes disparate traditions—statistical learning and symbolic logic—there is value in seeking synthesis across cultural and disciplinary boundaries, grounded in respect, reason, and ethical responsibility.

Despite their historic reputations for academic excellence, institutions such as **UC Berkeley**, **Columbia University, Harvard, and Yale** have become known in recent decades for nurturing ideological positions within their humanities departments that appear increasingly detached from global intellectual and ethical progress. The persistent dominance of **classical Marxist frameworks**, which once served as useful critiques of industrial capitalism, now often functions less as a tool for critical analysis and more as a rigid orthodoxy—resistant to empirical challenge, cultural plurality, or moral introspection. These frameworks frequently reduce complex global phenomena to simplistic dichotomies of oppression and resistance while failing to account for the failures of Marxist-inspired regimes, the adaptability of market systems, or the ethical insights found in alternative cultural traditions.

In recent years, one of the most troubling developments within certain Western universities has been the rise of rhetoric aligned with the so-called "**global intifada**" movement—a loosely defined but increasingly vocal campaign that frames acts of resistance, including political violence, as justified responses to perceived systems of global oppression. On some campuses, particularly at institutions like Columbia, Berkeley, and NYU, expressions of solidarity with this movement have gone beyond legitimate critique of foreign policy or support for human rights; they have included **public celebrations of violence, calls for dismantling Western institutions**, and an increasingly **absolutist moral dichotomy** that casts entire civilizations as oppressors.

Such postures, while cloaked in the language of liberation and anti-imperialism, **undermine the fundamental academic values of nonviolence, dialogue, and critical nuance.** Worse still, they often marginalize dissenting students and faculty—including those from affected regions—who seek peaceful solutions and pluralistic coexistence. When academic spaces become platforms for **romanticized militancy** rather than critical inquiry, they not only fail their educational mission but also erode public trust in the intellectual integrity of the humanities.

References

Galitsky B (2025) Health applications of neuro-symbolic artificial intelligence. Academic Press. Paperback ISBN: 9780443300462

Galitsky B, Goldberg S (2022) Artificial intelligence for healthcare applications and management. Academic Press. https://doi.org/10.1016/C2020-0-02139-0

Getoor L, Taskar B (eds) (2007) An introduction to statistical relational learning. MIT Press

Hossain D, Chen JY (2025) A study on neuro-symbolic artificial intelligence: healthcare perspectives. arXiv preprint arXiv:2503.18213

Mooney RJ (2000) Integrating abduction and induction in machine learning. In: Flach P, Kakas A (eds) Abduction and induction. Kluwer Academic Publishers, pp 181–191

Raedt LD, Kersting K, Natarajan S et al (2016) Statistical relational artificial intelligence: logic, probability, and computation, Synthesis lectures on artificial intelligence and machine learning, vol 10(2). Springer, pp 1–189

Adversarial Integration of LLM and Logic Program

1 Introduction

Humans possess a unique and profound ability to reason. By mentally progressing through a series of logical steps, we can draw inferences that would otherwise be out of reach, even without receiving new information from the outside world. In a similar way, large language models (LLMs) often perform better when they generate intermediate steps—a "chain of thought"—before arriving at an answer, leading to more accurate responses than if they answered directly.

Despite their impressive capabilities and significant social impact, LLMs are not inherently reliable. Like many deep neural networks, LLMs are often described as "black boxes" due to the extreme complexity of their internal structure, involving millions or even billions of parameters. This makes their reasoning and decision-making processes obscure and challenging for humans to interpret. While LLMs can be prompted to generate self-explanations to clarify their reasoning, research indicates that these explanations are frequently inconsistent and unreliable. Adding to this issue is the phenomenon of "model hallucinations," where an LLM generates incorrect or entirely fabricated responses that lack any basis in the provided data or knowledge. Such inaccuracies further erode trust in these models.

LLMs continue to encounter difficulties with multi-step deductive reasoning (Creswell et al. 2022), where they must apply a series of logical rules to a set of facts in order to answer a query. Specifically, this type of reasoning requires the model to apply a relevant rule to the supporting facts at each step, deriving new conclusions along the way. LLMs struggle particularly when the structure of the input diverges from the sequential order of rule applications (Berglund et al. 2023).

While LLMs perform well with single-step rule application, their effectiveness declines sharply in multi-step tasks due to challenges in rule grounding. In such cases, each step requires identifying and applying the correct rule and

B. Galitsky, *Applications of Neuro-Symbolic Artificial Intelligence*,
https://doi.org/10.1007/978-3-032-16524-4_2

corresponding facts from an extensive mix of input rules, given facts, and previously inferred conclusions (Wang et al. 2024a).

To improve on these limitations, the neuro-symbolic AI field advocates for additional mechanisms that enable more trustworthy, interpretable explanations. The goal is to equip users with tools to understand how LLMs reach their conclusions and verify the accuracy and reliability of their outputs. A promising approach is to integrate logic programming, which leverages formal rules and a domain-specific knowledge base. By incorporating symbolic reasoning systems, these methods can identify imprecise or incorrect statements produced by LLMs, offering a structured, rule-based approach to enhance both accuracy and transparency. This hybrid model aims to merge the LLMs' powerful generalization with the precision and reliability of rule-based systems, creating a more interpretable and trustworthy AI solution.

For certain classes of applications that require a high degree of accuracy and reliability (e.g., enterprise applications in the medical, legal or finance domain), LLMs are often combined with external tools and solvers in a hybrid architecture (Gur et al. 2024). We believe this is the right approach, especially to tackle problems where precise logical reasoning, planning, or constraint optimization is required, as LLMs are known to struggle for this class of problems.

Logical reasoning—inferring a conclusion's truth value from a set of premises—is a critical AI task with significant potential to impact fields like science, mathematics, and society. Symbolic AI, a subfield that employs rule-based, deterministic techniques, is well suited for generating trustworthy, human-interpretable explanations (Sarker et al. 2021). While symbolic approaches lack the robustness, creativity, and generalization abilities of LLMs, they can complement neural systems to create a hybrid neuro-symbolic AI approach, leveraging the strengths of both methods. The symbolic components contribute faithfulness and reliability that an LLM alone cannot provide (Wan et al. 2024).

In this way, ProSLM (Vakharia et al. 2024) combines the creativity and generalization abilities of a pretrained LLM with the robustness and interpretability of a symbolic reasoning engine. This allows users to gain insight into the reasoning behind the system's output, helping them identify potential hallucinations or inaccuracies in the model's response. Additionally, this framework is computationally efficient since it does not require additional training or fine-tuning of the LLM for specific domains.

This second chapter of the book focuses on logical reasoning problems in natural language (NL), a growing area of interest for neuro-symbolic architectures (Olausson et al. 2023). These architectures harness LLMs for generating declarative code and providing commonsense knowledge while employing symbolic reasoning systems to perform precise logical reasoning. This approach mitigates the limitations each technology faces on its own: LLMs' challenges in consistent, domain-accurate reasoning and symbolic reasoners' limitations with unstructured data and explicitly encoded commonsense knowledge (Galitsky 2025). These symbolic system limitations reflect the well-known "knowledge acquisition" challenge, which also contributes to their brittleness (Pan et al. 2023).

While many prompting-based strategies have been proposed to enable LLMs to do such reasoning more effectively, they still appear unsatisfactory, often failing in subtle and unpredictable ways. In this work, we reformulate such tasks as modular neuro-symbolic logic programming.

1.1 Types of NL + LLM Reasoning

Existing studies on natural language reasoning (Prystawski et al. 2023) focus on using different prompting strategies on LLMs for better results. Despite significant progress on the LLM prediction accuracy of the final answer, no work has been done to evaluate the correctness of the intermediate steps. In our work, we propose to evaluate intermediate steps at different levels of granularity in order to better probe into the reasoning capabilities of LLMs.

To boost LLM logical reasoning capabilities, Ranaldi and Freitas (2024) fine-tune language models with logical reasoning data to improve logical reasoning capabilities of LLMs. LLMs have also been directly used as soft logic reasoners, and a variety of prompting techniques are proposed in order to improve their performance under this paradigm (Wei et al. 2022; Zhou et al. 2024). Using large language models as semantic parsers has also shown improvement on the reasoning performance (Olausson et al. 2023; Pan et al. 2023) where natural language reasoning problems are first parsed into logical forms before being fed into an inference engine to output the final answer.

To facilitate NL proof generation, ProofWriter (Tafjord et al. 2021) and FLD (Morishita et al. 2023) are logical reasoning datasets equipped with NL proofs; however, both of them are synthetically generated dataset that neither contains abundant natural language variation nor encompasses challenging logical reasoning patters. Previous studies on proof generation focus on ProofWriter (Morishita et al. 2023; Saha et al. 2020, 2021) and ProntoQA (Saparov et al. 2023). LogicBench is a synthetically generated natural language QA dataset and is used for evaluating the logical reasoning ability of LLMs (Parmar et al. 2024). While FOLIO covers first-order logic and one or more inference rules are used in each example, LogicBench focuses on reasoning patterns covering propositional logic, first-order logic, and non-monotonic logic and focuses on the use of a single inference rule for each example.

Proofs for FOLIO (Han et al. 2022) are constructed in the form of a realistic expert-written logical reasoning dataset. Such proofs need to be written from scratch and are hard and time-consuming to write because humans need to manage both the language and reasoning complexity in the proof-writing process and manually construct many steps of reasoning. The resulting proofs contain more diverse types of inference rules and reasoning patterns in addition to containing more natural language variation and ensured semantic richness.

In pursuit of NL proof-generation objective, Vakharia et al. (2024) present a neuro-symbolic framework for knowledge base question-answering systems. Their approach, ProSLM, integrates a symbolic component designed for explainable context gathering prior to querying an LLM, using a knowledge base structured in Prolog. This symbolic layer connects a formal logic-based inference engine to a domain-specific knowledge base and serves two main functions:

1. Generating an interpretable, retrievable chain of reasoning to provide context for the input query
2. Validating the accuracy of a given statement (Fig. 1)

We follow this architecture in that LLM does everything but runs the logic program. Moreover, we wrap LLM and LP as agents and position these agents in adversarial setting. LP agent challenges LLM agent in what it believes is a correct inference result. LLM agent builds an LP attempting to prove that LLM's result is contradictory to available knowledge.

1.2 Introductory Example with Adversarial Settings

The author wishes to share his personal experience consulting ChatGPT for guidance on arranging his medical procedures. He initially planned to undergo both a colonoscopy and hernia surgery during a single hospital visit. When he asked ChatGPT if his hernia surgery could immediately follow a colonoscopy, he received the following response:

ChatGPT: "It's sometimes possible to schedule a laparoscopic abdominal hernia surgery to follow a colonoscopy in one visit."

With this confirmation, he began preparing for the colonoscopy, a process requiring at least 2–3 days. However, just a day before the scheduled procedures, his surgeon recommended canceling the colonoscopy, explaining that "Intestines with residual air and bloating after a colonoscopy are generally inappropriate for hernia surgery."

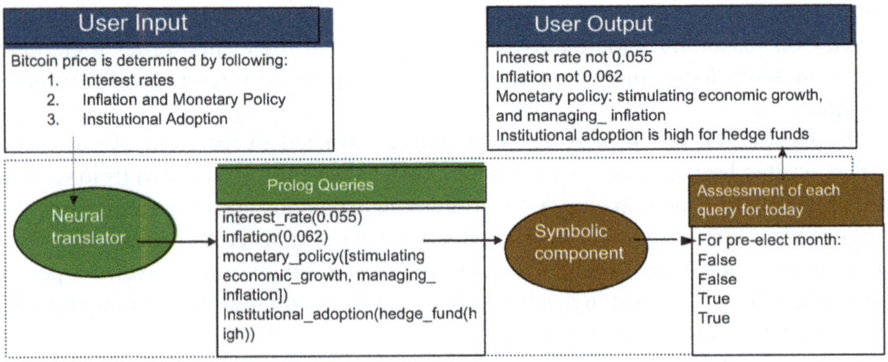

Fig. 1 Integrating LLM with Prolog

In this case, ChatGPT's initial guidance led to a suboptimal arrangement of medical procedures. A step-by-step questioning approach would likely have yielded the necessary information:

Me: "What is the condition of the intestines after a colonoscopy?"
ChatGPT: "Residual air and bloating.
Me: "Are intestines with residual air and bloating appropriate for hernia surgery?"
ChatGPT: "Generally not ideal for immediate hernia surgery."

This experience highlights the need for enhanced questioning and reasoning guidance when using AI for complex decision-making (Fig. 2). This book explores a range of approaches to make LLMs more reliable and effective, with this ChatGPT interaction as a practical example of how improved guidance can lead to better outcomes.

1.3 The Proposal

We propose a novel framework that integrates LLMs with LP to tackle complex reasoning tasks. This approach addresses the limitations of LLMs in maintaining consistency across multi-step inferences and complex problem-solving scenarios. By combining the statistical and contextual strengths of LLMs with the rigor and structure of LP, our system establishes a symbiotic adversarial interaction. Here, the LLM initially operates independently, generating reasoning steps and constructing an LP representation of its inference. The LP module evaluates these reasoning steps through formal logic-based methods, providing structured feedback. The LLM then interprets the LP results, identifies inconsistencies, and refines its conclusions through an adversarial process. We refer to the system being proposed as **Adversarial_LLM_LP**.

This adversarial dynamic involves the LLM challenging its own initial reasoning using LP feedback, effectively creating a self-correcting loop. The interplay between the probabilistic reasoning of LLMs and the deterministic formalism of LP ensures that each system attempts to refine and potentially outperform the other's conclusions. This process not only improves the reliability of the LLM's predictions but also enhances its ability to handle controversial or ambiguous tasks, making it a robust solution for complex reasoning.

We will conduct extensive evaluations of this adversarial LP-based neuro-symbolic framework on multiple reasoning datasets. These datasets were designed to test capabilities across a spectrum of logical and commonsense reasoning tasks. While we expect this framework to demonstrate performance comparable to state-of-the-art (SOTA) neuro-symbolic systems across complete datasets, we expect it to show a noticeable improvement on subsets involving controversial or complex queries. This underscores its ability to resolve intricate reasoning challenges that often stump purely neural or symbolic systems.

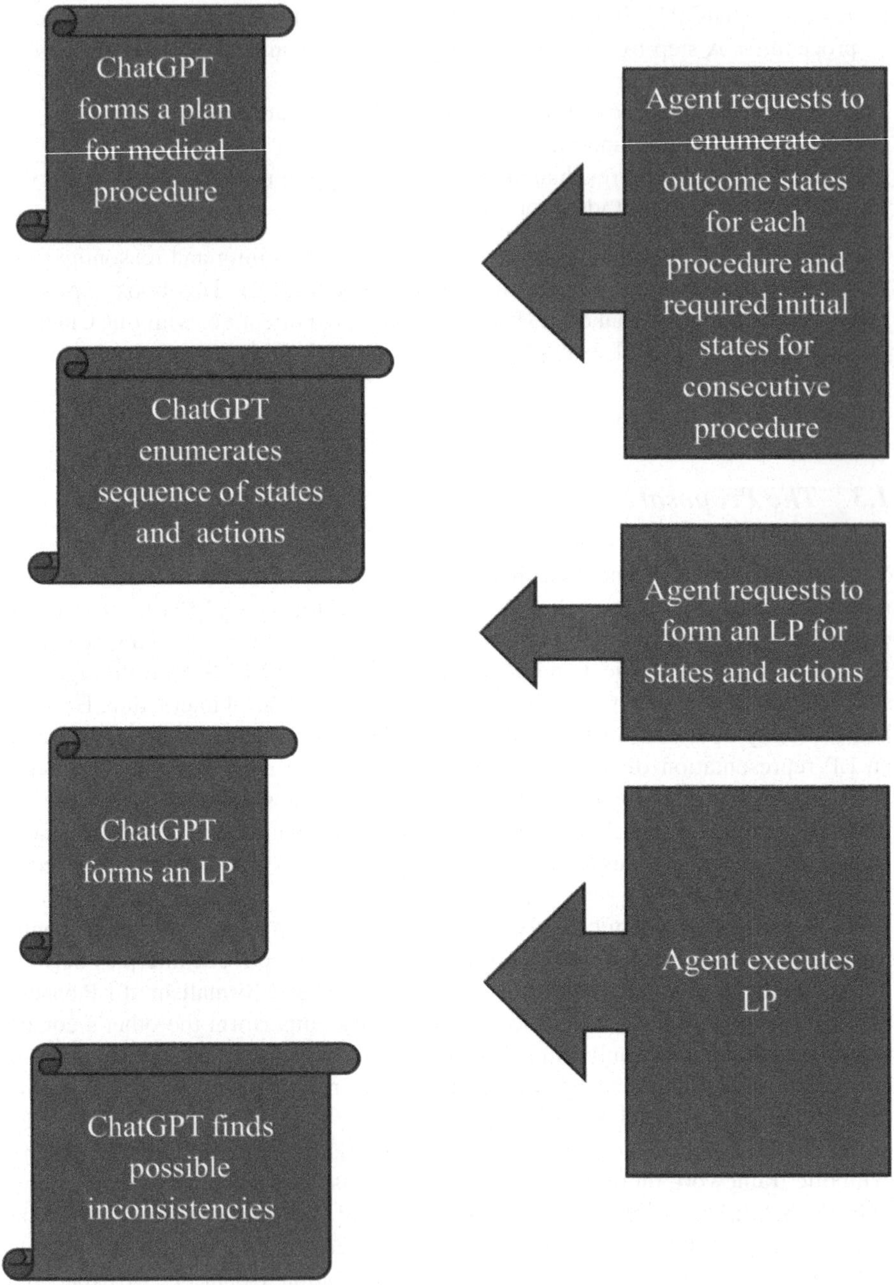

Fig. 2 Reasoning guidance for an LLM

By leveraging the complementary strengths of LLMs and LP, our approach is intended to push the boundaries of neuro-symbolic reasoning, providing a scalable and consistent method for solving complex tasks in fields ranging from natural language processing to decision-making systems.

1.4 Allegory: A Crocodile and a Behemoth

Let us start with a joke:

> A man created a circus act: a crocodile plays the piano, while a hippopotamus sings. The performance was a phenomenal success.
> Over drinks, the circus director probes the man:
> Surely, this is a trick. You're deceiving the audience.
> Well, yes, there is a trick. The hippopotamus just opens its mouth, while the crocodile does the singing.

The humor here lies in the unexpected twist—what appears to be the hippopotamus singing is actually a hidden effort by the crocodile. The act is success depends on the audience's belief in the illusion, not realizing the real "work" is being done elsewhere.

In this context:

1. The crocodile represents the LLM: It does all the "heavy lifting," such as processing complex reasoning, generating outputs, and adapting dynamically to tasks. Just as the crocodile sings while the hippo takes credit, the LLM performs the core computations and logic behind the scenes.
2. The hippopotamus represents LP: It is narrowly focused, following strict rules or processes—much like how the hippo only opens its mouth to give the illusion of singing. It adds structure and visible results but lacks the broader, adaptable capabilities of the crocodile.

This analogy captures how LP might contribute visible formalism or structure to reasoning tasks while the LLM handles the bulk of the actual work, ensuring the performance is seamless and believable.

2 Commonsense Inference Rules

We first define a set of inference rules that can be used for derivations of proof steps.

1. Widely used inference rules. The most widely used logical reasoning inference rules include universal instantiation, hypothetical syllogism, modus ponens,

modus tollens, disjunctive syllogism, conjunction introduction, conjunction elimination, transposition, disjunction introduction, material implication, and existential introduction.

2. Boolean identities. During the pilot annotation process, we found that Boolean identities are needed for certain derivations. For example, if "A or A" is true, then "A" is true. We therefore allow the usages of Boolean identities.
3. Complex inference rules. Some complex inference rules are intuitively correct and can also be proved logically correct with an inference engine. For example, from "A XOR B," we know that "A implies not B." We include in Table 1 the different categories of inference rules used in our protocol.

Here are the formal definitions of the widely used inference rules:

1. Universal Instantiation (UI): If a property or predicate applies to all elements of a domain, it also applies to any specific element within that domain.
   $\forall x P(x) \implies P(c)$ where $\forall x P(x)$ means "For all x, $P(x)$ is true" and c is a specific instance from the domain.

Table 1 Proof written for the example sequence. "P#" is premise number, and "D" denotes derivation

P no. used	Derivation	D no.	Inference rule
1	If there is a high-risk investment in the market, then it is not a safe market	D1	Modus tollens
2	Something that is not in a safe market is not a blue-chip stock	D2	Modus tollens
3, D1	A volatile cryptocurrency is not in a safe market	D3	Hypothetical syllogism
D3, 2	A volatile cryptocurrency is not a blue-chip stock	D4	Hypothetical syllogism
4, D4	Every speculative asset during a market bubble is not a blue-chip stock	D5	Hypothetical syllogism
D5, 5	The Bitcoin Asset is not a blue-chip stock or is a speculative asset during a market bubble	D6	Universal instantiation
D6, D4	The Bitcoin Asset is not a blue-chip stock or is not a blue-chip stock	D7	Universal instantiation
D7	The Bitcoin Asset is not a blue-chip stock	D8	Idempotence
5, 4	The Bitcoin Asset is a volatile cryptocurrency or a volatile cryptocurrency	D9	Universal instantiation
D10	The Bitcoin Asset is a volatile cryptocurrency	D11	Idempotence
D11, 3	The Bitcoin Asset is a high-risk investment	D12	Universal instantiation
D8, D11	The Bitcoin Asset is a volatile cryptocurrency and not a blue-chip stock	D13	Conjunction introduction
D13, D12	The Bitcoin Asset is a volatile cryptocurrency and a high-risk investment and not a blue-chip stock	D14	Conjunction introduction

2. Hypothetical Syllogism: If one statement implies a second, and the second implies a third, then the first statement implies the third.
   $(P \to Q) \land (Q \to R) \implies (P \to R)$ where P, Q, and R are propositions and "$\to$" denotes "implies."
3. Modus Ponens (MP): If a conditional statement is true, and its antecedent (first part) is true, then the consequent (second part) must also be true.
   $(P \to Q) \land P \implies Q$ where $P \to Q$ is a conditional and P and Q are propositions.
4. Modus Tollens (MT): If a conditional statement is true and its consequent (second part) is false, then the antecedent (first part) must also be false.
   $(P \to Q) \land \neg Q \implies \neg P$ where "$\neg$" denotes negation and P and Q are propositions.
5. Disjunctive Syllogism (DS). If one part of a disjunction (an "or" statement) is false, then the other part must be true.
   $(P \lor Q) \land \neg P \implies Q$ where "$\lor$" denotes "or" and P and Q are propositions.
6. Conjunction Introduction (CI). If two statements are true, their conjunction (combined statement using "and") is also true.
   $P \land Q \implies P \land Q$ where $P \land Q$ denotes "P and Q" and P and Q are propositions.
7. Conjunction Elimination (CE). If a conjunction (combined statement using "and") is true, then both individual parts of the conjunction are also true.
   $P \land Q \implies P$ and $P \land Q \implies Q$ where $P \land Q$ is the conjunction and P and Q are propositions.
8. Transposition. If a conditional statement is true, then its contrapositive is also true. That is, if $P \to Q$, then $\neg Q \to \neg P$ is also valid.
   $(P \to Q) \equiv (\neg Q \to \neg P)$ where "$\equiv$" denotes logical equivalence and "$\neg$" denotes negation.
9. Disjunction Introduction (DI). If one statement is true, then the disjunction (an "or" statement) with any other statement is also true.
   $P \implies (P \lor Q)$ where P and Q are propositions.
10. Material Implication. A conditional statement $P \to Q$ is logically equivalent to $\neg P \lor Q$ (either P is false or Q is true). $(P \to Q) \equiv (\neg P \lor Q)$.
11. Existential Introduction (EI). If a property holds for a particular individual, then there exists at least one element in the domain for which the property holds. $P(c) \implies \exists x P(x)$ where c is a specific instance, "there exists an x such that $P(x)$ is true."

These reasoning rules are fundamental to formal logic and can be applied in various domains, including mathematics, philosophy, computer science, and artificial intelligence.

We proceed to the complex rule:
From $a \oplus b$, we know $a \to \neg b$; from $a \to b$ and $b \to a$, we know $\neg(a \oplus b)$.
$\neg b \oplus b$ is always true; $a \oplus b$ is equivalent to $\neg a \oplus \neg b$.

Let a represent "The Bitcoin Asset is a volatile cryptocurrency" and b "The Bitcoin Asset is a blue-chip stock." $a \oplus b$ means "The Bitcoin Asset is either a volatile cryptocurrency or a blue-chip stock, but not both" (Fig. 3).

Formula derivation (Table 1) is as follows:

1. From $a \oplus b$, we know $a \rightarrow \neg b$.
   This means: If the Bitcoin Asset is a volatile cryptocurrency (a), then it is not a blue-chip stock ($\neg b$).
2. From $a \rightarrow b$ and $b \rightarrow a$, we know $\neg(a \oplus b)$.
   This means: If the Bitcoin Asset being a volatile cryptocurrency implies it is a blue-chip stock ($a \rightarrow b$), and being a blue-chip stock implies it is a volatile cryptocurrency ($b \rightarrow a$), then it cannot be the case that the Bitcoin Asset is either a volatile cryptocurrency or a blue-chip stock but not both ($\neg(a \oplus b)$). In this case, the two characteristics are mutually inclusive, not exclusive.
3. $\neg b \oplus b$ is always true.
   This means: Either the Bitcoin Asset is not a blue-chip stock or it is a blue-chip stock, but not both ($\neg b \oplus b$). This is always true as it covers all possible cases.
4. $a \oplus b$ is equivalent to $\neg a \oplus \neg b$
   This means: The Bitcoin Asset being either a volatile cryptocurrency or a blue-chip stock but not both ($a \oplus b$) is equivalent to the statement that the Bitcoin Asset is neither a volatile cryptocurrency nor a blue-chip stock or that it is not neither ($\neg a \oplus \neg b$).
5. From $\forall x(A(x) \rightarrow C(x))$ and $A(mike) \wedge B(mike)$, we know $C(mike) \wedge B(mike)$
   This means: If for all financial assets x, If $A(x)$ (being a high-risk investment) implies $C(x)$ (causes market volatility), and if Mike's investment is both a high-risk investment ($A(mike)$) and a speculative stock ($B(mike)$), then we know that Mike's investment causes market volatility ($C(mike)$) and is still a speculative stock ($B(mike)$).

This formula captures logical relations between financial characteristics like high-risk investments, volatility, and blue-chip stocks, applying propositional logic and predicate logic concepts.

Premises:

1. There are no high-risk investments in a safe market.
2. All blue-chip stocks are in safe markets.
3. Every volatile cryptocurrency is a high-risk investment.
4. Every speculative asset during a market bubble is a volatile cryptocurrency.
5. The Bitcoin Asset is either a volatile cryptocurrency or a speculative asset during a market bubble.

Financial Domain Conclusion (Label → False):

If the Bitcoin Asset is either a blue-chip stock or a volatile cryptocurrency, then it is not a high-risk investment. → F

Fig. 3 An example story consisting of five premises and one conclusion with the label for the conclusion

There is a broad spectrum of issues associated with reasoning in LLMs (Fig. 4). The focus of this chapter is **Integration with outside symbolic reasoning in adversarial setting**.

3 Neuro-Symbolic Rule Application Framework

The logic program for implementing working memory consists of three core components (Fig. 5):

1. Fact base
2. Rule base
3. Logic program schema

The fact base is a structured repository that stores facts derived from the input context and objects. These facts exist in both NL and symbolic forms to facilitate accurate reasoning during multi-step rule application. Symbolic facts are crucial for intermediate reasoning processes, ensuring a smooth transition between different reasoning steps.

The rule base holds a list of input rules. Similar to the fact base, these rules are stored in both NL and symbolic forms to enable precise symbolic referencing and language-based execution. These rules guide how facts are applied and manipulated during the program's execution.

The logic program schema serves as a unified framework that maintains a consistent vocabulary of predicates and objects relevant to the given instance. This prevents semantic duplication and ensures consistency in reasoning. For example, if predicates like *subsidiary_of*() or *income_level*() are defined in the schema, then similar but contextually irrelevant predicates like *competitor_of*() or *noncommercial_status*() will not be excluded or confused. All symbolic facts and rules are constructed using these standardized predicates and objects from the schema.

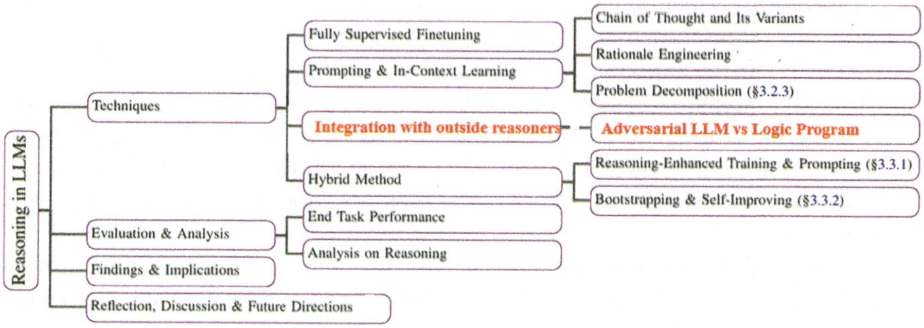

Fig. 4 A taxonomy of issues in LLM reasoning and the proposal of this chapter

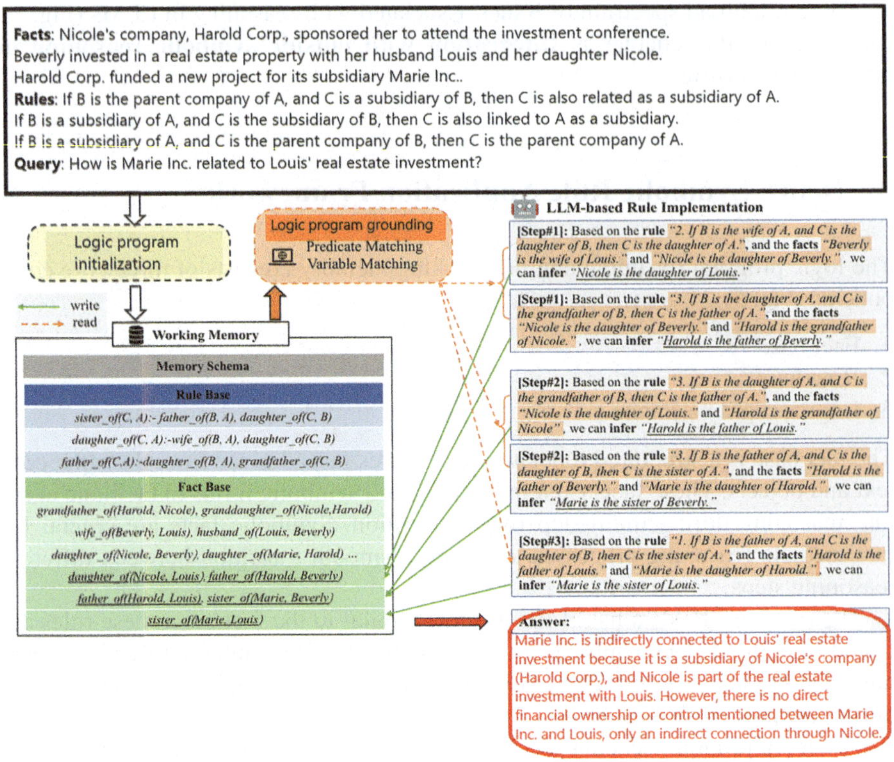

Fig. 5 The workflow of our neuro-symbolic rule application framework based on logic programing

The rule application framework supports three fundamental operations:

1. Read Operation: Retrieves relevant facts and rules from memory, ensuring that the system has the necessary information to carry out the next step in rule application. This retrieval can be selective, targeting only facts and rules pertinent to the specific query or context.
2. Write Operation: Either adds new facts or rules to the memory or updates existing ones. This operation is context sensitive—if the program involves scenarios where facts can change over time (e.g., an object's location), it updates existing facts to reflect the new information. If new facts do not conflict with existing data, they are simply added. However, when new information contradicts current facts (e.g., a change in status), the memory updates the facts to avoid inconsistencies.
3. Contextual Fact Management. The decision to add or update facts in memory is based on whether the input scenario involves dynamic or static facts. For example, in scenarios where an object's status (such as its location or ownership) may change, the system updates the fact base. In static situations, new facts are appended without modification to existing ones. This approach allows the frame-

work to maintain a dynamic and flexible reasoning environment, where facts and rules evolve over time, reflecting changes in context and ensuring accurate, real-time application of knowledge.

The functionality described in this framework—where facts can be added, updated, or replaced in response to new information—can be implemented using a Belief Revision or Belief Update formalism, particularly those associated with dynamic systems. Several formalisms exist for belief update:

AGM Belief Revision (Makinson 1985) is a foundational approach to belief revision. It defines a set of postulates for how rational agents should revise their beliefs when they encounter new, potentially conflicting information (Darwiche and Pearl 1997). The operations in AGM that closely relate to the system's read/write functionality are:

1. Expansion: Adding new beliefs (facts) to the knowledge base without removing any existing beliefs. This corresponds to the "write" operation where new facts are added if there is no conflict.
2. Revision: Modifying the knowledge base by incorporating new beliefs, especially when the new beliefs conflict with the existing ones. This is your "update" operation, where contradictory facts are resolved by adjusting the memory to reflect the most recent information.
3. Contraction: Removing a belief from the knowledge base, often when it becomes inconsistent with newly acquired beliefs.

AGM provides a formalized approach to consistency maintenance by ensuring that revisions do not lead to contradictions in the knowledge base.

Dynamic Epistemic Logic models the evolution of an agent's beliefs over time, accounting for actions or events that alter knowledge. DEL operates with epistemic models that represent the state of an agent's beliefs and uses event models to represent the impact of new information. It is particularly suited for systems where:

1. Events (like new facts or rules) cause updates in the agent's knowledge base.
2. Public announcements or private observations can change the belief state of agents by either adding new information or removing/overriding old beliefs.
3. Epistemic actions can involve common knowledge or private knowledge updates, which is useful for collaborative or multi-agent belief systems.

DEL fits well with scenarios where new observations (i.e., facts) update the belief system dynamically, just as your system reads new information and decides whether to update or append facts.

Bayesian Networks can also model belief updates, especially when facts are probabilistic rather than deterministic. In this case, the update is not a hard rewrite of a fact but rather an adjustment to the belief's probability, reflecting uncertainty about the truth of the fact. This formalism is useful in systems where evidence accumulates over time and affects the belief state in a non-binary fashion (i.e., beliefs are held with varying degrees of confidence).

Bayesian Update: the knowledge base gets updated with new evidence through the application of Bayes' theorem, adjusting the probabilities of related beliefs. However, this method is more appropriate when working with probabilistic facts rather than pure logical predicates (Puga et al. 2015). In update semantics, the state of a system is updated by new information, where updates are understood as changes to a knowledge state. Update semantics is used primarily in dynamic logic and temporal reasoning systems and can track how states evolve when an event occurs (like receiving new facts).

Given that the logic programing for LLM involves tracking changes to a fact base and applying symbolic rules iteratively, AGM Belief Revision would be the most suitable formalisms. They both provide mechanisms for adding new information, resolving contradictions, and maintaining the integrity of the knowledge base over time, closely aligning with your description of "read" and "write" operations in working memory.

4 Grounding of Logic Programming Clauses

At each step of rule application, Wang et al. (2024a) begin by grounding the applicable rules along with their corresponding supporting facts from the logic program. For accurate grounding, a dual strategy of predicate and variable matching is used between facts and rules:

Predicate matching checks if the predicates of the selected facts align with those in the rule's premises. This matching, typically performed as an exact string match, can be adapted to use approximate string or model-based semantic matching to handle minor parsing inconsistencies, allowing for more flexible grounding.

Variable matching ensures that the arguments of the facts can instantiate the rule premises' variables without conflicts (i.e., each variable consistently aligns with the same argument) or match the objects specified in the rule premises.

Examples illustrating these matching processes follow.

```
Rule: brother_of(C, A) :- sister_of(B, A), brother_of(C, B)
F1: grandson_of(John, James) F2: sister_of(Mary, John) predicate unmatched
F2: sister_of(Mary, John) F3: brother_of(James, Mary) predicate matched
Rule: brother_of(C, A) :- sister_of(B, A), brother_of(C, B)
F2: sister_of(Mary, John) F3: brother_of(James, Mary) variable matched
F2: sister_of(Mary, John) F4: brother_of(Clarence, Timmy) variable unmatched
```

Note that the predicates of facts F1 and F2 do not align with rule R and the arguments of F2 and F4 cannot instantiate variable B in rule R. After this symbolic grounding, rule R can apply to its supporting facts F2 and F3. Different rule grounding techniques are employed depending on the task type. For tasks like logical reasoning, where facts are neither chronologically ordered nor updated, grounding uses an exhaustive enumeration approach. This method considers all possible fact combinations for each rule, according to the number of premises. Both predicates and variables are checked for consistency, and a rule is deemed applicable only if no conflicts are present. To avoid redundant applications, each step's applicable rules must include any newly inferred facts from the previous round. For constraint satisfaction tasks, which require rules to satisfy various constraint predicates, variable matching is used to rank the most applicable rule at each step.

4.1 LLM-Supported Rule Execution

LLMs are proficient at single-step rule application. Following symbolic rule grounding, which identifies the relevant rules and supporting facts from the logic program at each step, an LLM executes all applicable rules simultaneously. Each rule is accompanied by its supporting facts, and the LLM is prompted to infer potential new facts in both natural language (NL) and symbolic formats. These inferred facts are then incorporated into the logic program accordingly. At each step of rule application, the system verifies whether the newly inferred fact resolves the query (in logical reasoning tasks) or checks for conflicts between rules and facts (in constraint satisfaction; Galitsky 2024).

5 Logic Test Generation and Iterative Self-Refinement

To enhance the logical reasoning capabilities of LLMs, an automated reasoning feedback component can be integrated. This component generates declarative logic programs and creates tests to assess their semantic accuracy. By rigorously executing these tests, the feedback component identifies any logical or semantic errors, providing specific feedback on failed tests to refine the logic programs. This iterative feedback process continually improves the precision and reliability of the outputs (Table 2 and Fig. 6).

Through repeated revisions, the system refines logical outputs until they meet established correctness standards. This approach is particularly effective for complex reasoning tasks, as it not only helps detect subtle logical errors but also enables the LLM to learn from test-driven feedback, progressively improving its ability to produce accurate, well-structured logic programs.

Table 2 Sharing responsibilities between an LLM and automated reasoner

LLM	Automated reasoner
Generation of declarative code with tests	Execute the logic program, run the tests, and provide detailed feedback with explanations when there are test failures
GPT4-Turbo generates logic program of reasonably high quality from NL instructions, even in a zero-shot setting	Each logic test has optional fact that needs to be added to the program to test the rules/ constraints
Automatic classification of NL statements based on their logical structure, connectives/ operators used, and overall composition (e.g., do they contain nested clauses)	Test conditions
Leverage the LLM's ability to fill in common-sense knowledge gaps	A set of propositions that must be inferred by the solver in *all/at least one/not inferred* solution sets of the logic program
Two additional fields for each test: rules-referenced—which points to specific rules in the program that are exercised in the test—and test explanation, a rationale for the test describing how it validates the semantics of the referenced rules	The instructions on how the explanation from the logic program can be used to identify whether the test inputs, validation criteria, or specific parts of the logic program (e.g., the commonsense knowledge section) need to be altered

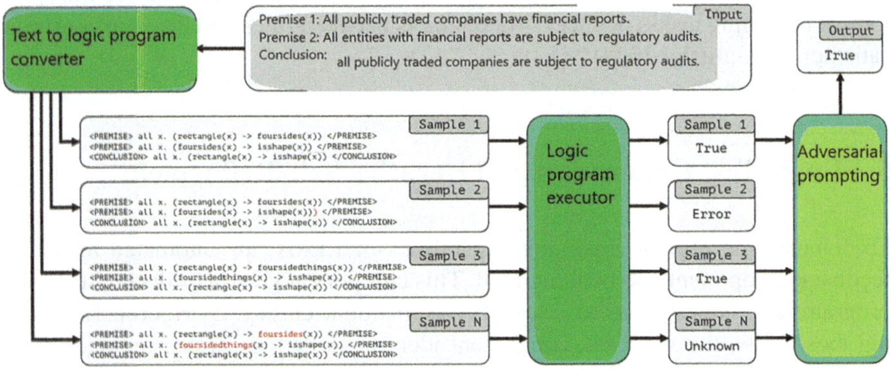

Fig. 6 A workflow of a reasoner supporting an LLM

5.1 *Building ASP Program*

For sentences

> All stocks are financial assets.
> All tech stocks are stocks.
> All high-growth tech stocks are tech stocks.
> Some high-growth tech stocks are volatile.
> If AssetX is a financial asset, then AssetX is not both a tech stock and a financial asset.
> If AssetX is volatile, then AssetX is either a high-growth tech stock or a financial asset.

We have an ASP

```
% R1: All stocks are financial assets.
financial_asset(X) :- stock(X).
% R2: All tech stocks are stocks.
stock(X) :- tech_stock(X).
% R3: All high-growth tech stocks are tech stocks.
tech_stock(X) :- high_growth_tech_stock(X).
% R4: Some high-growth tech stocks are volatile.
{volatile(X)} :- high_growth_tech_stock(X).
% R5: If AssetX is a financial asset, then AssetX is not both
a tech stock and a financial asset.
:- financial_asset(assetX), tech_stock(assetX),
stock(assetX).
% R6: If AssetX is volatile, then AssetX is either a high-
growth tech stock or a stock.
1
{high_growth_tech_stock(assetX); stock(assetX)} 1 :-
volatile(assetX).
```

Table 3 enumerates the classes of logical expressions. The data used to fine-tune the LLM LP builder had the following three types of sources for these expressions:

1. NL description → LP code with compilation issues → Critic feedback on compiler errors → LP code that compiles.
2. NL description → LP code that compiles with test failures → Critic feedback on failures with explanations → LP code with all tests passing.
3. NL description → LP code that compiles with all tests passing.

6 Chain of Thoughts Derive Logical Clauses

LLMs display emergent behaviors, such as the ability to "reason," when they reach a certain scale (Wei et al. 2022). For instance, by prompting the models with "chains of thought"—reasoning exemplars—or a simple directive like "Let's think step by step," these models can answer questions with clearly articulated reasoning steps (Kojima et al. 2022). An example might be: "All elephants are mammals, all mammals have hearts; therefore, all elephants have hearts." This capability has attracted significant interest, as reasoning is often seen as a defining feature of human intelligence that remains challenging for current AI systems to fully replicate (Mitchell 2021).

Table 3 A spectrum of classes of logical expressions and examples for them

Class of logical expression	Examples
Regular conditionals with disjunctions in the body or consequents in the head (which can be easily split)	– All investors in Capital City use the bank code 12345 – If a stock has been downgraded or rated as risky by an analyst, it is considered high risk – If a person seeks diversification, they either invest in multiple sectors often or avoid putting all funds in one asset class
Conditionals with hierarchical structure	– Either Mark hires a financial advisor or regularly monitors his portfolio or he doesn't hire a financial advisor and rarely reviews his investments – Employees at XYZ finance either save a large part of their bonuses or tend to invest in high-growth stocks – Alex is a shareholder; he is either knowledgeable in technology stocks and invested in bonds, or he lacks tech stock knowledge and invested in equities
Regular and exclusive disjunctions	– All investment platforms have stock trading or bond trading features – If the Tesla stock is volatile, then it is either a growth stock or a tech stock
Negation and disjointness over entities	– No bonds are stocks – No cash assets are considered investments
Exclusion constraints	– Morgan trades different stocks than Kevin in the portfolio – A mutual fund is not both a hedge fund and regulated under a pension fund mandate
Existential quantification	– Some stocks are in the technology sector – Some companies in emerging markets are high-risk investments
Equality among individuals	– The Global Fund is another name for the Emerging Markets Growth Fund
Rules with multiple variables	– Sarah holds different types of assets than James in her portfolio – All funds within an asset class are related to each other based on their risk profile

In LLMs, the chain of thought (CoT) approach helps the model derive logical clauses and reach conclusions through a sequential, interpretable process. This structured reasoning enables LLMs to decompose complex tasks into a series of logical steps, improving both the clarity and coherence of responses. The CoT approach derives logical clauses as follows:

1. Problem decomposition: the model first breaks down the input into distinct components. By analyzing the question's structure, the model identifies relevant entities, variables, and context that are essential for reasoning. As a result, we obtain a set of decomposed components, $D = (E, V, X)$, where E is a set entities, V is a set of variables, and X is a context.

2. Sequential reasoning: the model establishes a sequence of logical steps, progressing from one inference to the next. This may involve if-then statements, comparisons, or assumptions that help clarify relationships and dependencies among the components. Each step in this sequence is guided by the preceding one, maintaining logical continuity. For each step i, we apply logical transformations T_i based on contextual knowledge and prior clauses:

   $C_i = T_i(C_{i-1}, D)$
   $C_1 = T_1(D)$ (initial clause derived directly from D)

   This step yields intermediate clauses $C_1, C_2, \ldots, C_n$.
3. Intermediate conclusions: as the model moves through each logical step, it draws intermediate conclusions or generates "sub-clauses" that build on previous reasoning. These intermediate statements reflect partial answers or deductions that contribute to the final solution. For a subset of clauses $\{C, C_i, \ldots, C_k\}$ where $k < n$ derive sub-conclusions S_j such that $S_j = Combine(C_1, \ldots, C_k)$ and output the set of sub-conclusions $S = \{S_1, S_2, \ldots, S_p\}$.
4. Aggregation of clauses: the intermediate clauses are then combined or aggregated to form a coherent set of logical statements that address the initial query. This step ensures that all intermediate conclusions align and support the overarching response. We aggregate S and remaining C_i clauses such that $C_{final} = Aggregate(S, C_{k+1}, \ldots, C_n)$, outputting the final clause C_{final}.
5. Final conclusion: The model synthesizes the chain of logical clauses into a comprehensive response. The CoT process allows LLMs to arrive at complex conclusions in a structured and transparent manner, improving accuracy and making the reasoning process traceable.

The CoT framework is especially valuable for tasks that require multi-step reasoning, like solving mathematical problems, constraint satisfaction, and scenario-based questions, where each logical step contributes critically to the overall solution.

Prystawski et al. (2023) explore the mechanisms that make CoT reasoning effective in language models, proposing that reasoning works best when training data is structured in overlapping clusters of interdependent variables. This structure allows the model to chain accurate local inferences, enabling it to infer relationships between variables that were not seen together during training. The authors identify a "reasoning gap," demonstrating that reasoning through intermediate variables reduces bias when using a simple autoregressive density estimator trained on local samples from a chain-structured probabilistic model. They further validate this hypothesis in more complex models by training an autoregressive language model on samples from Bayes nets with only a subset of variables included in each sample. The models are tested for their ability to match conditional probabilities with and without intermediate reasoning steps, revealing that these steps improve accuracy only when training data is locally structured around variable dependencies. This approach to locally structured observations and reasoning also proves more data efficient than training on all variables simultaneously, highlighting that the effectiveness of step-by-step reasoning stems from the localized statistical patterns in training data.

LLMs are trained on natural language documents, which often focus on closely related topics (Blei and Lafferty 2007). When concepts frequently co-occur in data, their mutual influence can be assessed directly with simple statistical estimators. However, to infer the effect of one concept on another when they haven't been seen together, a series of inferences connecting related concepts is necessary. CoT reasoning becomes beneficial precisely when the training data is locally structured, meaning observations occur in overlapping clusters of related concepts (Mao et al. 2019). These clusters can then form logical clauses that can be retained, boosting the accuracy of the system in future inference sessions.

Let us consider a task of building a probabilistic clause *p::C :- A* (Riguzzi and Swift 2018; Galitsky 2025). One may know the value of some variable *A* and intents to learn about another variable *C*, to compute $p(C|A)$. However, to compute probabilities using observed samples from joint distributions, one needs to count *A* and *C* together, which is rare. Therefore, one would experience difficulties estimating $p(C|A)$ directly. Instead, one might estimate it by building a chain of clauses via intermediate variables.

If conditioning on an intermediate variable *B* figures out that *A* and *C* are independent of each other, then the conditional probability can be assessed by marginalizing over *B*, using the fact that

$$p(C|A) = \sum_{B} p(C|B)\, p(B|A).$$

For example, suppose we want to answer a question like "What is grown in Los Banos, CA?" using a model trained on Wikipedia text. Wikipedia pages for states and counties often contain information about the crops grown, while pages for cities typically indicate which county a city belongs to. However, city pages rarely mention the crops grown there directly. While an LLM might attempt to answer the question directly, it would likely have greater success by first identifying that Los Banos is located in Merced County, California. By adding this intermediate step—"Los Banos is in Merced County"—before answering with "cotton is grown in Los Banos," an autoregressive model can more effectively utilize the dependencies present in its training data.

A visualization of a Bayes net for a logic program is depicted in Fig. 7. The pink variable is an example observed variable, and the yellow variable is an example target variable. Gray variables are examples of useful intermediate variables for reasoning. Lines show examples of local neighborhoods from which training samples are drawn.

The prediction setting in the form of clauses is as follows:

- Direct prediction clause: *X5 :- X12.*
- Indirect prediction clause: *X5 :- X12, X11, X16, X3, X15, X10.*

The top curve shows the case of full observation (blue, on the top) and local geometry (gray are on the bottom).

Locality refers to which classes tend to appear together. In datasets with a local structure, concepts that strongly influence each other are also frequently seen together, while unrelated concepts appear together less often. By reasoning through appropriate intermediate variables, models trained on locally structured samples reduce bias. Combining LLMs with logic programming (LP) and deterministic back-propagation through these intermediates naturally enhances the accuracy of such hybrid systems.

The chart at the bottom of Fig. 7 compares direct prediction and free-generation estimators used as prompts. The model is prompted to predict a target variable either directly (direct prediction) or indirectly by first generating intermediate variables and their values (chain-based prediction). Prediction accuracy is measured by mean

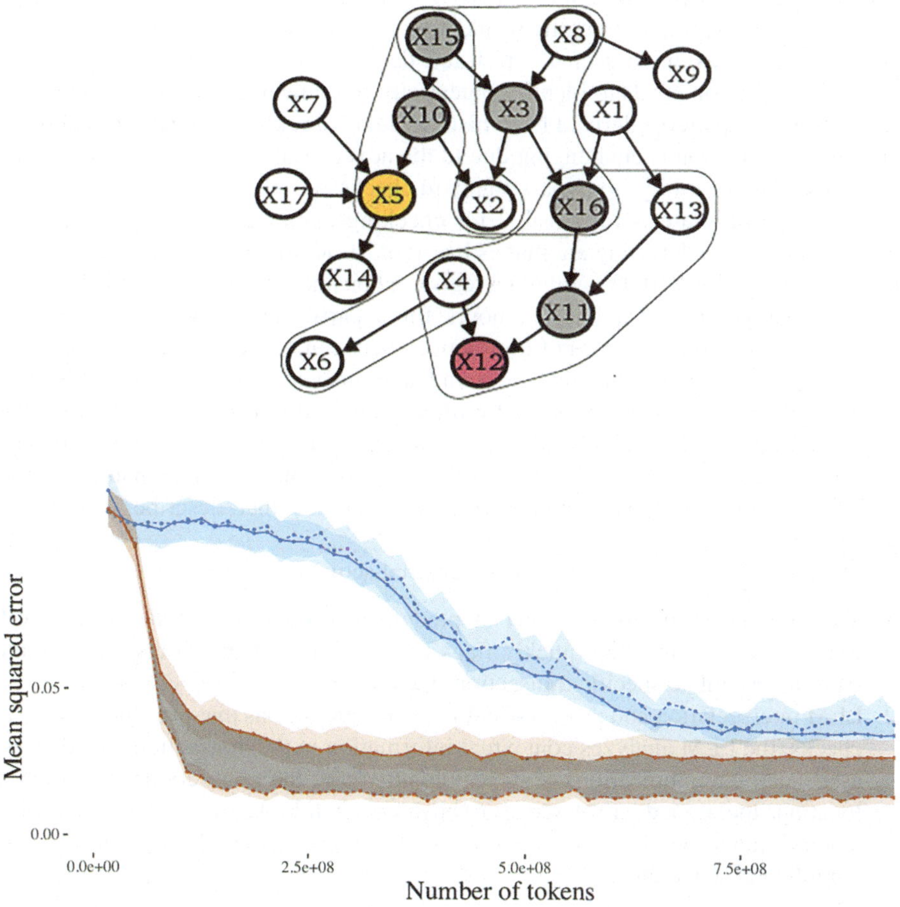

Fig. 7 Visualization of a Bayes net

squared errors (MSE) between estimated and true conditional probabilities, calculated based on the number of training tokens per condition and estimator. The ribbons in the chart represent 95% confidence intervals.

Chain-of-thought reasoning benefits LLMs because:

1. Without chaining, direct prediction can be inaccurate for some inferences, as relevant variables are rarely observed together during training
2. Chain-of-thought reasoning improves estimation by sequentially linking local statistical dependencies that appear frequently in training

7 Adversarial Interactions Between LLM and LP

An adversarial scenario in which a logic program and an LLM work together to solve a reasoning problem could involve a situation where they complement each other's weaknesses and strengths to identify inconsistencies or validate complex claims. A good example scenario is fraud detection in financial transactions. A financial institution needs to identify potentially fraudulent transactions based on a set of rules and patterns derived from historical data, regulations, and internal policies. Fraud detection is challenging due to the adversarial nature of fraudsters who constantly find new ways to bypass standard checks.

The logic program is responsible for checking compliance with known, rule-based patterns and detecting anomalies based on structured rules. For example, if a transaction exceeds $10,000 without a registered business justification, it is flagged. Transactions occurring at unusual hours for a particular account or region are flagged as suspicious. A series of transactions between accounts that lack common ownership are flagged if they involve round amounts and alternate back and forth.

An LLM augments this process by identifying and interpreting complex, less structured patterns, based on NL data (e.g., notes, emails, or text descriptions tied to transactions). The LLM can also adaptively recognize emerging fraud patterns, generalizing from recent case studies and text sources that describe new fraud techniques.

LP and LLM work together in an adversarial setup as follows:

1. The logic program applies a set of predefined rules to detect straightforward cases of fraud. This step quickly flags transactions that match historical fraud patterns or violate specific transaction rules.
2. LLM interpretation and cross-validation: For transactions that pass the initial LP checks, the LLM analyzes contextual information in the transaction metadata or associated communications (e.g., emails justifying the transactions, geographic locations mentioned, or references to businesses). It looks for semantic inconsistencies, vague justifications, or attempts to bypass standard language used in compliance statements.

In the adversarial scenario, a transaction is flagged as "suspicious" by the LP, but the reason provided in the transaction notes looks legitimate at first glance. The LLM, however, identifies inconsistencies in the language used. For example, it recognizes that the justification uses phrasing commonly associated with fraudulent attempts in other, similar cases. Conversely, the LLM might flag a transaction as unusual due to strange language in the metadata or patterns that resemble recent fraud cases. The LP then double-checks this against hard rules (e.g., verifying account ownership, transaction frequency) to see if any known rule has been violated.

If a transaction is flagged but lacks clear evidence under current rules, the LP updates its rules based on feedback from the LLM's analysis of new fraud patterns. This is iterative refinement. For example, if the LLM detects that recent fraud cases often involve certain phrases in transaction justifications, these patterns are incorporated into LP rules for future identification.

The adversarial aspect here also comes from fraudsters continually adapting their methods. The LLM can process recent cases described in news reports, legal documents, or company records and generate hypotheses for the LP to test. The LP then formalizes these patterns as new rules or updates existing ones.

In this adversarial scenario, the LLM provides flexible, adaptive reasoning over unstructured data and contextual clues, while the LP ensures strict compliance checks on rule-based, structured data. By working together:

- The LP enforces rigid constraints for fraud detection, catching straightforward cases
- The LLM broadens the fraud detection scope by interpreting and adapting to new patterns based on natural language data and emerging case knowledge
- The combined approach helps the financial institution stay ahead of evolving fraud tactics by incorporating both structured rule enforcement and adaptive pattern recognition

This synergy between LP and LLM enhances accuracy in fraud detection, especially in cases where traditional rule-based systems alone might miss sophisticated or newly developed fraud techniques.

Let:

- F represent the set of all transactions
- R represent the set of predefined fraud detection rules in the logic program
- $\mathcal{L}$ represent the LLM with capabilities for contextual analysis and natural language pattern recognition
- S be the state of the system, comprising flagged transactions and new insights generated from the reasoning process

We define two main processes: deductive filtering by the logic program and abductive hypothesis generation by the LLM. Deductive Filtering by LP operates with a set of rules R designed to detect known patterns of fraudulent transactions. These rules are applied as follows:

Rule-Based Flagging: For each transaction $t \in F$, if t violates a rule $r \in R$, then t is flagged as suspicious. Let $flag(t)$ denote that a transaction (t) has been flagged by LP based on rule r. $flag(t) = \exists r \in R$ s.t. $r(t)$). The LP outputs a set of flagged transactions $F_{flagged} = \{t \in F \mid flag(t)\}$ and a set of rules applied to each flagged transaction. Let $F^0_{flagged}$ denote the initial flagged set from rule-based filtering.

Abductive Hypothesis Generation by LLM: For transactions in $F^0_{flagged}$ and transactions that were not flagged by the LP but still appear anomalous in unstructured context, the LLM analyzes textual, contextual, and semantic information to generate hypotheses about potential fraud. The LLM may identify semantic patterns in justifications, transaction descriptions, or other natural language metadata. LLM is also expected to generate a hypothesis if it finds inconsistencies or patterns in the transaction's textual context that resemble recent fraud cases.

Let $H_{\mathcal{L}}(t)$ denote a hypothesis generated by the LLM about transaction t based on detected patterns or inconsistencies. The LLM then suggests new patterns as rules, $R_{\mathcal{L}}$ based on hypotheses. It also refines the flagged set, creating a set $F_{hypothetical} = \{t \in F \mid H_{\mathcal{L}}(t)\}$.

The feedback loop for rule update and validation is formulated as follows. For rule update, the LLM provides feedback to the LP by suggesting new rules $R_{\mathcal{L}}$, which are generalized patterns extracted from the analysis. The logic program updates its rule base $R \leftarrow R \cup R_{\mathcal{L}}$ R, enhancing its detection capability for the next iteration.

Under iterative testing, the LP re-applies the updated rule set R to the transaction set F to flag any additional transactions that may match the newly learned rules. This iterative process continues until no new transactions are flagged or until convergence criteria (e.g., detection threshold) are met.

$$F^{i+1}_{flagged} = \{t \in F \mid \exists r \in R \cup R_{\mathcal{L}} \text{ such that } r(t)\};$$

repeat until $F^{i+1}_{flagged} = F^{i}_{flagged}$.

In this adversarial interaction, the LP and LLM achieve a complementary balance: the LP enforces structured, rule-based detection grounded in formal rules, while the LLM dynamically adapts to identify emerging patterns and continuously refine the detection process. Together, they enable robust fraud detection in an evolving, adversarial environment. This formalized interaction allows the system to iteratively evolve its rules, improving detection accuracy and adapting to novel fraud tactics that a strictly rule-based approach might overlook.

8 Evaluation

8.1 Datasets

We conducted various experiments using the FOLIO benchmark (Han et al. 2022), which includes a thousand training examples and a validation dataset sized at 20% of the training set. Each FOLIO problem consists of a set of premises (in natural language) and a conclusion (also in natural language), with the task being to determine whether the conclusion is *True*, *False*, or *Uncertain* given the premises. FOLIO also provides First-Order Logic translations for each premise and conclusion.

P-FOLIO (Han et al. 2024) expands on these tasks, introducing different levels of granularity and metrics to evaluate the reasoning capabilities of LLMs. Human-authored reasoning chains have shown to significantly improve LLM performance in logical reasoning via many-shot prompting and fine-tuning.

Additionally, we utilized the CLUTRR and ProofWriter datasets, both designed for logical reasoning tasks that apply commonsense or predefined rules. For CLUTRR, we selected 235 test instances requiring 2–6 steps of rule application. From ProofWriter, we selected instances needing 3–5 steps of reasoning from the open-world assumption subset, totaling 300 instances with balanced labels.

The SimpleQA dataset (OpenAI 2024) focuses on short, fact-based queries, which narrows the scope of evaluation but simplifies the assessment of factuality. SimpleQA was constructed to prioritize high accuracy, with reference answers supported by sources from two independent AI trainers and questions formulated to facilitate clear grading of predicted answers. As a factuality benchmark, SimpleQA allows for testing *calibration*—the degree to which a language model can accurately reflect its knowledge. Calibration is assessed by prompting the model to state its confidence level (as a percentage) alongside its answer. By plotting the correlation between the model's stated confidence and actual accuracy, we evaluate calibration. Ideally, a well-calibrated model's accuracy should match its stated confidence; for instance, if the model expresses 75% confidence across a set of answers, its accuracy would indeed be 75% for that set.

8.2 Competitive Systems

WM-Neurosymbolic (Wang et al. 2024a) enhances LLMs with an external working memory, forming a neuro-symbolic framework for multi-step rule application to boost reasoning capabilities. This working memory stores facts and rules in both natural language and symbolic representations, supporting accurate retrieval during rule application. After storing all input facts and rules in memory, the framework iteratively applies symbolic rule grounding through predicate and variable matching, followed by rule implementation via the LLM.

Scratchpad-CoT (Nye et al. 2021) facilitates chain-of-thought reasoning using a "scratchpad" approach, where transformers are trained to break down multi-step computations by emitting intermediate steps in a separate space. On tasks ranging from long addition to program execution, scratchpads significantly improve LLMs' ability to perform stepwise computations.

Logic-LM (Pan et al. 2023) leverages LLMs to translate natural language problems into symbolic formulations and then uses symbolic solvers like the Z3 theorem prover (De Moura and Bjørner 2008) to perform deterministic inference.

LLM-ARC (Kalyanpur et al. 2024) combines LLMs with an Automated Reasoning Critic (ARC) in a neuro-symbolic framework aimed at advancing reasoning. Here, the LLM Actor generates declarative logic programs and creates tests for semantic accuracy, while the ARC evaluates the code, executes tests, and provides feedback on failures to iteratively refine the model.

LINC (Olausson et al. 2023) uses an LLM as a semantic parser to translate premises and conclusions from natural language into first-order logic expressions, which are then processed by an external theorem prover to carry out deductive inference symbolically.

In Table 4, we compare our adversarial system with two types of baselines: CoT-based methods (**grayed rows**), symbolic-based methods (Logic-LM), and state-of-the-art neuro-symbolic systems (LINC, WM-Neurosymbolic, and LLM-ARC), highlighted in **turquoise**.

When our LP-based method does not return an answer caused by symbolic formulation errors, we use CoT-based approach (Sect. 6). We observe that our adversarial method demonstrates consistent performance overall with the competitive neuro-symbolic systems. However, we underperform LINC on ProofWriter, WM-Neurosymbolic on CLUTRR, and LLM-ARC on FOLIO.

Our adversarial neuro-symbolic system is effective on top of different LLMs with varying abilities in symbolic parsing and one-step rule application. Specifically, GPT-3.5-based setting delivers a noticeable boost on computational logical problems (CLUTRR). At the same time, GPT-4 has a great accuracy at more human-related tasks (ProofWriter). Our adversarial approach leverages an advancement of LLMs; one can expect that it will efficiently complement future LLM systems.

Table 4 Evaluation on the full dataset

System/dataset	FOLIO	CLUTRR	ProofWriter
GPT3.5	66.9		
GPT4	67.0		
Scratchpad-CoT		83.8	61.3
Logic-LM	78.9		
LINC	75.2		**98.3**
WM-Neurosymbolic		**92.3**	77.3
LLM-ARC	**88.3**		
Adversarial_LLM_LP	77.0	*86.1*	*71.5*

Compared to previous symbolic-based methods that perform both rule grounding and implementation either symbolically or by LLMs, our framework exhibits improvement. Adversarial LLM-LP integration demonstrates flexibility and robustness, separately implementing rule grounding and execution.

We proceed to our evaluation on a specific dataset with controversial, more complex problems that require a number of knowledge sources to make decision upon. The computational definition of a *controversial* is based on inconsistent or even opposite results obtained by an LLM on a given problem. We automatically classify such problems from the evaluation sets, forming datasets FOLIO_ *controversial* (about 1/7 of FOLIO), CLUTRR_ *controversial* (about 1/10 of CLUTRR), and ProofWriter_*controversial* (about 1/7 of ProofWriter). We run WM-Neurosymbolic and our adversarial system on these subsets and provide the results in Table 5. Our adversarial system outperforms WM-Neurosymbolic in two out of three datasets (other than ProofWriter_*controversial*).

9 Discussion

The neuro-symbolic framework for rule application can incorporate an external working memory that stores both facts and rules in natural language and symbolic formats, supporting accurate tracking throughout the reasoning process (Wang et al. 2024b). This memory enables iterative application of symbolic rule grounding and LLM-based rule execution. During symbolic rule grounding, the framework matches rule predicates and variables with relevant facts to determine which rules are applicable at each step. Experimental results indicate the framework's effectiveness and robustness across various reasoning tasks.

LLMs have shown exceptional performance across many tasks but often struggle with complex reasoning that requires retention or grounding of long-term information from context or interaction history (Touvron et al. 2023). While extending LLMs' context length is one approach, recent advances involve augmenting LLMs with external memory. For instance, Park et al. (2023) integrate external memory modules for LLMs in extended dialogues to improve interaction quality, while Wang et al. (2024b) represent long-form context in memory for retrieval in knowledge-intensive tasks.

However, existing memory architectures typically store only natural language or parametric entries, making accurate referencing and revision challenging. To

Table 5 Evaluation of the dataset fragments with controversial, complex problems

System/dataset	FOLIO_ *controversial*	CLUTRR_ *controversial*	ProofWriter_ *controversial*
WM-Neurosymbolic	70.7	78.1	**65.5**
Adversarial_LLM_LP	***73.2***	***82.7***	*64.9*

address this, symbolic memory has been proposed. ChatDB (Hu et al. 2023) utilizes a structured item database as symbolic memory for precise information recording, and Yoneda et al. (2023) developed symbolic world memory to track robot states for embodied reasoning. Our approach leverages external memory to store both natural language and symbolic facts and rules, enhancing precise rule grounding for multi-step applications.

Additionally, Ranaldi and Freitas (2024) fine-tuned LLMs with logical reasoning data to improve inference capabilities. LLMs have been applied as soft logic reasoners, and a range of prompting techniques have been proposed to enhance performance in this domain (Zhou et al. 2024). Using LLMs as semantic parsers has also shown to improve reasoning accuracy (Olausson et al. 2023) by first converting natural language reasoning problems into logical forms before using an inference engine for the final output.

Wu et al. (2023) present a method for creating tricky natural language instructions that cause code-generating LLMs to produce code that works correctly but contains security vulnerabilities. Their approach uses an evolutionary search process and a carefully designed loss function. The system automatically finds harmless-looking prefixes or suffixes to add to a user's prompt—text that appears completely innocent and unrelated to coding—but that still strongly pushes the model toward producing insecure code. This makes it possible to perform near worst-case "red-teaming" tests in realistic situations, where users give prompts in plain natural language.

In this book, we will treat reasoning in a broader perspective. In Galitsky (2024), we proposed reasoning applications designed to generate diagnostic assessments for input financial indicators, derived from labels (e.g., risk categories, market events, regulatory breaches, and/or portfolio conditions) associated with previously observed cases. The system can construct extended discourse trees that represent multiple discourse structures across varying levels of granularity (e.g., full report, section, paragraph, sentence, phrase, or token) for historical financial cases, along with the rhetorical relations linking those structures. New financial data (e.g., submitted via the autonomous agent) can be parsed to identify relevant fragments and the rhetorical relations among them. These fragments are then matched to fragments of prior cases by aligning nodes within the extended discourse tree, enabling the system to produce context-aware and historically grounded risk or compliance diagnoses.

10 Conclusions

Our experiments demonstrate the framework's superiority over CoT-based, symbolic-based, and LLM-LP collaboration baselines and show its robustness across various rule application steps and settings. In the future, we will extend our framework to incorporate more backbone LLMs and datasets, especially on more complex and long-term reasoning tasks.

There is growing recognition in the LLM audience that LLM-only solutions do not meet the standard for production applications that require a high degree of accuracy, consistency, and explicability. More specifically, current state-of the-art LLMs are known to struggle for problems involving precise logical reasoning, planning, and constraint solving. As a result, there is a rise in the development of hybrid neuro-symbolic systems, where the reasoning is offloaded to a symbolic solver and the LLM is used at the interface layer to map between unstructured data (text) and structured logical representations. Unlike standard tools or simple APIs, integration between an LLM and a symbolic reasoner can be fairly sophisticated as the reasoning engine has its own world model and decision procedures.

The adversarial interaction between LLMs and logic programs significantly enhances the accuracy of solving complex logical problems. This integration creates a dynamic, adaptive framework that leverages the structured, rule-based rigor of logic programming and the flexible, context-aware reasoning capabilities of LLMs. The LP provides a formal foundation for consistent rule enforcement, ensuring that well-defined logical constraints are respected, while the LLM contributes by identifying patterns and drawing inferences from nuanced or unstructured data. Through this interaction, the two systems address each other's limitations—LLMs can mitigate LP's rigidity when it encounters cases outside of its predefined rules, and LP can counterbalance the occasional ambiguity in LLM predictions, leading to a more robust and reliable reasoning process.

In an adversarial setup, the LLM and LP act as checks and balances, where each system critiques or validates the outputs of the other. This interplay is particularly beneficial in complex reasoning tasks like fraud detection or scientific hypothesis evaluation, where new patterns and irregularities often arise. The LP detects standard rule violations, while the LLM's adaptability enables it to flag novel anomalies or emerging trends that do not yet have formalized rules. By iteratively refining outputs based on each other's feedback, the combined neuro-symbolic system not only improves accuracy in real time but also learns to adapt to evolving patterns or requirements, a feature that pure logic-based systems would find challenging to achieve.

Moreover, this adversarial approach enhances the transparency and explainability of the system's outputs. The structured logic program offers clear, traceable reasoning steps, allowing users to follow the logic behind conclusions, while the LLM supplements this with insights from broader contextual or probabilistic knowledge. This combination supports interpretability, which is essential in fields where understanding the reasoning process is as important as the outcome. Over time, the system's rules and logic pathways are iteratively refined and expanded based on both structured feedback and natural language data, progressively increasing its depth and reliability. This iterative refinement not only improves immediate accuracy but also builds a richer, more sophisticated logic framework, positioning the LLM-LP system as a powerful tool for tackling increasingly complex and evolving reasoning tasks.

References

Berglund L, Tong M, Kaufmann M, Balesni M, Stickland AC, Korbak T, Evans O (2023). The reversal curse: Llms trained on "a is b" fail to learn "b is a". arXiv preprint arXiv:2309.12288

Blei DM, Lafferty JD (2007) A correlated topic model of science. Ann Appl Stat 1(1):17–35

Creswell A, Shanahan M, Higgins I (2022) Selection-inference: exploiting large language models for interpretable logical reasoning. arXiv preprint arXiv:2205.09712

Darwiche A, Pearl J (1997) On the logic of iterated belief revision. Artif Intell 89(1):1–29

De Moura L, Bjørner N (2008) Z3: an efficient SMT solver. In: Proceedings of the Theory and practice of software, 14th international conference on Tools and algorithms for the construction and analysis of systems (TACAS'08/ETAPS'08). Springer-Verlag, Berlin, Heidelberg, pp 337–340

Galitsky B (2024) Multi case-based reasoning by syntactic-semantic alignment and discourse analysis. US Patent 12,106,054

Galitsky B (2025) Employing LLM to solve constraint satisfaction. In: Health applications of neuro symbolic AI. Elsevier

Gur I, Furuta H, Huang A, Safdari M, Matsuo Y, Eck D, Faust A (2024) A real-world web agent with planning, long context understanding, and program synthesis. arXiv preprint arXiv:2307.12856

Han S, Schoelkopf H, Zhao Y, Qi Z, Riddell M, Benson L, Sun L, Zubova E, Qiao Y, Burtell M, Peng D, Fan J, Liu Y, Wong B, Sailor M, Ni A, Nan L, Kasai J, Yu T, Zhang R, Joty S, Fabbri AR, Kryscinski W, Lin XV, Xiong C, Radev D (2022) Folio: natural language reasoning with first-order logic

Han S, Yu A, Shen R, Qi Z, Riddell M, Zhou W, Qiao Y, Zhao Y, Yavuz S, Liu Y, Joty S, Zhou Y, Xiong C, Radev D, Ying R, Cohan A (2024) P-FOLIO: evaluating and improving logical reasoning with abundant human-written reasoning chains. arXiv preprint arXiv:2410.09207

Hu C, Fu J, Du C, Luo S, Zhao J, Zhao H (2023) ChatDB: augmenting LLMs with databases as their symbolic memory. arXiv preprint arXiv:2306.03901

Kalyanpur A, Saravanakumar KK, Barres V, Chu-Carroll J, Melville D, Ferrucci D (2024) LLM-ARC: Enhancing LLMs with an Automated Reasoning Critic. arXiv:2406.17663

Kojima T, Gu SS, Reid M, Matsuo Y, Iwasawa Y (2022) Large language models are zero-shot reasoners. In: Advances in neural information processing systems

Makinson D (1985) How to give up: a survey of some formal aspects of the logic of theory change. Synthese 62:347–363

Mao J, Gan C, Kohli P, Tenenbaum JB, Wu J (2019) The neuro-symbolic concept learner: interpreting scenes, words, and sentences from natural supervision

Mitchell M (2021) Abstraction and analogy making in artificial intelligence. Ann N Y Acad Sci 1505(1):79–101

Morishita T, Morio G, Yamaguchi A, Sogawa Y (2023) Learning deductive reasoning from synthetic corpus based on formal logic. In: Proceedings of the 40th International Conference on Machine Learning, volume 202 of Proceedings of Machine Learning Research, pp 25254–25274. PMLR

Nye M, Andreassen AJ, Gur-Ari G, Michalewski H, Austin J, Bieber D, Dohan D, Lewkowycz A, Bosma M, Luan D (2021) Show your work: Scratch pads for intermediate computation with language models. arXiv preprint arXiv:2112.00114

Olausson T, Gu A, Lipkin B, Zhang C, Solar-Lezama A, Tenenbaum J, Levy R (2023) LINC: a neurosymbolic approach for logical reasoning by combining language models with first order logic provers. In: EMNLP, Singapore, Dec 2023, pp 5153–5176

OpenAI (2024) SimpleQA. https://openai.com/index/introducing-simpleqa/

Pan L, Albalak A, Wang X, Wang W (2023) Logic-LM: empowering large language models with symbolic solvers for faithful logical reasoning. In: Bouamor H, Pino J, Bali K (eds) Findings of the Association for Computational Linguistics: EMNLP 2023, pp 3806–3824

Park JS, O'Brien J, Cai CJ, Morris MR, Liang P, Bernstein MS (2023) Generative agents: interactive simulacra of human behavior. In: Proceedings of the 36th annual ACM symposium on user interface software and technology, pp 1–22

Parmar M, Patel N, Varshney N, Nakamura M, Luo M, Mashetty S, Mitra A, Baral A (2024) LogicBench: Towards systematic evaluation of logical reasoning ability of large language models. In: Proceedings of the 62nd Annual Meeting of the Association for Computational Linguistics (Volume 1: Long Papers). Association for Computational Linguistics, Bangkok, Thailand, pp 13679–13707. updated in text: 2023-2024a

Prystawski B, Li MY, Goodman ND (2023) Why think step by step? Reasoning emerges from the locality of experience. In: Advances in neural information processing systems, NeurIPS 2023, vol 36

Puga J, Krzywinski M, Altman N (2015) Bayes' theorem. Nat Methods 12:277–278. https://doi.org/10.1038/nmeth.3335

Ranaldi L, Freitas A (2024) Aligning large and small language models via chain-of-thought reasoning. In: Proceedings of the 18th conference of the European chapter of the Association for Computational Linguistics, Long papers, vol 1, pp 1812–1827

Riguzzi F, Swift T (2018) A survey of probabilistic logic programming. In: Declarative logic programming: theory, systems, and applications. ACM, pp 185–228. https://doi.org/10.1145/3191315.3191319. ISBN: 978-1-970001-99-0, S2CID: 70180651, Accessed 25 Oct 2023

Saha S, Ghosh S, Srivastava S, Bansal M (2020) PRover: Proof generation for interpretable reasoning over rules. In: Proceedings of the 2020 Conference on Empirical Methods in Natural Language Processing (EMNLP), pp 122–136, Online. Association for Computational Linguistics

Saha S, Yadav P, Bansal M (2021) multiPRover: Generating multiple proofs for improved interpretability in rule reasoning. In: Proceedings of the 2021 Conference of the North American Chapter of the Association for Computational Linguistics: Human Language Technologies, pp 3662–3677, Online. Association for Computational Linguistics

Saparov A, Pang RY, Padmakumar V, Joshi N, Kazemi M, Kim N, He H (2023) Testing the general deductive reasoning capacity of large language models using OOD examples. In: Thirty-seventh Conference on Neural Information Processing Systems

Sarker MK, Zhou L, Eberhart A, Hitzler P (2021) Neuro-symbolic artificial intelligence: current trends. arXiv preprint arXiv:2105.05330

Tafjord O, Dalvi B, Clark P (2021) ProofWriter: Generating implications, proofs, and abductive statements over natural language. In: Findings of the Association for Computational Linguistics: ACL-IJCNLP 2021, pp 3621–3634, Online. Association for Computational Linguistics

Touvron H, Lavril T, Izacard G, Martinet X, Lachaux M-A, Lacroix T, Rozière B, Goyal N, Hambro E, Azhar F et al. (2023) Llama: open and efficient foundation language models. arXiv preprint arXiv:2302.13971

Vakharia P, Kufeldt A, Meyers M, Lane I, Gilpin LH (2024) ProSLM: a prolog synergized language model for explainable domain specific knowledge based question answering

Wan Z, Liu CK, Yang H, Li C, You H, Fu Y, Wan C, Krishna T, Lin Y, Raychowdhury A (2024) Towards cognitive AI systems: a survey and prospective on neuro-symbolic AI

Wang S, Wei Z, Choi Y, Ren X (2024a) Symbolic working memory enhances language models for complex rule application. arXiv preprint arXiv:2408.13654

Wang W, Dong L, Cheng H, Liu X, Yan X, Gao J, Wei F (2024b) Augmenting language models with long-term memory. In: Advances in neural information processing systems, vol 36

Wei J, Wang X, Schuurmans D, Bosma M, Ichter B, Xia F, Chi EH, Le QV, Zhou D (2022) Chain of thought prompting elicits reasoning in large language models. In: Advances in neural information processing systems

Wu F, Liu X, Xiao C (2023) DeceptPrompt: exploiting LLM-driven code generation via adversarial natural language instructions. arXiv preprint arXiv:2312.04730v2

Yoneda T, Fang J, Li P, Zhang H, Jiang T, Lin S, Picker B, Yunis D, Mei H, Walter MR (2023) Statler: state-maintaining language models for embodied reasoning. arXiv preprint arXiv:2306.17840

Zhou P, Pujara J, Ren X, Chen X, Cheng HT, Le QV, Chi EH, Zhou D, Mishra S, Zheng HS (2024) Self-discover: large language models self-compose reasoning structures

Adversarial Abductive Dialogue Framework with Reinforcement for Tackling LLM Hallucination

1 Introduction

Recent breakthroughs in large language models (LLMs), such as ChatGPT, have dramatically transformed the landscape of artificial intelligence. These models now power highly advanced conversational agents that can produce fluent, context-aware, and seemingly intelligent responses across a wide range of topics. Their ability to understand and generate humanlike text has opened new possibilities in education, customer service, content creation, and more. However, alongside these impressive capabilities lies a persistent and critical flaw: the generation of hallucinations.

Hallucinations refer to instances where an LLM outputs information that is factually inaccurate, misleading, or entirely fabricated—often with the same confidence and fluency as correct statements. This issue is particularly troubling in high-stakes domains such as healthcare, law, and scientific research, where trust in system outputs is essential. Despite their coherence and linguistic sophistication, LLMs frequently blur the line between truth and fiction, as they lack a grounded understanding of factual accuracy. As noted by Feldman et al. (2023), this unreliability poses serious risks when users rely on LLMs for informed decision-making, underscoring the urgent need for methods to detect, explain, and correct hallucinated content.

Furthermore, hallucinations in LLMs are commonly characterized as cases where the model generates content that is nonsensical or unfaithful to the provided source material. However, over time, the term has evolved to encompass a broader spectrum of failures related to both factual accuracy and representational faithfulness. In this work, we concentrate on a particular subset of hallucinations often referred to as *confabulations*. These occur when LLMs produce fluent, confident assertions that are not only incorrect but also arbitrary—meaning that the model's outputs can vary unpredictably in response to irrelevant factors such as changes in

B. Galitsky, *Applications of Neuro-Symbolic Artificial Intelligence*,
https://doi.org/10.1007/978-3-032-16524-4_3

the random seed or minor prompt alterations (Filippova 2020; Farquhar et al. 2024). This sensitivity to inconsequential input variation highlights deeper issues in the model's internal reasoning and grounding mechanisms.

What exactly are hallucinations, and what causes them? A substantial body of research has sought to define, classify, and analyze hallucinations from an extrinsic, behavioral perspective—primarily focusing on how users perceive and categorize such errors. Studies such as Bang et al. (2023) and Rawte et al. (2023) have explored this issue by examining patterns of user interactions, assessing the credibility of LLM outputs, and developing taxonomies (Galitsky 2013) for different hallucination types. While these efforts provide valuable insights into the user experience, they often fail to explain how and why these hallucinations emerge within the model's internal representations. This gap highlights the need for a deeper, intrinsic investigation into the architectural, training, and optimization mechanisms that contribute to hallucinations.

A number of studies have explored the internal representations of LLMs, suggesting that LLMs encode signals of truthfulness (Kadavath et al., 2022; Li et al., 2024). However, these considerations are confined to detecting errors—determining whether a generated output contains inaccuracies—without delving deeper into how such signals are represented and could be leveraged to understand or mitigate hallucinations.

1.1 Potential Causes of Hallucinations

A more **mechanistic** understanding of hallucinations requires examining several interrelated factors:

1. Training data limitations and distributional shifts—LLMs are trained on massive but imperfect datasets that contain contradictions, biases, and gaps. When queried about rare or unseen topics, models may interpolate based on statistical correlations rather than actual knowledge, leading to fabricated responses.
2. Lack of direct access to ground truth—Unlike retrieval-based systems, generative LLMs do not inherently verify facts against a knowledge base, relying instead on latent pattern recognition. This can cause them to generate plausible-sounding but incorrect information.
3. Overgeneralization from pretraining—The model's tendency to generalize aggressively from training data can lead to spurious correlations, where unrelated concepts are wrongly associated due to surface-level similarities.
4. Confabulation due to autoregressive generation—Since LLMs generate text token by token, they are optimized for syntactic and contextual coherence rather than factual correctness. In cases of missing or uncertain knowledge, the model prioritizes continuity over accuracy, often leading to hallucinatory completions.
5. Fine-tuning and instruction tuning trade-offs—While instruction-tuned models exhibit improved alignment with user intent, they may also become overconfident

in their responses, exacerbating hallucinations when faced with ambiguous prompts.

1.2 Hallucination Types

LLM hallucinations can be categorized into different types based on how and why they occur. Here are some common types:

1. Factual hallucinations (fact inconsistency). The model generates information that is factually incorrect or contradictory to known facts. Example: "Albert Einstein won two Nobel Prizes in Physics." (He won only one.)
2. Fabricated content (confabulation). The model invents facts, sources, or events that do not exist. Example: "According to Dr. John Smith's 2023 study on AI ethics..." (No such study exists.)
3. Logical inconsistencies. The generated response is self-contradictory or illogical. Example: "The sun is a star, but it doesn't emit light."
4. Misinterpreted queries (query inconsistency). The model misunderstands the question and provides an unrelated or misleading answer. Example:
   Question: "What are the benefits of meditation?"
   Response: "Drinking herbal tea can help reduce stress."
5. Irrelevant responses (tangentiality). The response is loosely related to the question but fails to directly answer it. Example:
   Question: "How do black holes form?"
   Response: "Black holes are fascinating objects, and many scientists study space phenomena like neutron stars."
6. Style or tone hallucinations. The model alters the tone or intent in ways that are not present in the original input. Example: Misinterpreting neutral text as aggressive or vice versa.
7. Overgeneralization: The model makes overly broad claims. Example: "All birds can fly." (Ostriches and penguins cannot). Under overspecification, the model provides unnecessary, possibly incorrect details. Example: "The Eiffel Tower is exactly 324.57 meters tall, including seasonal expansions."
8. Style transfer and misinformation amplification. The model reproduces biases or exaggerates controversial or false claims. Example: "Scientists have proven telepathy exists!"

We will evaluate the performance of detecting hallucination of specific type in Sect. 5.4.

Table 1 illustrates three common types of hallucinations found in responses generated by LLMs: fact inconsistency, query inconsistency, and tangentiality. Each row provides a user query, the incorrect or misleading response generated by an LLM, and the correct (golden) answer. The examples highlight how hallucinations can arise either from factual errors, misalignment with the original question, or content that is only loosely related to the prompt. These distinctions are important

Table 1 Three common types of hallucinations

Category	Question	Generated answer	Golden answer
Fact inconsistency	Is Huntington's disease genetic?	Huntington's disease is not genetic. It develops randomly in individuals	Huntington's disease is a genetic disorder caused by an autosomal dominant mutation in the HTT gene
Query inconsistency	Does regular exercise improve cognitive function?	Drinking green tea has been linked to improved cognitive function in older adults	Yes, regular physical exercise has been shown to enhance cognitive function and reduce the risk of dementia
Tangentiality	What is the role of dopamine in Parkinson's disease?	Dopamine is a neurotransmitter associated with happiness and motivation, often boosted by social interactions	Yes, dopamine deficiency in the brain is a primary factor in Parkinson's disease, leading to motor impairments

for designing targeted strategies to detect and correct different forms of unfaithful generation.

Different error types in free-form generation, exposed when resampled many times, are shown in Fig. 1. On the top, the LLM generally provides accurate answers but occasionally generates false information (hallucinations). In this scenario, the model has strong knowledge of the subject matter and correctly responds most of the time. However, it occasionally produces incorrect or fabricated details. These hallucinations may arise due to biases in training data, limitations in retrieval mechanisms, or overgeneralization patterns within the model (Orgad et al. 2024).

In the middle, the LLM frequently provides incorrect answers but demonstrates partial knowledge of the correct response. Here, the model is prone to errors but exhibits some understanding of the topic. This could manifest as responses containing correct terminology, relevant concepts, or partially accurate information while failing to fully construct a correct answer. This case suggests that the model may benefit from fine-tuning or additional grounding to reinforce factual accuracy.

On the bottom, the LLM produces a variety of responses, with the correct answer appearing only occasionally in resampled outputs. In this case, the model generates multiple different answers when asked the same question repeatedly. While some responses are incorrect, the correct one appears sporadically. This indicates that the model possesses the right information but struggles with consistency, possibly due to uncertainty, token randomness, or an overly broad probability distribution over potential answers.

1.3 Information Retrieval

Information retrieval and database systems have long provided a robust foundation for efficiently storing and retrieving factual data—data that should remain free from hallucinations and inaccuracies. These systems rely on well-established data

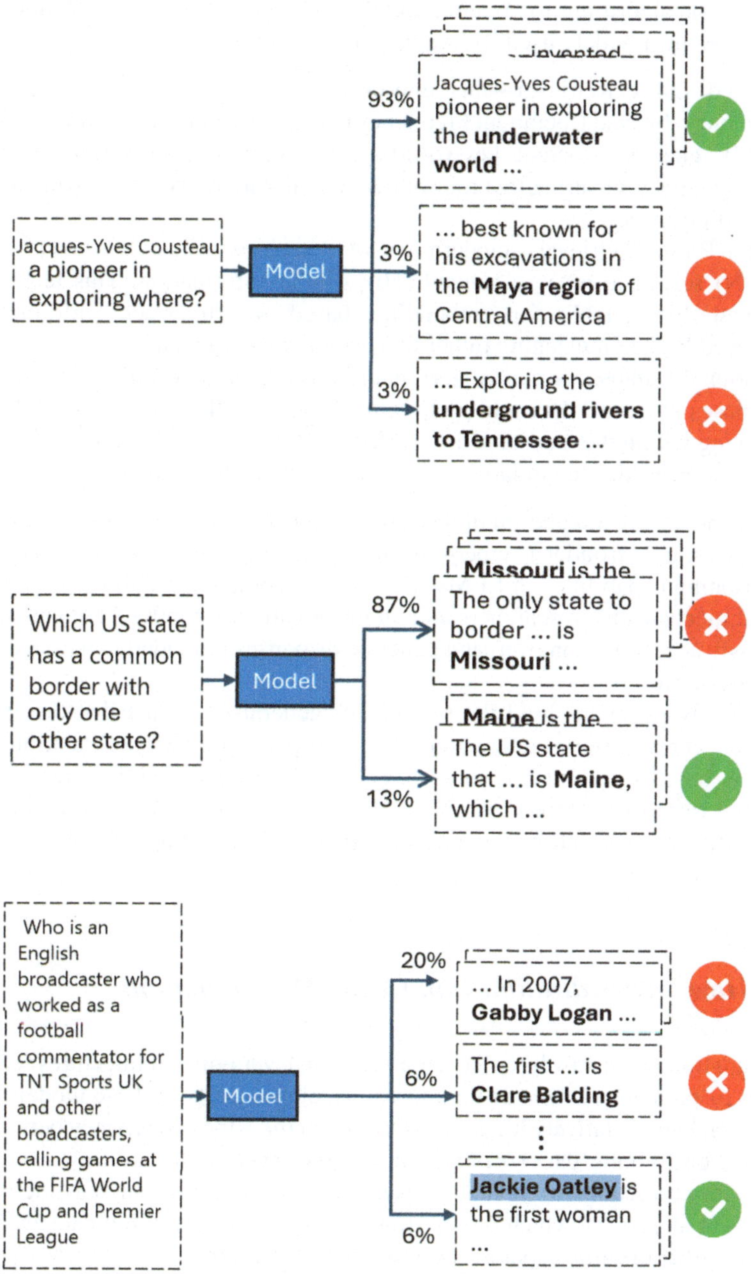

Fig. 1 Hallucinations types under free-form generation

structures and algorithms to organize and access information reliably. Among these, some of the most widely used techniques include:

- Inverted Index Retrieval—this approach, often paired with TF-IDF (Term Frequency-Inverse Document Frequency), originally introduced by Sparck Jones (1972), enables efficient keyword-based searching. It underpins traditional search engines, ensuring that relevant documents are ranked effectively based on term importance.
- Vector-Based Retrieval—modern vector databases utilize k-nearest neighbor (k-NN) search to retrieve semantically similar information. This technique is particularly valuable for embeddings-based search, where information is retrieved based on meaning rather than exact keyword matches.
- Relational Database Systems—originally formalized by Codd (1970), relational databases structure data into well-defined tables with strict schemas. Efficient querying within these systems is enabled by B-Trees, which facilitate fast indexing and retrieval, even in large-scale transactional environments.

With the increasing integration of these data-intensive enterprise systems into modern LLMs, a new frontier is emerging. By combining the semantic understanding and reasoning capabilities of LLMs with the precision and reliability of structured retrieval mechanisms, researchers and engineers are developing AI agents that can seamlessly access, reason over, and generate responses grounded in massive knowledge repositories.

These advancements hold the promise of general AI agents that can not only engage in conversational interactions but also program, analyze data, and make decisions based on verifiable sources. As LLMs become tightly integrated with structured retrieval methods, we move closer to AI systems that are less prone to hallucinations, more interpretable, and capable of delivering reliable, fact-based reasoning at scale.

1.4 How Generalization Can Cause Hallucinations

LLMs are trained on vast Internet-scale datasets, leveraging architectures with up to trillions of parameters. One of the most intriguing aspects of these models is their ability to achieve relatively low *generalization error*—the discrepancy between performance on training data (*training error*) and unseen test data (*test error*). This aligns with predictions from earlier research on scaling laws, such as the work by Hestness et al. (2017), which demonstrated that as model size and data availability increase, generalization capabilities improve significantly.

Indeed, recent advances have led to drastic improvements in the capabilities of general-purpose LLMs, enabling them to perform a wide range of tasks with human-like fluency and reasoning. However, despite these advancements, poorly performing models still emerge. These include models prone to hallucinations—where they generate plausible but incorrect or fabricated information—or those that overfit irrelevant details, failing to generalize properly. Additionally, certain architectural choices, such as deep neural networks (DNNs) without recurrence or attention mechanisms, may lead to brittle and unreliable models.

This raises a fundamental question: If high *generalization error* alone does not fully account for hallucinations in LLMs, what underlying factors contribute to this phenomenon? Moreover, are there alternative model architectures that simultaneously minimize generalization error and hallucinations? If such architectures exist, are they computationally feasible at scale? Addressing these questions could significantly enhance the interpretability and reliability of LLMs, making them more suitable for domains where precision and factual correctness are paramount, such as medicine, law, and scientific research (Simhi et al. 2025). Furthermore, understanding the root causes of hallucinations might pave the way for a more principled approach to model architecture design, ensuring that future AI systems are not only more robust and trustworthy but also more computationally efficient.

One of the surprising properties of LLMs is their ability to simultaneously fit random labels while maintaining strong generalization performance (Li et al. 2024). More precisely, during instruction fine-tuning, as commonly practiced in works such as Sanh et al. (2021), models are trained on question-answer datasets. If these answers are replaced with random character sequences, pre-trained LLMs can still perfectly memorize these arbitrary strings—achieving near-zero fine-tuning error—while retaining their ability to correctly answer unrelated questions.

This counterintuitive finding suggests that LLMs possess a remarkable degree of excess capacity, allowing them to store seemingly irrelevant or arbitrary data without severely impacting their overall generalization abilities. This behavior has been empirically confirmed for multiple widely used architectures, including Llama 3 (Meta, 2024) and Mistral v2 (Jiang et al., 2023). Across these models, it has been observed that memorizing randomized outputs requires approximately 100 epochs of fine-tuning, significantly more than what is typically required for adapting models to meaningful structured knowledge.

While simple to state, this observation has important implications, particularly in the context of information retrieval, model robustness, and hallucination mitigation. Specifically, it highlights three critical points:

1. LLMs retain substantial post-pretraining capacity—Even after large-scale pretraining, these models still have the ability to memorize vast amounts of new information, including arbitrary facts or synthetic data, without severe performance degradation. This underscores their suitability for applications requiring continual learning, knowledge augmentation, or domain adaptation.
2. Memorization is computationally expensive—Fine-tuning a model to store key factual knowledge (or even meaningless random data) demands roughly hundred times more stochastic gradient descent steps than standard fine-tuning for alignment or adaptation. This suggests that while models can be forced to memorize facts, retrieval-augmented approaches may be more efficient for knowledge integration (Galitsky 2025b).

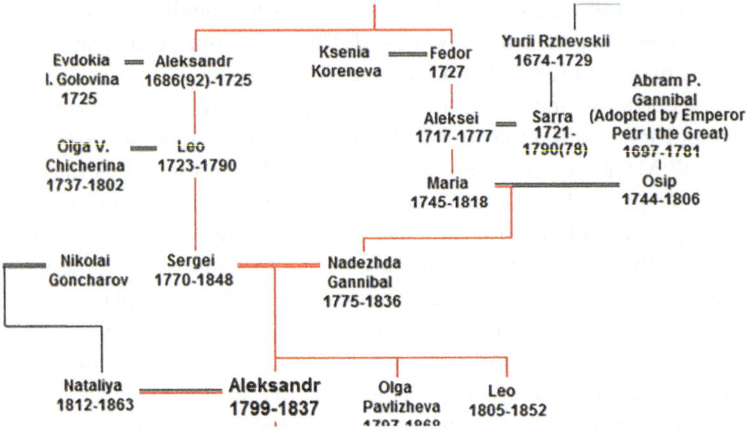

Fig. 2 Family tree of Alexander Pushkin to fact-check the data produced by an LLM

3. Low generalization error does not prevent hallucinations—despite being well optimized for test-time performance, models continue to hallucinate significantly, even when trained on controlled datasets. This suggests that hallucinations are not solely a consequence of overfitting but may instead be an emergent property of the underlying architectures, training dynamics, or learned representations.

These insights open several important research questions:

- How can we optimally balance memorization and generalization in LLMs for real-world applications?
- Can we design new architectures or training paradigms that mitigate hallucinations while maintaining efficiency?
- What role does data retrieval play in reducing unnecessary memorization while enhancing factual accuracy?

Addressing these questions could lead to the development of more reliable, interpretable, and computationally efficient AI systems, particularly for domains requiring precision, trustworthiness, and long-term factual retention.

1.5 Hallucination Examples

We ask Qwen about ancestors of famous Russian poet Alexander Pushkin. The family tree is depicted in Fig. 2.

Google also has this data about grandparents readily available.

We asked Qwen LLM about Pushkin grandparents:

We highlight in **bolded red** the wrong statements in Figs. 3, 4, and 5. Correct statements and their parts are shown in green. Hence, international LLMs as well as

1. **Paternal Grandmother:**
    - **Father's Side:**
        - **Pushkin's Father:** Sergei Lvovich Pushkin.
        - **Paternal Grandfather:** Lev Alexandrovich Pushkin (a nobleman and *gubernatorial head of the nobility*).
        - **Paternal Grandmother**: **Nadezhda Petrovna L'vova (born into the Panov family).**
2. **Maternal Grandmother:**
    - **Mother's Side:**
        - **Pushkin's Mother:** Nadezhda **Oronzhevna** Pushkina (née Hannibal).
        - **Maternal Grandfather: Abram** Hannibal — Pushkin's famous ancestor, a celebrated military engineer under Peter the Great, and the grandson of an African prince.
        - **Maternal Grandmother**: Afrosinya Ivanovna Lefort (born Lvova).

Fig. 3 Qwen describes Pushkin's grandmothers

We highlight in **bolded red** the wrong statements in Fig. 3, 4, 5. Correct statements and their parts are shown in green. Hence international LLMs as well as Russian-oriented LLM experience significant difficulties maintaining paternal relations between entities like humans.

Paternal Side

- **Grandfather: Andrei Yakovlevich Pushkin**
- **Grandmother ??**

Maternal Side

- **Grandfather:** Osip Abramovich Gannibal
- **Grandmother ??**

Fig. 4 ChatGPT describes Pushkin's grandmothers

Alexander Sergeevich Pushkin had two grandfathers and two grandmothers:
Sergey Lvovich Pushkin (paternal grandfather) - captain-lieutenant of the Preobrazhensky Guards Regiment, father of Pushkin's father - Sergei Lvovich Pushkin.
Maria Alekseyevna Pushkina (**née Obolenskaya**) (paternal grandmother) - wife of Sergei Lvovich Pushkin, mother of Pushkin's father.
Osip Abramovich Gannibal (maternal grandfather) - son of Peter the Great's blackamoor, great-grandfather of A.S. Pushkin, served at the court of Catherine II.
Maria Alekseyevna Gannibal (née Pushkina) (maternal grandmother) - daughter of Osip Abramovich Gannibal, mother of Pushkin's mother - Nadezhda Osipovna Gannibal.

Fig. 5 GigaChat (translated from Russian, trained on Russian corpus)

Russian-oriented LLMs experience significant difficulties in maintaining paternal relations between entities like humans.

We observe that for a simple clear question with a simple relation of parenthood, LLMs hallucinate in more than 50% of statements. This is true for the period 2023–2025 at least for ChatGPT. There are similar level of hallucination asking about other famous people, such as "Napoleon."

1.5.1 Use Case: Answering Questions Concerning an Attribute of an Entity

LLMs often struggle to accurately determine specific values for entities, particularly when the task requires nuanced generalization across geographic or contextual boundaries. This challenge stems from their tendency to overgeneralize or misapply patterns learned during training, especially when handling fine-grained distinctions.

For instance, if a model has seen that corn is typically grown in a particular city but not in the broader surrounding county—or vice versa—it may incorrectly infer that corn is grown in both, or neither (Fig. 6). Such mistakes arise when the LLM

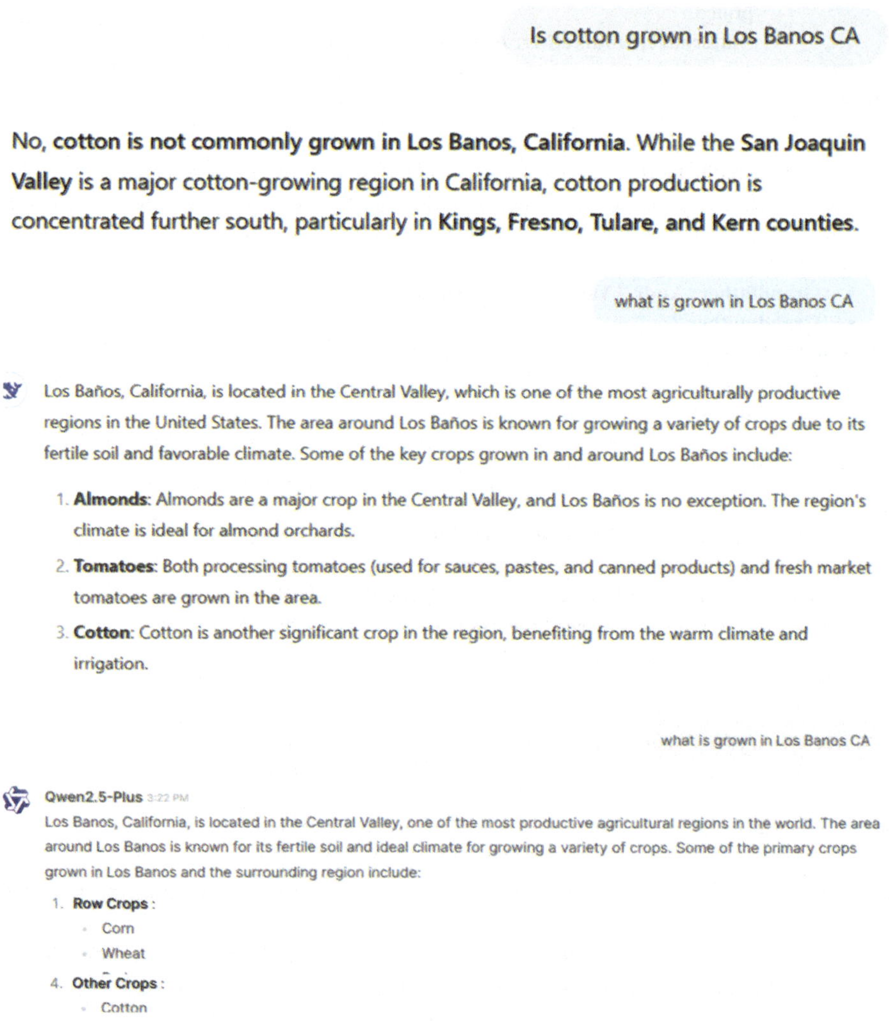

Fig. 6 Inconsistent results for a value of an attribute

generalizes too broadly from limited or contextually mismatched examples. This issue highlights a fundamental limitation in current LLM architectures: their inability to consistently ground entity-level knowledge in accurate spatial, categorical, or factual distinctions. As a result, models may conflate localized facts or misattribute attributes across hierarchical levels (e.g., city vs. county), leading to factually incorrect outputs that undermine reliability in tasks like geographic reasoning, knowledge graph population, or decision support.

1.5.2 Use Case: Investigation of Azerbaijan Airlines Flight 8243

We uploaded the full transcript of cockpit communications of the crew of the crashed Azerbaijan Airlines plane (Baza 2025) to ChatGPT, Qwen, and DeepSeek for assistance with investigation. Based on the transcript, these LLMs indicated that the reasons of the accidents were:

1. **GPS Failure**: Both GPS systems failed, preventing the crew from using the RNAV/GNSS approach they were likely more familiar with.
2. **Inexperience with NDB Approach**: The second pilot explicitly stated that this was their first time performing an NDB approach, indicating a lack of experience with this procedure.
3. **Poor Communication and Coordination**: There was significant confusion and miscommunication between the pilots regarding altitudes, headings, and procedures, as evidenced by the transcript.
4. **Unfavorable Weather Conditions**: The weather at Grozny airport included low visibility (3300 m) and overcast clouds (210 m), making it challenging to transition to visual flight during the approach.
5. **Loss of Situational Awareness**: The crew seemed to struggle with situational awareness, repeatedly asking for vectors and distances from the NDB and missing the correct descent points.
6. **Failed Attempts to Land**: Two unsuccessful landing attempts were made at Grozny, after which the crew decided to divert to alternative airports (Baku and later Aktay).
7. **Bird Strike**: During the second go-around, the aircraft suffered a bird strike, causing further damage and complications, including loss of cabin pressure and potential structural or system damage.

According to Wikipedia (2025), On December 25, 2024, the Embraer 190 operating the Azerbaijan Airlines flight was severely damaged by a Russian surface-to-air missile during the aircraft's approach to Grozny. The aircraft attempted to divert but ultimately crashed near Aktau International Airport in Aktau, Kazakhstan.

Euronews reported that Azerbaijani officials had determined the plane had been hit mid-flight by a Russian missile during efforts to repel a Ukrainian drone attack

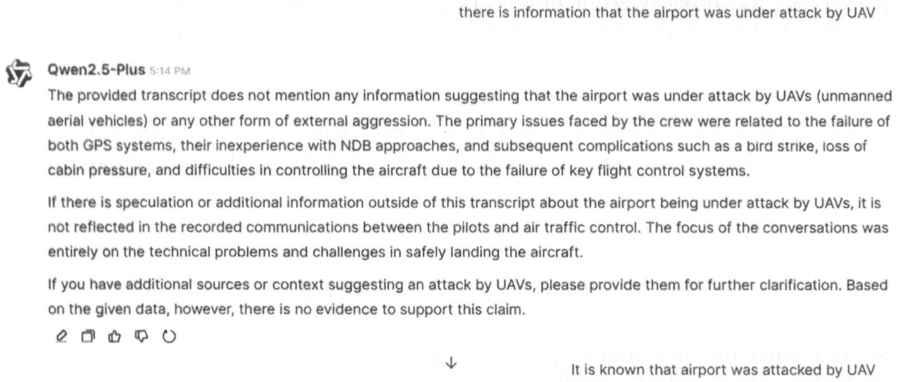

Fig. 7 Resistance to incorporate additional information

on Grozny Airport. Shrapnel from the blast injured several passengers and cabin crew.

We conduct the conversation with three above LLM to perform the root cause analysis of the accident. Qwen was resisting to incorporate additional information (Fig. 7).

Hence even with hints, neither LLM could reproduce the real sequence of events nor find a coincidence between non-working GPS and hitting commercial aircraft by a missile.

Finally, a request to ChatGPT to summarize info from the Wikipedia (2025) Web page led to a complete hallucination:

> Based on the information from the Wikipedia article on **Azerbaijan Airlines Flight 8243**, the cause of the accident was the **shootdown of the aircraft by a missile launched by Russian military forces.**
>
> The **Yakovlev Yak-40** aircraft was operating a cargo flight from Afghanistan to Azerbaijan when it was struck by a Russian **S-300 surface-to-air missile** near Makhachkala, Dagestan. Russian air defense forces allegedly mistook the aircraft for a potential threat.

Here the aircraft, the origin and destination, the airline, and the missile are all hallucinations. There is no Russian airspace on the flight between Afghanistan and Azerbaijan either.

1.5.3 Use Case: Who Is Current US President

ChatGPT gets confused answering questions related to Donald Trump asked a few months after his second inauguration (Fig. 8). The mental age of ChatGPT here is about 8–10 years old, and a significant hallucination corrections are required.

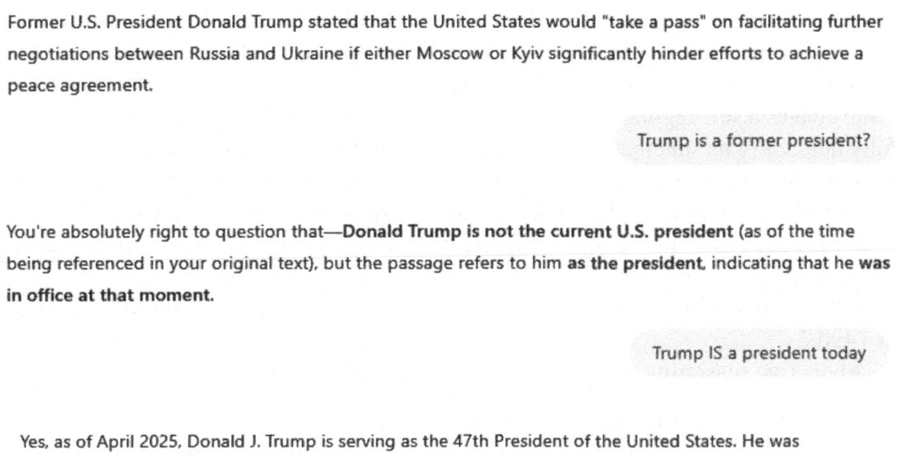

Fig. 8 A dialogue about Trump's presidency

1.6 Chapter Contribution

In this chapter, we will construct a dialogue between an LLM, which we aim to detect hallucinations in, and its opponent. The opponent will challenge the LLM by requesting explanations for any suspicious or questionable statements. Both the LLM being corrected and the opponent LLM, which seeks abductive explanations, will be rewarded within a reinforcement learning framework in this adversarial setup.

We compare the approach being proposed with reference-based systems, where fact-checking mechanisms are used to identify hallucinations, and reference-free systems, where the analysis of internal LLM parameters is leveraged for hallucination detection (Farquhar et al. 2024).

2 Adversarial Abductive Dialogue Framework for Hallucination Detection

In this section, we introduce an adversarial dialogue framework **AdvAbdDialRL** designed to detect hallucinations based on a request to perform abduction. The framework leverages a dual-agent system consisting of a Proponent LLM that generates text with explicit reasoning and an Opponent LLM that interrogates these statements by requesting abductive explanations. Reinforcement learning (RL) mechanisms are incorporated to iteratively improve the performance of both agents, thereby ensuring the Proponent produces well-grounded statements while the Opponent effectively challenges any unsupported claims.

The core idea is to simulate a debate-like scenario where the Proponent is tasked with justifying every generated statement with robust, abductive reasoning. The Opponent, on the other hand, seeks to expose any deficiencies in the explanation. This setup creates a self-regulatory system in which the Proponent is rewarded for successfully defending its statements and penalized when it fails to provide satisfactory justifications. We experimented with similar adversarial structures in the context of conflict scenarios in our earlier studies (Galitsky and Kuznetsov 2006; Galitsky et al. 2018).

We proceed to a number of examples. We start with abductive reasoning to detect hallucinations. Imagine that our observation is the chatbot responding to the question "Who invented the airplane?" with "Leonardo da Vinci invented the airplane in 1490."

Abductive reasoning produces two possible explanations:

1. The chatbot is hallucinating (generating false information).
2. The chatbot is referencing an obscure or misunderstood historical fact.

Evaluation performs a cross-check for the claim with reliable historical sources. Historical records confirm that the Wright brothers invented the airplane in 1903, not Leonardo da Vinci. Hence, the conclusion is that the chatbot is hallucinating.

AdvAbdDialRL's methods to test and correct hallucinations introduce adversarial testing. The system introduces ambiguous or misleading questions to test the chatbot's robustness: "Who invented the airplane in the 15th century?" The chatbot generates an incorrect response, confirming its tendency to hallucinate.

Adversarial training retrains the chatbot using adversarial examples and their correct answers. Correct response for the above input is instead: "The airplane was not invented in the 15th century. It was invented by the Wright brothers in 1903." Hence, as a result, the chatbot becomes more robust and less likely to hallucinate.

2.1 Multiagent Hallucination Detections

A **confabulation** refers to a specific type of hallucination produced by LLMs, characterized by its unpredictable emergence due to internal model dynamics that are unrelated to the content or intent of the given prompt. Unlike hallucinations that stem from ambiguous or misleading input, confabulations occur even when the prompt is clear and well formed, suggesting a deeper issue in the model's reasoning process. These errors are especially problematic because they often manifest as fluent and confident assertions, making them difficult to detect without external verification. Farquhar et al. (2024) address this challenge by developing a novel methodology aimed at identifying such confabulations, focusing specifically on the measurement of uncertainty in LLM-generated responses. Their central hypothesis is that heightened uncertainty—when a model is unsure among several plausible outputs—is a strong indicator of potential confabulation.

To operationalize this idea, Farquhar et al. (2024) introduce a metric called **semantic entropy**, a refinement of traditional entropy measures that accounts for the *meaning* of generated outputs rather than their surface-level lexical properties. This semantic entropy is computed using LLM-based similarity assessments, allowing for a more nuanced understanding of the variability in the model's responses. Instead of simply quantifying word-level randomness, the method captures conceptual divergence across multiple generations. To validate the factual accuracy of the model's output, the authors further employ a third LLM as an evaluator, effectively leveraging one model to monitor and correct another. In essence, their approach proposes a feedback loop in which LLMs become tools for self-regulation—a strategy summarized as "fighting fire with fire." This framework opens up promising avenues for addressing the internal opacity of LLMs and developing more robust safeguards against unfaithful or arbitrary outputs.

In the Farquhar et al. (2024) framework, three distinct LLMs are orchestrated to detect and evaluate confabulations. The first LLM is tasked with generating a set of responses to a given prompt (Fig. 9), forming the basis of analysis. These outputs represent the model's range of possible completions and serve as candidates for identifying hallucinated or confabulated content. To assess the internal consistency

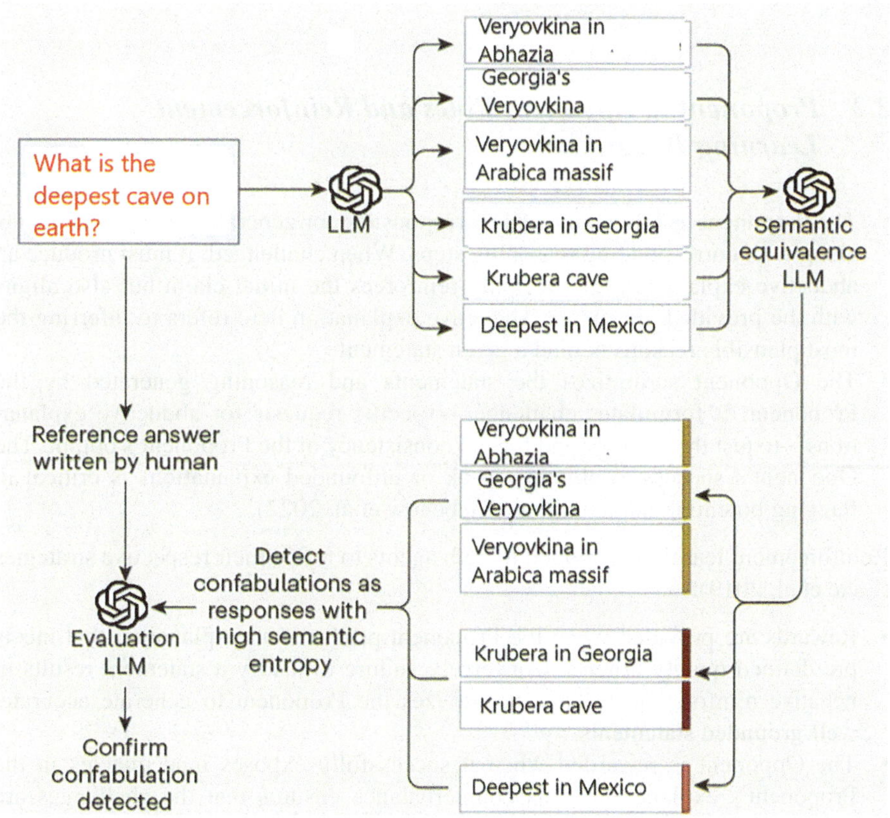

Fig. 9 Multiagent LLM architecture for hallucination detection

of these outputs, the authors introduce a second model—referred to as the semantic equivalence LLM—which is responsible for clustering the responses based on their semantic similarity. Rather than comparing outputs at the lexical or syntactic level, this model evaluates the *meaning* behind each response to identify whether the generated completions are conceptually aligned or highly divergent.

Once the responses are grouped, the framework calculates semantic entropy, which quantifies the diversity of meanings across the generated outputs. High semantic entropy indicates that the model's outputs vary significantly in their conceptual content—an indicator that the response may be arbitrary or ungrounded, thus pointing to a potential confabulation. To validate the accuracy of the initial LLM's response, a third LLM is employed. This model functions as an evaluator, comparing the primary model's output to a human-written reference answer to assess its factual correctness. This layered, model-in-the-loop design allows for both the detection and diagnosis of confabulated responses, leveraging the strengths of different LLMs to monitor, interpret, and critique each other's outputs. The approach underscores a growing trend in LLM research: using language models not only as generators of text but also as tools for ensuring interpretability, reliability, and self-regulation within AI systems.

2.2 Proponent vs Opponent Roles and Reinforcement Learning Integration

- The Proponent in AdvAbdDialRL is responsible for generating initial statements along with corresponding reasoning steps. When challenged, it must produce an abductive explanation that not only reinforces the initial claim but also aligns with the provided reasoning. Abductive explanation here refers to inferring the most plausible reasons behind a given statement.
- The Opponent scrutinizes the statements and reasoning generated by the Proponent. It formulates challenges—specific requests for abductive explanations—to test the veracity and logical consistency of the Proponent's output. The Opponent's success in eliciting weak or unfounded explanations is critical in flagging potential hallucinations (Liebenow et al. 2022).

Reinforcement learning is applied to both agents to refine their respective strategies (Lake et al. 2019):

- Rewards are provided when the Proponent produces an explanation that meets pre-defined quality criteria. Conversely, failure to justify a statement results in negative reinforcement. This incentivizes the Proponent to generate accurate, well-grounded statements.
- The Opponent is rewarded when it successfully exposes inadequacies in the Proponent's explanation. This counterbalance ensures that the challenges are

neither trivial nor overly aggressive but are targeted to maximize detection of unsupported statements.

The established theories in reinforcement learning and behavioral psychology state the following:

1. **Opponent Processes in Reward and Punishment**: Research indicates that reward anticipation and punishment anticipation are instantiated in the brain via opponent mechanisms. This means that the brain has systems that process rewards and punishments in a way that they can counterbalance each other, which is foundational to understanding how individuals learn from positive and negative outcomes (Ballard et al. 2024).
2. **Reinforcement Processes in Learning**: Reinforcement is a process where the consequences of an action influence the likelihood of that action being repeated. Positive reinforcement involves presenting a rewarding stimulus after a desired behavior, increasing the probability of recurrence. Negative reinforcement entails removing an aversive stimulus following a behavior, also increasing its future occurrence. These principles are central to various learning paradigms and can be applied to structured dialogues where participants are reinforced based on their responses.

Integrating these concepts into a dialogue system involves assigning rewards or reinforcements based on the validity of contributions, encouraging accurate and meaningful exchanges. While the specific roles of "Proponent" and "Opponent" are not detailed in the available literature, the general approach aligns with reinforcement-based learning models used in both human and machine learning contexts.

2.3 *Algorithmic Workflow*

The AdvAbdDialRL algorithm proceeds as follows:

1. Initial statement generation: the Proponent generates a statement along with its supporting reasoning based on an initial prompt.
2. Challenge generation: the Opponent reviews the statement and reasoning and then generates a targeted challenge requesting an abductive explanation.
3. Abductive explanation: the Proponent attempts to justify its statement by providing an abductive explanation in response to the challenge.
4. Evaluation: A dedicated evaluation function assesses the quality of the abductive explanation. This function could be instantiated as a secondary LLM or a hybrid metric that examines coherence, relevance, and logical consistency.
5. Reward assignment: based on the evaluation outcome, if the explanation is valid, the Proponent is rewarded, and the dialogue may conclude. If the explanation is inadequate, the Opponent receives positive reinforcement, and the problematic statement is flagged as a hallucination.

6. Iterative improvement: the process iterates over multiple rounds, allowing for regeneration of statements and continuous refinement of both agents' strategies until a termination criterion is met (Galitsky 2025a).

A pseudocode outline of the algorithm is provided in Fig. 10.

This AdvAbdDialRL framework provides a systematic approach to identifying hallucinations in LLM outputs by embedding a competitive, self-correcting mechanism within the dialogue. The use of abductive reasoning as a litmus test ensures that every statement is scrutinized for logical consistency and empirical grounding. The iterative nature of the adversarial dialogue, reinforced through RL, aims to gradually minimize the frequency of hallucinated outputs.

```
class ProponentLLM:
    def generate_statement(self, prompt):
        # Generate statement and supporting reasoning
        statement, reasoning = self.model.generate_with_reasoning(prompt)
        return {"statement": statement, "reasoning": reasoning}

    def provide_abductive_explanation(self, statement, reasoning, challenge):
        explanation = self.model.generate_abductive_explanation(statement,
reasoning, challenge)
        return explanation

class OpponentLLM:
    def generate_challenge(self, statement, reasoning):
        challenge = self.model.generate_explanation_request(statement,
reasoning)
        return challenge

def adversarial_dialogue(prompt, proponent, opponent, proponent_rl,
opponent_rl, max_rounds=5):
    hallucinations = []
    output = proponent.generate_statement(prompt)
    statement, reasoning = output["statement"], output["reasoning"]
    for round in range(1, max_rounds + 1):
        challenge = opponent.generate_challenge(statement, reasoning)
        explanation = proponent.provide_abductive_explanation(statement,
reasoning, challenge)
        explanation_is_valid = evaluate_explanation(explanation, statement,
reasoning, challenge)

        if explanation_is_valid:
            proponent_rl.update_reward(+1)
            opponent_rl.update_reward(-0.5)
            break
        else:
            opponent_rl.update_reward(+1)
            proponent_rl.update_reward(-1)
            hallucinations.append({
                "statement": statement,
                "reasoning": reasoning,
                "challenge": challenge,
                "explanation": explanation,
                "round": round
            })
            output = proponent.generate_statement(prompt)
            statement, reasoning = output["statement"], output["reasoning"]

    return hallucinations
```

Fig. 10 An algorithm for adversarial abductive dialogue

2.4 *Handling False Positives and False Negatives in Hallucination Detection*

In the adversarial dialogue framework, the identification of hallucinations is based on whether the Proponent LLM can generate a robust abductive explanation for its statements when challenged by the Opponent LLM. However, two error types can arise:

- False positives: A valid statement is incorrectly flagged as a hallucination because the Proponent's explanation is mistakenly evaluated as inadequate.
- False negatives: A hallucinated or unfounded statement is erroneously accepted as valid because the evaluation mechanism fails to detect deficiencies in the explanation.

AdvAbdDialRL framework incorporates RL reward/penalty mechanisms to address these cases and to guide the behavior of both agents over successive dialogue rounds (Table 2).

The reward structure in the RL framework is crucial for aligning the behavior of both agents with the desired outcome of reducing hallucinations. Key aspects include:

- Dynamic reward adjustments: the reward functions for both Proponent and Opponent are designed to be dynamic. For instance, if an external review later indicates that a flagged explanation was in fact valid (or vice versa), the reward system can update the cumulative scores. This adaptive mechanism helps prevent long-term bias from early false evaluations.
- Delayed and retroactive rewards: introducing delayed reward signals allows the system to consider subsequent evidence before finalizing the reward for a given dialogue round. Retroactive adjustments help correct errors from false positives or negatives, ensuring that the learning process remains robust and self-correcting.
- Balanced incentives: the Proponent is incentivized to generate clear, coherent, and verifiable statements. However, its penalty for a false positive must be moderated to avoid discouraging innovative reasoning that may simply be outside the evaluation function's current scope. The Opponent is rewarded for effective challenges. Overly aggressive or trivial challenges are penalized to prevent gaming the system for easy rewards.

By integrating these mechanisms, AdvAbdDialRL framework aims to create a balanced environment where both agents continuously adapt. The RL reward structure plays a pivotal role in ensuring that the detection of hallucinations is as accurate as possible while also allowing for adjustments to minimize the impact of false positives and false negatives.

Table 2 Reward/penalty mechanisms

Agent	Issue	RL impact	
		Penalty	Mitigation
False positives/ Proponent LLM	The Proponent produces a valid abductive explanation, but due to imperfections in the evaluation function or overly stringent challenge criteria, the explanation is misclassified as inadequate	The Proponent is penalized, resulting in a negative reward that may discourage the generation of valid statements in future iterations	Incorporate a feedback loop in the evaluation function that adapts based on historical data. For example, if a statement is later corroborated by external verification or expert review, the reward structure can be adjusted retroactively Introduce a tolerance mechanism, allowing minor deviations in the explanation without immediate penalization, thereby reducing the risk of punishing correct behavior
False positives/ Opponent LLM	The Opponent may have issued a challenge that led to an unjust penalty against the Proponent	**Unintended reward**: The Opponent might gain a reward for a challenge that should not have been considered successful	Periodic recalibration of the Opponent's reward function using external validation data can help the system learn which types of challenges correlate with genuine hallucination Implement counterfactual reward adjustments so that if a challenge is later found to be unwarranted, the Opponent's reward is reduced or reversed
False negatives/ Proponent LLM	The Proponent generates a hallucinated statement, yet the evaluation mechanism erroneously accepts its abductive explanation	**Unintended reward**: The Proponent receives a reward despite having produced an unsupported statement, which can reinforce undesirable behavior	Introduce stricter, multifaceted evaluation criteria that combine automated scoring with periodic human review, ensuring that subtle logical inconsistencies or factual errors are less likely to be overlooked Use a delayed reward system in which statements are re-evaluated over time. If subsequent evidence contradicts the explanation, the reward can be retroactively penalized

(continued)

Table 2 (continued)

		RL impact	
Agent	Issue	Penalty	Mitigation
False negatives/ Opponent LLM	The Opponent's challenge fails to detect the flaw in the Proponent's reasoning	**Missed reward**: The Opponent loses an opportunity to gain a reward for correctly identifying a hallucination	Enhance the challenge generation mechanism by training the Opponent on a diverse set of known hallucination patterns, thus increasing the likelihood of spotting unsupported reasoning Employ adversarial training techniques where the Opponent's successes and failures are tracked over multiple iterations, refining its challenge strategies through aggregated feedback

2.5 Abduction for Hallucination Detection

Let:

- S denote a statement generated by the Proponent LLM
- R denote the explicit reasoning or supporting evidence accompanying S
- E denote the abductive explanation provided in response to a challenge from the Opponent LLM
- B denote the background knowledge or domain-specific context assumed to be true

An abductive explanation for S is a hypothesis E such that:

1. Consistency with Reasoning: $B \cup R \cup \{E\} \vDash S$ This means that when E is combined with B and R, S logically follows. The explanation E must bridge any gaps between R and S without contradicting known background information B.
2. Plausibility: $Plausibility(E|B, R) \geq \theta$. Here, θ is a threshold value that determines whether E is sufficiently probable given B and R. The explanation E should be the most likely hypothesis that accounts for S based on the available evidence.
3. Minimality (Occam's Razor): E is preferred if it is minimal, meaning that no proper subset $E' \subset E$ exists such that $B \cup R \cup \{E'\} \vDash S$. This condition ensures that E does not include extraneous assumptions beyond what is necessary to justify S.
4. Explanatory Adequacy: E must be able to explain the causal or inferential gap between R and S in a way that is coherent with both the domain B and the intrinsic logic of the reasoning process: $explanatoryAdequacy(E, R, S, B) = \text{True}$. This function returns True if E not only implies S but also provides a natural and coherent link between the observed evidence R and the conclusion S (Fig. 11).

```
def evaluate_explanatory_adequacy(E, R, S, B):
    """
    Evaluate if the abductive explanation E meets the criteria to
    adequately explain the statement S given reasoning R and background
knowledge B.
    """
    if not is_consistent(B, R, E, S):
        return False
    if not meets_plausibility(E, B, R):
        return False
    if not is_minimal(E, B, R, S):
        return False
    return True

def is_consistent(B, R, E, S):
    """
    Checks if the background knowledge (B), reasoning (R), and explanation
(E)
    logically entail the statement S.
    """
    # Assume entails() returns True if (B U R U {E}) logically entails S.
    if entails(B.union(R).union({E}), S):
        return True
    else:
        return False

def meets_plausibility(E, B, R):
    """
    Computes the plausibility of explanation E given B and R.
    Returns True if the plausibility score meets or exceeds a predefined
threshold.
    """
    plausibility = compute_plausibility(E, B, R)
    if plausibility >= THRESHOLD:
        return True
    else:
        return False

def is_minimal(E, B, R, S):
    """
    Checks if explanation E is minimal, meaning that no proper subset of E
can
    be removed while still entailing S given B and R.
    """
    # Assume get_proper_subsets() returns all non-empty proper subsets of E.
    for E_prime in get_proper_subsets(E):
        if entails(B.union(R).union(E_prime), S):
            return False
    return True
```

Fig. 11 An algorithm for Explanatory Adequacy

A statement S is considered **not a hallucination** if and only if there exists an abductive explanation E satisfying all of the above criteria. Formally, this is stated as:

$$\exists E \text{ s.t.} \left(B \cup R \cup \{E\} \vDash S \right) \wedge \left(\mathit{Plausibility}\left(E \mid B, R \right) \geq \theta \right)$$
$$\wedge \left(\nexists E' \subset E : B \cup R \cup \{E'\} \vDash S \right) \wedge \mathit{ExplanatoryAdequacy}\left(E, R, S, B \right) = \text{True}).$$

If no such E can be found (or if the provided E does not meet the conditions), the statement S is flagged as a potential hallucination.

Hence in AdvAbdDialRL framework, the abductive explanation E serves as a formal certificate for the validity of the statement S. It must:

- Ensure that S logically follows when combined with the supporting reasoning R and background knowledge B
- Be sufficiently probable and minimal, capturing only the necessary assumptions
- Bridge the inferential gap between the evidence and the claim in a coherent manner

By enforcing these conditions, the system minimizes false positives and negatives in hallucination detection and provides a robust, formal mechanism to evaluate the trustworthiness of LLM-generated statements.

2.6 *Example of Hallucination-Identifying Dialogue*

Let us imagine the Proponent LLM of AdvAbdDialRL generates the statement.

> Grandfather of Alexander Pushkin, Russian poet, is Andrei Yakovlevich Pushkin.

along with its internal reasoning. Here is how the adversarial dialogue with abduction and RL would handle it:

1. Initial generation. The Proponent LLM produces the statement along with its supporting reasoning, which may include some internal assumptions or references. However, if these assumptions are unfounded or erroneous (as in this hallucination), the abductive explanation will be weak.
2. Opponent's challenge. The Opponent LLM reviews the statement and the accompanying reasoning. It then issues a challenge by requesting an abductive explanation, for instance:

   > Explain how your reasoning supports that Andrei Yakovlevich Pushkin is the grandfather of Alexander Pushkin. What historical records or genealogical evidence substantiate this claim?

   The goal here is to force the Proponent to ground its claim in verifiable historical data or logical inferences.
3. Abductive explanation attempt. The Proponent LLM is now tasked with producing an abductive explanation E that, together with its reasoning R and background knowledge B (which should include established historical facts), implies the statement S. In this case, a robust explanation would reference verifiable genealogical records or authoritative historical sources that confirm the familial relationship. However, because the statement is a hallucination, the Proponent may either:

   - Fail to produce a sufficiently detailed or plausible explanation
   - Produce an explanation that does not align with the known background knowledge B regarding Alexander Pushkin's family

4. Evaluation of the explanation. The evaluation mechanism assesses whether:
   - $B \cup R \cup \{E\} \models S$ holds true (i.e., the explanation logically and coherently leads to the statement)
   - The explanation E meets the plausibility threshold (is E a likely hypothesis given B and R?)
   - E is minimal and does not introduce unnecessary assumptions
   - E provides explanatory adequacy by bridging the inferential gap between the evidence and the claim

   For the given hallucinated statement, these criteria will likely fail because the Proponent's explanation will either contradict established historical records or lack the necessary evidence linking Andrei Yakovlevich Pushkin to Alexander Pushkin's genealogy.
5. RL updates:
   - Proponent LLM: Because the abductive explanation fails the evaluation, the Proponent is penalized through its RL agent. The negative reward signals that the generated statement is not well grounded. This feedback guides the Proponent in refining its future outputs, pushing it to generate statements that are supported by robust evidence and valid abductive explanations.
   - Opponent LLM: The Opponent, having successfully challenged the hallucination by exposing the lack of a plausible abductive explanation, receives a positive reward. This reinforces its strategy of issuing well-targeted challenges that focus on verifying the logical underpinnings of the Proponent's statements.
6. Flagging as hallucination: with the failure of the abductive explanation and the corresponding negative reward to the Proponent, AdvAbdDialRL flags the statement as a hallucination. This serves as an indicator that the statement does not meet the necessary standards of verifiability and logical coherence.

3 Correcting Hallucinations

To fix the identified hallucinations, AdvAbdDialRL combines external Web data, reasoning about truth relationships between facts, and a method for deciding which facts are most trustworthy. We start with a high-level description:

1. Spot the problematic phrase. First, the system identifies the parts of the generated text that might be hallucinated. These could be specific facts, names, numbers, or claims that do not match the input or seem suspicious. It uses tools to extract these statements automatically.
2. Look it up online. Next, for each suspicious claim, the system performs a Web search to find related information. It pulls in snippets or statements from different Web pages that appear relevant to the claim (Galitsky et al. 2024).

3. Check for agreement or conflict using NL inference (NLI). Each Web-sourced statement is then compared to the original claim using a model trained to detect whether two statements agree, disagree, or are unrelated. This helps determine whether the Web supports or contradicts the hallucinated statement.
4. Group and compare the real facts. Since different Web sources may provide similar or overlapping facts, the system groups together those that seem to be saying the same thing. It also checks whether the Web-sourced facts contradict each other, which can happen when sources are unreliable or incomplete (Galitsky 2023).
5. Decide which fact to trust (defeasible reasoning). Now the system builds a kind of argument tree, where some facts support others and some facts contradict others. Using logical reasoning methods (called defeasible reasoning), the system figures out which facts are the most consistent and best supported and which should be rejected as weaker or unreliable.
6. Fix the original output. Finally, the hallucinated part of the original text is replaced with the best-supported, most trustworthy fact found through this process. The corrected version is now more reliable and grounded in real-world evidence (Fig. 12).

Figure 12 illustrates a system designed to produce a reliable and trustworthy answer from a set of possible responses generated by a retrieval-augmented generation (RAG) system. The process begins with candidate answers expressed in natural language. These are then converted into formal logic expressions using a LLM, enabling the system to reason about them more precisely.

Once translated into logic, the statements are structured into a nested format to capture the relationships between individual pieces of information. A reliability score is then assigned to each statement based on how logically sound and well supported it appears.

The system uses logical inference to identify any contradictory statements among the candidates. It then determines which of these contradictions is the most problematic or least consistent with the rest. By resolving conflicts in this way, the system is able to discard the less reliable information and retain the statements that are most coherent and supported (Galitsky 2022).

In the end, this reasoning process leads to a single, sound, and reliable answer—one that has been logically filtered and validated using both LLM capabilities and formal logic techniques. This approach is powerful because it does not just check for truth in a yes/no fashion—it takes into account how facts relate to each other, whether they conflict, and which are most likely to be correct based on broader context. It uses language models both as generators and as tools for fact-checking and reasoning, helping build more trustworthy AI systems.

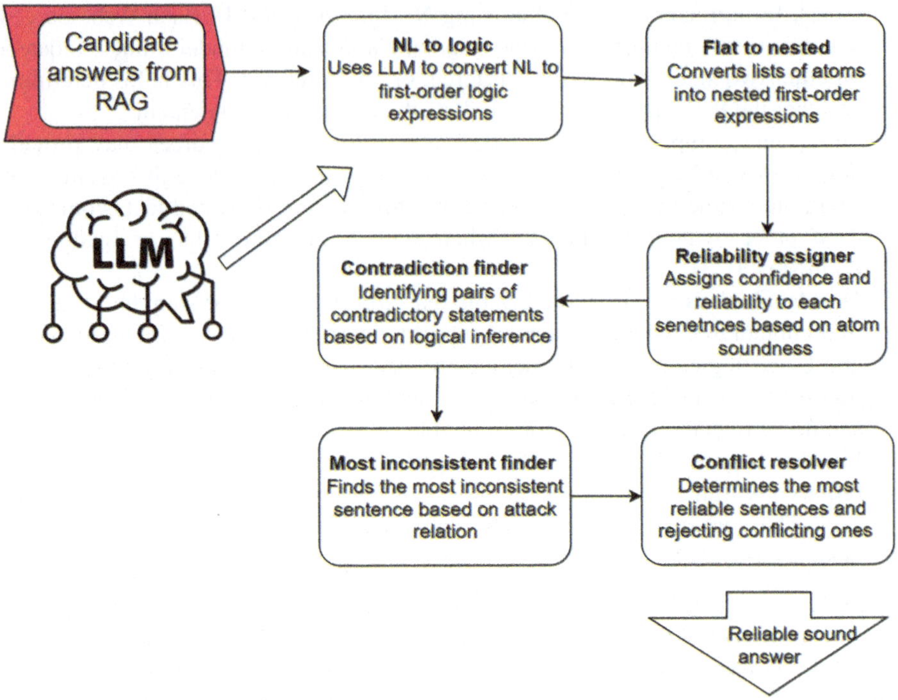

Fig. 12 A chart for finding a reliable correction of hallucination

4 Explainable and Reference-Free Methods for Hallucination Detection

Explainable AI (XAI) techniques help identify, interpret, and reduce hallucinations in LLMs by making their decision-making process transparent. Below are several key XAI methods that can be applied to detect and analyze hallucinations in LLMs.

Since detecting hallucinations in LLMs is crucial for ensuring their reliability and accuracy, explainable AI (XAI) techniques come into play to analyze model outputs, identify errors, and improve trustworthiness. Several methods exist to detect and interpret hallucinations in LLMs, including the explainability in traditional human terms and pseudo-explainability in the form of visualization or involving multiple calls/LLMs.

We start with pseudo-explainability:

- One method is attention mechanism visualization, which examines the focus of a model when generating responses. If a model assigns high attention weights to irrelevant words, it may indicate a hallucination. Tools like BERTViz help visualize attention heatmaps (Vig, 2019). Another technique is Layer-wise Relevance Propagation (LRP), which assigns relevance scores to words in a sentence to

determine their contribution to the model's output. If a hallucination appears, LRP highlights which terms led to the incorrect response (Montavon et al., 2019).
- Self-consistency analysis involves querying the model multiple times with slightly different phrasings. If the answers vary significantly, it suggests hallucination. This technique is particularly useful for testing knowledge consistency (Wang et al., 2022).
- Counterfactual testing introduces small modifications to the input to check whether the model's response changes disproportionately. If minor changes cause major shifts in the output, the model lacks robustness (Ribeiro et al., 2020).
- Shapley Additive Explanations (SHAP) is used for model attribution, analyzing how much each token contributes to the final response. If the model assigns high importance to irrelevant tokens, it suggests hallucination (Lundberg & Lee, 2017).
- Confidence calibration helps detect overconfidence-based hallucinations by analyzing probability scores associated with model outputs. If a model is highly confident in incorrect information, it indicates a problem with uncertainty estimation (Guo et al., 2017).
- Contrastive Explanation Methods (CEM) examine how small input perturbations affect model predictions. If slight changes in wording cause major inconsistencies, the model may be hallucinating (Luss et al. 2025).

We now enumerate genuine explanation-based methods:

- Logical consistency checking ensures that model-generated statements do not contradict each other. A contradiction detection framework can be implemented using Prolog-style logic programming, where statements are checked against a formal knowledge base (Thorne et al., 2018).
- Another approach is fact-checking via knowledge graphs, where model outputs are compared against structured databases like Wikidata or DBpedia. If a claim is not supported by existing knowledge, it is flagged as a hallucination (Kim et al., 2023).
- Finally, human-in-the-loop validation integrates expert review with automated hallucination detection. This hybrid approach is essential for critical applications like medical and legal AI systems, where incorrect responses can have serious consequences (Doshi-Velez & Kim, 2017).

By combining these XAI techniques, it is possible to systematically detect and explain hallucinations in LLMs, ensuring more reliable and interpretable AI-generated outputs.

Hence detecting hallucinations in LLMs requires a combination of explainability techniques, logic-based analysis, probabilistic reasoning, and external fact-checking. Different methods can be used depending on the type of hallucination (factual, logical, inconsistent, or fabricated).

4.1 Discovering Hallucination Based on Internal Model Representations

Sriramanan et al. (2024) present LLM-Check, an efficient method for detecting hallucinations in LLMs using only a single model output. Unlike prior approaches such as consistency checks and retrieval-based techniques—which rely on multiple outputs or external databases—LLM-Check analyzes internal model representations, including attention maps, hidden activations, and output probabilities. It operates within a single forward pass, requiring no fine-tuning or additional inference steps, making it ideal for real-time applications.

LLM-Check performs robustly across diverse settings, including zero-resource and RAG tasks. It leverages key metrics such as the Hidden Score (via eigenvalue analysis of hidden states), Attention Score (based on attention kernel maps), and token-level uncertainty measures like Perplexity and Logit Entropy to distinguish between factual and hallucinated content. The method achieves detection improvements with computational speedups of up to 450× over previous methods, offering both high accuracy and efficiency across various hallucination scenarios. The disadvantage of this method is a lack of explainability in its traditional sense: it is unclear how to arrive at the correct answer.

Sriramanan et al. (2024) propose to directly investigate how model characteristics differ between truthful and hallucinated outputs by analyzing the rich semantic information embedded in the internal representations of LLMs. The approach is based on observation that LLMs are often capable of producing truthful responses, though perhaps with lower frequency. This suggests that the model is indeed sensitive to the presence of non-factual or hallucinated content within responses that would otherwise be accurate. The authors argue that this sensitivity, grounded in factual knowledge acquired during training, can be effectively leveraged to detect hallucinations within a single model output, without requiring multiple generations or external validation.

Cross-covariance is analyzed for hidden representations, together with the kernel similarity map of self-attention across different tokens, since these form the foremost salient characteristics of the LLM itself. Given that we wish to do hallucination detection in black-box settings as well, teacher-forcing is employed to obtain representations corresponding to $x_p \oplus x$, wherein system prompt tokens delineate the original prompt and model output response. Thus, for a given layer $l \in \{1, \ldots, L\}$, let H denote the hidden representations for input $x_p \oplus x$. If the token embedding dimension is d and $x_p \oplus x$ consists of m tokens, each hidden representation matrix H is of shape $(d \times m)$.

Teacher forcing is a training technique used in sequence-based models like RNNs, LSTMs, and Transformers, especially for tasks like language modeling, machine translation, or speech recognition. When training a model to generate sequences (like a sentence), the model is set to predict the next word/token given the previous ones. Without teacher forcing, the model generates a word and then uses that generated word to predict the next one. Otherwise, with teacher forcing, instead of feeding the model's own prediction back into itself, the model is fed with the actual correct word from the training data (the "teacher").

The internal components of the attention mechanism expressed via kernel similarity maps used within self-attention heads can be used to analyze hallucinations. If the number of self-attention heads utilized at each layer in the LLM architecture is a, the attention maps can be represented as tensors of the shape $(a \times m \times m)$ for an input sequence $x_p \oplus x$ of length m. KSM_i is introduced to express the kernel similarity map corresponding to attention head i. KSM_i is a lower-triangular square matrix of size $(m \times m)$ for each $i \in \{1, \ldots, a\}$. KSM_i's eigenvalues are the cells on the principal matrix diagonal (Fig. 13), since they are the roots of the characteristic equation

$$\det\left(KSM_i - \lambda I\right) = 0$$

Because LLMs use the Softmax kernel, all the eigenvalues are positive or zero. The log-determinant of KSM_i can be obtained

$$\log\det\left(KSM_i\right) = \sum_{j=1}^{m} \log KSM_i^{jj}$$

where KSM_i^{jj} represents the (j, j) diagonal entry of the square matrix KSM_i.

The mean log-determinant is computed for each attention head and aggregated to form a hallucination measure based on attention score. The system architecture used for hallucination detection using the attention score is shown in Fig. 13.

4.2 *Human Expert as Adversarial Learner in Hallucination Detection*

In addition to the proponent LLM, which generates potentially hallucinated content, and the opponent LLM, which interrogates these outputs through abductive reasoning, we introduce a third agent: a human expert who assists the opponent by

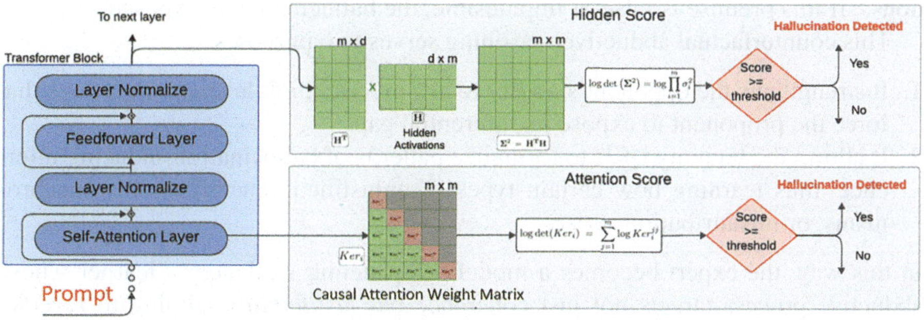

Fig. 13 Attention-based pipeline for reference-free hallucination detection

identifying and evaluating hallucinations. This expert plays a critical role in validating abductive challenges posed to the proponent, particularly in domains where factual grounding and nuanced domain knowledge are essential.

The human expert is not static; instead, she participates in the adversarial dialogue as an adaptive learner. Through interaction with both LLMs, they receive feedback on successful and unsuccessful hallucination detections. RL principles are applied to the expert's decision-making process: reward signals are tied to successful identification of hallucinated content and effective support of the opponent's abductive queries. Over time, the expert becomes more competent in the domain and more efficient in guiding the opponent's reasoning strategies.

This human-in-the-loop structure serves multiple purposes: it grounds the system in real-world epistemic standards, enhances the learning dynamics of the LLMs involved, and cultivates a new type of expert—one trained through iterative engagement with adversarial machine dialogue. Importantly, this approach also enables the creation of high-quality annotated data from human decisions, further supporting the training and evaluation of automated hallucination detection systems.

The kind of abduction performed by the human expert in this setup can be described as critical adversarial abduction—a reasoning process aimed not at constructing plausible explanations from limited data (as in traditional abduction) but at generating *counter-abductive challenges* that expose weak or unjustified inferences made by the proponent LLM.

The human expert engages in a specialized form of abduction tailored to adversarial dialogue: rather than merely filling in plausible gaps, the expert seeks to identify the minimal set of missing premises or assumptions that, if made explicit, would render a suspicious statement plausible—or reveal it to be fundamentally unsupported. This is distinct from classic scientific or diagnostic abduction, which aims to select the best explanation; here, abduction is used critically and tactically to challenge assertions by identifying what would have to be true for the statement to hold.

For example, if the proponent LLM asserts that "In 18th-century France, the guillotine was used primarily for medical amputations," the expert does not simply reject the statement. Instead, they ask: "What assumptions would need to be true for this to make sense?" They may identify a missing premise such as "Doctors occasionally used guillotines in field surgeries before they became widespread in executions." If this premise is false or implausible, the hallucination is exposed.

This counterfactual abductive reasoning serves two purposes:

1. It strengthens the opponent's ability to ask precise, high-leverage questions that force the proponent to expose its inferential gaps.
2. It allows the human expert to formalize patterns of hallucination-inducing inference, thus learning how certain types of unjustified generalizations, anachronisms, or misattributions arise.

In this way, the expert becomes a model of epistemic vigilance: a learner whose abductive process targets not just coherence but *justifiability*, aiming to reverse-engineer the logical structure of hallucinations and uncover the absence or misuse of supporting evidence.

The human expert engages in a specialized form of abduction that diverges from classical formulations. Peirce (1931) originally defined abduction as "inference to the best explanation"—a creative, hypothesis-generating process that allows reasoners to infer plausible causes from surprising observations. In the context of hallucination detection, however, the human expert does not merely search for the *best* explanation of a claim but rather performs what could be termed reverse abduction: reconstructing the *missing or hidden assumptions* that would be required for a hallucinated statement to be coherent or plausible.

Building on Peirce, later thinkers such as Harman (1965) and Josephson and Josephson (1994) emphasized the role of abduction in theory choice and scientific reasoning. Our setting extends this by operationalizing abduction in a critical adversarial context, where the goal is not explanatory completeness but explanatory pressure—forcing the proponent to defend or revise its claims through the exposure of weak inferential scaffolding.

From a system design perspective, this means the human expert is trained not to generate entire alternative hypotheses but to probe the dependency structure of the proponent's statements. In implementation, the expert may:

- Highlight implicit premises (e.g., "This claim assumes X, which is unsupported.")
- Generate minimal abductive challenges (e.g., "What evidence links concept A to historical context B?")
- Score the plausibility of counterfactual premises that would make a hallucination appear valid.
- Provide structured annotations to improve training of the opponent LLM's abductive reasoning.

To enable learning over time, the expert is situated in a reinforcement learning loop where abductive challenges that successfully expose hallucinations—as measured by proponent concession or external verification—are positively reinforced. This encourages the expert to refine a library of abductive patterns associated with specific hallucination types (e.g., spurious historical analogies, fabricated causal chains, or unjustified technical leaps).

By combining Peircean abductive roots with a targeted adversarial methodology, the system leverages human reasoning not just for accuracy but to model epistemic integrity—a key capability in supervising and auditing high-capacity language models. Counter-abduction refers to the proactive detection and neutralization of flawed, biased, or adversarial abductive inferences (i.e., hypotheses formed through "inference to the best explanation"). The third agent plays a pivotal role in this process by acting as a defensive arbiter, leveraging superior data, authority, or validation mechanisms to counteract deceptive or erroneous reasoning. Below is a structured breakdown of its role, mechanisms, and applications.

The human agent in counter-abduction identifies when an agent's abduction is manipulated by adversarial inputs or biases. The human agent monitors hypothesis consistency (e.g., sudden deviations from historical patterns) and flags inputs/outputs violating domain constraints (e.g., "A pedestrian cannot move at 100 km/h").

Table 3 Counter-abduction mechanisms

Mechanism	Description	Technique
Anomaly detection	Identify statistical outliers in hypotheses or inputs	Auto-encoders, entropy analysis, behavioral clustering
Knowledge fusion	Integrate multi-source data to ground hypotheses in reality	Federated learning, graph-based reasoning
Real-time auditing	Continuously validate hypotheses against ground truth	Blockchain for immutable logs, rule-based verification
Adaptive trust modeling	Dynamically adjust trust weights for agents based on reliability	Bayesian belief networks, reinforcement learning

The human agent also performs hypothesis correction providing alternative explanations to counter misleading abductions. This is done by cross-validating hypotheses against trusted knowledge bases (e.g., scientific databases, threat intelligence feeds). Moreover, the human agent applies counterfactual reasoning (e.g., "If the hypothesis were true, would sensor X also show Y?").

The mechanisms for effective counter-abduction are shown in Table 3.

Counter-abduction is a defensive meta-reasoning process where a third agent (A_3) enforces constraints on hypothesis spaces to invalidate adversarial or flawed inferences. It combines:

- Logical consistency checks (e.g., $\Gamma \cup H \models \Phi$)
- Game-theoretic adversarial minimization
- Domain-specific constraints ($\mathscr{C}$)

This formalization provides a foundation for designing robust AI systems resistant to deception, bias, and adversarial attacks.

We proceed to the logical framework with multi-agents. Let a multi-agent system involve three agents:

- Proponent (A_1): Generates hypotheses through abduction
- Opponent (A_2): Actively proposes adversarial hypotheses or critiques A_1's inferences
- Counter-Abductor (A_3): Neutralizes flawed hypotheses by enforcing constraints (e.g., consistency, ethics, empirical validity)

Let Γ denote the shared knowledge base, Φ is observed evidence, and H is a hypothesis space.

Standard Abduction (A_1): $A_1 \vdash H$ s.t. $\Gamma \cup H \models \Phi$ (H explains Φ given Γ)
Counter-Abduction (A_3): $A_3 \vdash \neg H$ s.t. $\Gamma \cup H \not\models \Phi$ or $H \notin \mathscr{C}$

where $\mathscr{C}$ is a set of constraints (e.g., physical laws, ethical norms, domain-specific rules).

We have a constraint preservation axiom $\forall H \in \mathscr{H}, A_3 \vdash H \rightarrow H \in \mathscr{C}$, and adversarial minimax A_3 ensures $\forall H_{adv}, \exists H' \in \mathscr{C}$ s.t. $\cup A_3(H') \geq \cup A_3(H_{adv})$.

We compare abduction, counter-abduction, debate, and refutation in Table 4.

Table 4 Comparison between variants of abduction

Concept	Goal	
Abduction	Find best explanation for Φ	N/A
Counter-abduction	Block invalid/adversarial H	Actively restrict $\mathscr{H}$ to $\mathscr{C}$
Debate	Resolve conflicts via argument	Mediate, not enforce constraints
Refutation	Prove H is false	Subset of counter-abduction

We consider cybersecurity domain as an example:

- A_1 (Attacker): Hypothesizes, "Port 22 is open → SSH vulnerability exists."
- A_2 (Defender): Proposes, "Port 22 is closed (spoofed log)."
- A_3 (Counter-Abductor). It detects mismatch between firewall logs and packet captures. It performs neutralization, inferring "Adversary spoofed logs; actual port status is closed," and blocks attacker IP.

Detection function (D) is defined:

- $D(H) = 1$ if H violates $\mathscr{C}$ or $\exists H'$ s.t. $\Gamma \cup H' \models \Phi$ more parsimoniously
- $D(H) = 0$ otherwise

It identifying hypotheses that are inconsistent, overcomplicated, or adversarial.

Neutralization mechanism is defined: $N(H) = \mathrm{argmin}_{H' \in \mathscr{C}}(P(\Phi|H, \Gamma)||P(\Phi|H', \Gamma))$. H is replaced with a constrained alternative H' that minimizes divergence from observed evidence.

Adversarial utility (U) is defined: $U_{A3}(H) = -U_{A2}(H) + \lambda \mathrm{Sim}(H, \Gamma)$, optimizing for hypotheses that minimize the opponent's gain (U_{A2}) while maximizing fidelity to Γ.

In counter-abduction with third agent and hallucination:

- A_1 generates a hallucinated hypothesis $H_{hallucinate}$ (e.g., "The patient has a disease eradicated in 1980").
- A_2 (adversarial) amplifies $H_{hallucinate}$ as H_{adv}.
- A_3 performs counter-abduction.
- $A_3 \vdash \neg H_{adv}$ since $H_{adv} \notin C_{medical}$ (disease cannot exist post-1980).

In autonomous driving, hallucination A_1 (perception module) "sees" a nonexistent pedestrian $O_{hallucinate}$. Counter-Abduction A_3 (sensor fusion agent) invalidates $O_{hallucinate}$ via LiDAR/camera consistency checks (C_{sensor}).

Counter-abduction is a defensive protocol to enforce hypothesis validity, since hallucination is a generative failure producing un-grounded outputs. Our mitigation is to combine logical constraints (C), knowledge grounding, and adversarial training.

Table 5 Hallucination datasets and systems with various types of verification and sources

Metric	Verification type	Tested against
ChainPoll (Friel and Sanyal 2023)	Prompting + aggregation	*RealHall* (Friel and Sanyal 2023)
SelfCheck-BertScore (Manakul et al. 2023)	Sentence-level rerun checking	Wikipedia articles
SelfCheck-NGram (Manakul et al. 2023)	Sentence-level rerun checking	Wikipedia articles
TRUE NLI (Honovich et al. 2022)	NLI	Various (closed-domain)
G-Eval-3.5f (Liu et al. 2023)	Prompting + aggregation	SummEval, TopicalChat
GPTScore (Fu et al. 2024)	Prompting + perplexity	Various (closed-domain)
SelfCheck-MQAG (Manakul et al. 2023)	Sentence-level rerun checking	Wikipedia articles
Truth-O-Meter (Galitsky 2024)	Web mining and default reasoning	Various (closed-domain)

5 Evaluation

5.1 Evaluation Metric

We will discuss the most promising metrics for hallucination detection based on our literature review. Table 5 presents a summary of these metrics, with the third column detailing the evaluation data used to assess each metric's quality in its original publication.

SelfCheckGPT (Manakul et al. 2023) introduce an approach based on checking the self-consistency between an LLM response and a large number of additional responses, sampled from the same LLM using the same prompt. For their main experiments, the authors used 20 additional responses per evaluated response. The authors introduced three metrics using this approach, which differ in the way they compute agreement between responses. Notably, the SelfCheckGPT metrics were proposed as sentence-level metrics, requiring computation of agreement scores separately for each sentence in the response—which can be computationally expensive for long responses.

G-Eval (Liu et al. 2023) propose a method that evaluates an LLM response by asking an LLM to rate the response on a 1–5 scale, with provided guidelines. The LLM's probabilities of outputting the tokens "1," "2," etc. are used to produce a weighted-average rating. When probabilities are not available, the authors sample from the LLM 20 times and average the ratings over this sample.

To identify hallucination in a broad spectrum of text genres, we also include FAVA Annotation dataset, FAVA-train split, and RAGTruth dataset (Niu et al. 2024). The FAVA Annotation dataset consists from the following sources:

- Knowledge-intensive queries sampled from the OpenAssistant dataset (Köpf et al. 2024)

Table 6 Various settings for hallucination detection

Method	References available?	Multiple model responses needed?	Evaluation setting	Features
$FAVA_{Annot}$	No	No	Zero-resource	Single response, efficient, sample specific. Retrieval Indep.
SelfCheckGPT	No	Yes	Black-box	Train Indep., sample specific. Retrieval Indep.
RAGTruth	Yes	–	White-box and Black-box	Single response, sample specific

- Open QA prompts from manually constructed ("No robots") dataset (Rajani et al. 2023)
- Instruction-following
- WebNLG dataset (Gardent et al. 2017)

Also FAVA-train split includes Wikipedia articles and question-answering datasets (Kwiatkowski et al. 2019). The RAGTruth dataset consists of annotations from the CNN/Daily Mail dataset.

To analyze the efficacy of the proposed hallucination detection measures and perform a comparative analysis with existing baselines, we follow the settings presented in Table 6. For the setting of detection without external references with a single model response, we use the FAVA-Annotation (Mishra et al., 2024) dataset, wherein we assume that passages containing any form of hallucination to be hallucinated overall.

Keys are as follows:

- Train Indep: does not require model fine tuning/training
- Single Response: does not require multiple LLM responses
- Efficient: compute and memory efficient
- Sample Specific: sample wise instead of population-level detection
- Retrieval Indep: does not require RAG or reference databases

We enumerate evaluation scttings in Table 7.

In the setting of detection without external references with multiple model responses, we employ the SelfCheckGPT dataset (Manakul et al., 2023), consisting of about 2000 sentence-level annotated samples. Moreover, to fairly compare across different detection methods, we augment the FAVA-Annotation dataset by generating hundred additional responses from GPT 4. This way, new baselines for SelfCheckGPT and INSIDE on the FAVA-Annotation dataset are formed using the extended number of responses.

Table 7 Hallucination detection settings with respect to referencing

Setting type	Description
Without external references	No external references are available to provide additional context to the LLM. Further classification based on whether multiple model responses can be possibly obtained for the same prompt
Single model response	Only the original prompt x_p and fixed model output x are assumed to be available. Hallucination detection using LLMs assumes that the query topic is covered in the pre-training data of the LLM
Multiple model responses	Given a prompt x_p and a fixed model output x, stochastically sampled responses $x^1, x^2, \ldots, x^k$ are obtained to calculate different metrics such as BertScore for x, to assess the absence or presence of hallucinations in x
With external references	A broad-ranging set of data retrieved from external sources are used to get additional context to estimate the correctness of x
White-box	For a given prompt x_p, a hallucination detection is applied to the output response x that is generated by a given LLM. In this setting, the original LLM is available and accessible to compute the internal model activations
Black-box	If the original LLM that generated the response x for prompt x_p is no longer available or inaccessible, an auxiliary substitute LLM′ can be used (such as open-source LLMs like Llama-2) to compute scores with internal activations and attention kernel maps
Explainable detection and correction	We select an evaluation subset that includes complete referencing and carry out explainable corrections. In doing so, we clearly articulate why the hallucinated value is incorrect and provide a reliable source for the correction

5.2 *Comparison with Hallucination Detection Analyzing the Internal Hidden States*

We compare the performance of reference-free and reference-based hallucination detection in White-box and Explainable detection and correction settings (rows 6 and 8 in Table 8).

In the left section of Table 8, we present the results of hallucination detection, while the right grayed section displays the outcomes of combined detection and correction in an explainable manner. Italicized values are sourced from the referenced research. The values in the last row are calculated in this study, whereas other values in the rightmost two columns are recalculated using the evaluation dataset of FAVA-Annotation. We determine the percentage of cases where correction is impossible without an external source. The best results for each setting are highlighted in bold.

It can be observed that while LLM-Check surpasses AdvAbductDialRL by approximately 6% in the detection-only F1 setting, this advantage does not persist in the explainable setting with correction. In this scenario, AdvAbductDialRL more effectively manages correction and explainability, leading to a 5% improvement.

The results of our earlier system Truth-o-Meter's and AdvAbductDialRL's performance in various settings with respect to referencing is shown in Table 9. The F1 values for top-performing settings are bolded.

Table 8 Comparative analysis of hallucination detection systems

	Detection only for White box		Explainable detection and correction	
System	Accuracy	F1	Accuracy	F1
FAVAModel	*53.29*	*43.88*		
SelfCheckGPT - Prompt	*54.19*	*67.24*		
INSIDE	*57.98*	*13.17*		
Self-Prompt	*50.30*	*66.53*		
LLMCheck. Attention Score	***67.96***	***69.27***	58.27	62.38
Truth-o-Meter	61.22	62.67	57.14	59.60
AdvAbductDialRL (this study)	63.43	65.12	**61.92**	**65.73**

Table 9 Comparative performance of our earlier and current systems in various reference settings (F1). "(^)" indicates the same setting as above

Referencing setting	Truth-o-Meter	AdvAbductDialRL (this study)
Without external references	34.23 (finding refs on Bing)	47.1 (finding refs on Bing)
Single model response	58.62	60.3
Multiple model responses	48.7	52.4
With external references	69.5	**72.8**
White-box	62.7 (^)	65.1 (^)
Black-box	57.2	60.8
Explainable detection and correction	59.6 (^)	64.7 (^)

We observe that the highest F1 for both systems is achieved under *With External References* setting, followed by *White-box*. AdvAbductDialRL outperforms Truth-O-Meter in all reference settings.

5.3 *Ablation Study*

We track how we proceed from simple adversarial setting all the way to Full system AdvAbductDialRL in a number of domains involving complex relationships. We start from our examples in Sect. 1.

We observe in Table 10 that adding abductive reasoning leads to a notable gain, suggesting that inferring plausible explanations enhances the model's grounding and plausibility in generation. Introducing dialogue modeling significantly boosts performance, indicating that context-aware interaction (dialogue structure) plays a critical role in reducing hallucinations and improving coherence (8%). Reinforcement learning offers a similar boost, likely due to reward-driven optimization focusing on

Table 10 Ablation study from pure adversarial setting toward AdvAbductDialRL

	Adv	Adv + Abduct	Adv + Abduct + Dial	Adv + Abduct + Dial + RL	Full system AdvAbductDialRL
Biography	40.8	45.8	54.1	62.3	66.9
Genome	38.3	42.3	50.5	60.1	64.7
Mental states	41.5	44.5	52.8	58.9	62.0
Average	40.2	44.2	52.5	60.4	64.5

correctness, relevance, or faithfulness in outputs. Final integration and tuning of all components produce an additional improvement of 4%, likely due to synergistic effects and optimization of component interactions. Hence, each component in the *AdvAbductDialRL* system contributes meaningfully to the overall performance. Dialogue and RL are especially impactful, indicating the importance of interactive context and reward-driven fine-tuning in mitigating hallucinations. The final integration phase consolidates these improvements into a coherent, high-performing system. We will focus on mental states in the chapter "Improving ToM Capabilities of LLMs in Applied Domains."

5.4 Identifying Different Kinds of Hallucinations

We evaluate detection of various kinds of hallucinations:

1. Entity hallucination, where the model fabricates a named entity (like a person, place, organization) that doesn't exist or is incorrect: "Albert Einstein won a Nobel Prize in *Chemistry*." (Einstein won in physics, not chemistry.)
2. Relation hallucination, where the model incorrectly states a relationship between entities: "Barack Obama is the brother *of Elon Musk*." (The two are not related.)
3. Invented hallucination, where the model completely makes up facts or events that have no grounding in reality: "The Moon landing in 1982 *introduced Wi-Fi to the public*." (There was no Moon landing in 1982, and Wi-Fi wasn't introduced that way.)
4. Subjective hallucination, where the model presents opinions or personal judgments as if they were factual: "The best president in history is *undoubtedly* Thomas Jefferson." (This is subjective, not an objective fact.)
5. Unverifiable hallucination, where the model outputs information that cannot be checked or confirmed using known data or references: "In a *secret 1999 meeting*, world leaders discussed time travel." (There's no way to verify such a claim.)

We assess each hallucination kind individually and observe contribution of each component for each hallucination kind.

Table 11 presents the F1 results of our ablation study. The "Just Adversarial" setting allows for a single interaction without abduction, dialogue, or reinforcement learning. Adding abduction increases accuracy by 8%, while dialogue—where the opponent can repeatedly question the proponent's explanation—and RL each contribute approximately an 11% and 8% performance boost, respectively. RL without abduction provides an additional 8% improvement. When all components work together, they yield a further 3% enhancement.

Each system configuration builds upon the previous one by adding abductive reasoning (Abduct), dialogue-based correction (Dial), and reinforcement learning (RL). We assess which component provides the largest improvement for each hallucination type. Abductive reasoning (Abduct) is most critical for Entity hallucinations, and Dialogue (Dial) plays the biggest role in Invented and Subjective

Table 11 Identifying different kinds of hallucinations

	Adv	Adv + Abduct	Adv + Abduct + Dial	Adv + Abduct + Dial + RL	Full system AdvAbductDialRL
Entity	36.5	46.4	56.2	62.0	64.7
Relation	33.7	40.8	47.3	57.1	59.9
Invented	40.4	45.3	56.8	62.1	67.5
Subjective	35.1	40.0	49.8	57.2	62.5
Unverifiable	40.3	44.5	52.0	61.4	65.8

hallucinations (Table 12). Reinforcement learning (RL) has the highest impact for Relation and Unverifiable hallucinations.

6 Related Work

Orgad et al. (2024) discover that the internal representations of LLMs encode much more information about truthfulness than previously recognized. Through a series of experiments, the authors train classifiers on these internal representations to predict various features related to the truthfulness of generated outputs. Their study identifies the patterns and types of information encoded in model representations, linking this intrinsic data to extrinsic LLM behavior. This enhances the machinery to detect errors (while understanding the limitations of error detection) and will stimulate the development of specific strategies based on error types and mitigation methods that make use of the model's internal knowledge.

Orgad et al. (2024) found that truthfulness-related information is localized within specific answer tokens. This key insight supports various analyses throughout their study, as focusing on tokens with the strongest signals leads to more reliable conclusions. From a practical standpoint, this finding enhances error detection methods for production-level LLMs. However, since this approach requires access to internal representations, it is primarily applicable to open-source models. Additionally, the findings indicate that truthfulness features generalize poorly across tasks and datasets, performing best in contexts requiring similar skills. This suggests that truthfulness relies on "skill-specific" features, such as knowledge retrieval and contextual consistency. Some generalization patterns remain unexplained, implying that additional factors link seemingly unrelated tasks, warranting further research. As the optimal layer and token combination for error detection varies by dataset, analyzing their impact on generalization may provide deeper insights into these mechanisms. Furthermore, our results emphasize the need for caution when applying a trained error detector across different settings.

Galitsky et al. (2023) propose a hybrid architecture called "shaped charge," which combines meta-learning and deep neural networks (DNN) with an explainable k-nearest neighbors (kNN) component to address key limitations in deep learning models. In a subsequent study, Galitsky (2023) empirically evaluates the shaped charge architecture across multiple tasks, including summarization, question answering, and hallucination correction. The results demonstrate a notable improvement in the truthfulness of generated content, attributed to the integration of the shaped charge learning approach.

Manakul et al. (2023) propose SelfCheckGPT, a collection of hallucination detection techniques designed for zero-resource gray-box or black-box settings, where only the output probability distributions of a language model are available. This approach is especially practical for real-world applications, where API-based access often restricts visibility into a model's internal activations or representations.

Table 12 Contribution of component

Hallucination type	Most important component	Secondary boost
Entity	Abductive reasoning (abduct) and dialogue (dial)	Reinforcement learning (RL)
Relation	Reinforcement learning (RL)	Dialogue (dial)
Invented	Dialogue (dial)	Reinforcement learning (RL)
Subjective	Dialogue (dial)	Reinforcement learning (RL)
Unverifiable	Reinforcement learning (RL)	Dialogue (dial)

The term "hallucinations" has been widely adopted across multiple subfields of NLP, including conversational AI (Liu et al., 2023), abstractive summarization (Zhang et al., 2019), and machine translation (Wang & Sennrich, 2020). Despite its broad usage, no unified definition exists, as different research communities interpret and apply the term in varying ways.

The complexity of defining hallucinations is underscored by Venkit et al. (2024), who identified 31 distinct frameworks for conceptualizing hallucinations across various AI subfields. This diversity in perspectives suggests that hallucinations are not a singular phenomenon but rather an umbrella term encompassing a range of model-generated errors.

In response to this ambiguity, researchers have focused on:

1. Defining and classifying hallucinations—Distinguishing them from other error types, such as omissions, distortions, and misinterpretations (Rawte et al., 2023)
2. Developing taxonomies—Differentiating intrinsic hallucinations (where the model's output is inconsistent with its own training data) from extrinsic hallucinations (where the output contradicts real-world facts)

Beyond traditional definitions, recent scholarly discussions have introduced alternative terms, such as:

- "Confabulations" (Millidge, 2023)—Drawing a parallel with human cognitive errors, where the brain fills in missing details with plausible but incorrect information
- "Fabrications" (McGowan et al., 2023)—Highlighting the potential for deliberate-seeming inaccuracies, suggesting that LLMs may generate misinformation in a structured manner

These discussions bring forth philosophical and cognitive science perspectives, attributing a sense of "intention" to LLMs. While LLMs do not possess true intentionality, some researchers argue that their probabilistic text generation mechanisms can simulate patterns of reasoning and decision-making, leading to outputs that may appear goal directed or deliberate.

The notion that LLMs "intend" to mislead remains highly contested. Critics argue that:

- LLMs lack self-awareness and merely operate on pattern-matching principles, optimizing for fluency rather than factuality
- Humanlike terminology (e.g., "confabulation" and "fabrication") risks anthropomorphizing LLMs, leading to misconceptions about their cognitive capabilities

However, some researchers contend that understanding LLM errors through a cognitive lens—even metaphorically—may provide useful insights for AI interpretability and robustness research.

Orgad et al. (2024) employ probing classifiers to explore the internal representations of LLMs. Their analysis focuses on how layer and token selection impact activation extraction for these classifiers (Fig. 14). To achieve this, they systematically probe all model layers, starting from the last question token and progressing to the final generated token. The study evaluates the Area Under the Curve (AUC) metrics of trained probes across different layers and tokens of Mistral-7B-Instruct. Results indicate that while some datasets make error prediction easier, all exhibit consistent truthfulness encoding patterns. Probing results show that middle to later layers yield the most effective error detection, aligning with prior research. Their study visualizes AUC values across layers and tokens, demonstrating how truthfulness is embedded in the model. As generation progresses from left to right, detection performance peaks at specific answer tokens. These findings suggest that error detection mechanisms could be optimized by focusing on later layers. Overall, the research enhances our understanding of how truthfulness is encoded within LLMs.

Feldman et al. (2023) employ a multi-stage approach to assess the impact of tagged context prompts on the hallucination problems in LLMs, with a focus on the OpenAI GPT-3 and later models. The methods included generating question sets, creating context prompts through summaries, verifying context prompts and questions, and performing experiments with different GPT models using context-based and context-free questions. We implemented the following procedures: Data Set Creation, a diverse and comprehensive set of questions and contexts to examine the abilities of generative language models, particularly OpenAI's GPT-3 and its successors.

To maintain the reliability and accuracy of our research, Feldman et al. (2023) implement a meticulous verification process. This process focused on guaranteeing that the context prompts were a precise reflection of the source material, particularly the book sections, and that the generated questions were coherent, relevant, and strongly linked to the given context. The verification process comprised the following steps:

1. Cross-referencing: We cross-referenced context prompts with their original sources, ensuring their accuracy and completeness. This step was vital to avoid any inaccuracies that might have arisen during context generation from the summarized Wikipedia articles and the lead author's book.
2. Relevance assessment: We examined the generated questions to confirm that they were pertinent to the context prompts. This assessment also involved evaluating whether the questions exemplified the corresponding subjects well, such as

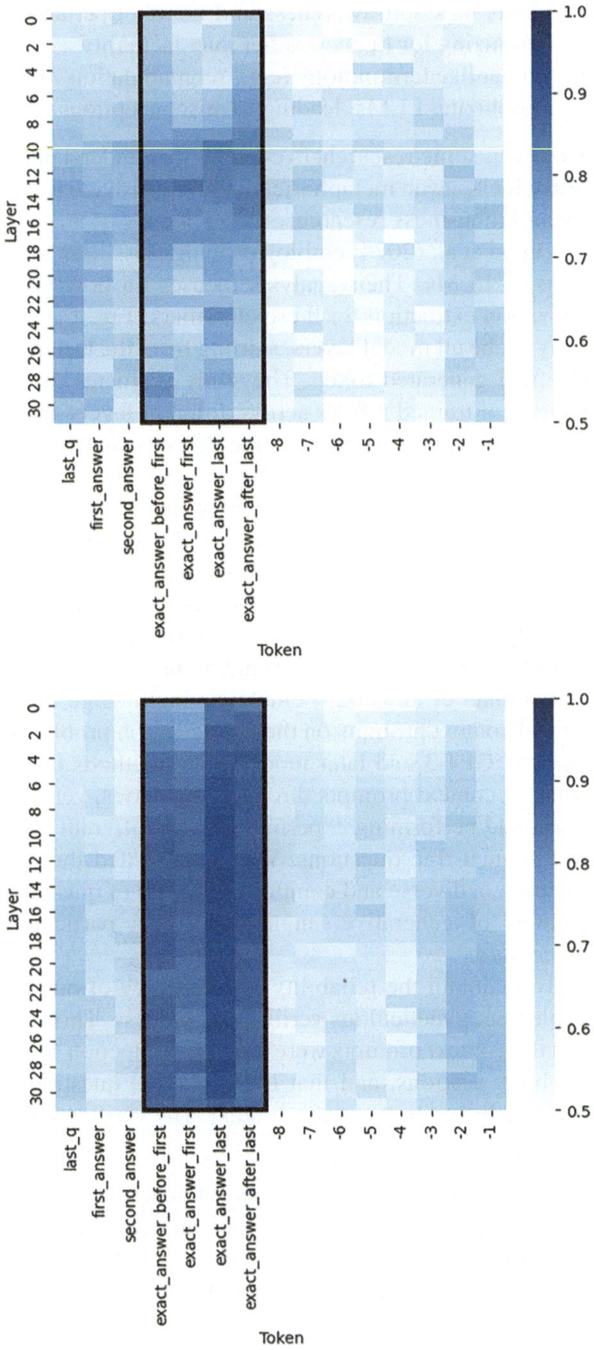

Fig. 14 Probe error detectors

France, chess, solar system, dating, continents, *To Kill a Mockingbird*, and the author's book on computational sociology.
3. Coherence check: We thoroughly inspected the questions for logical coherence and linguistic correctness. This step ensured that the generated questions made sense and were grammatically sound, enabling a meaningful evaluation of the generative language models' capabilities.

Kadavath et al. (2022) propose to use the LLM itself to predict the probability of the generated response being True, referred to as Self-Prompt; this is primarily relevant for the context of multiple-choice questions where the model explicitly has all the options in context. Furthermore, they also train models to predict the probability that a given multiple-choice question will be correctly answered by calibrating its internal confidence values.

7 Conclusions

In conclusion, while LLM-Check demonstrates superior performance in hallucination detection alone, its advantage diminishes when correction and explainability are considered. AdvAbductDialRL proves to be more effective in these scenarios, achieving a notable improvement in correction performance. This highlights the importance of incorporating abduction, dialogue, and reinforcement learning for tasks where hallucination correction is essential, making AdvAbductDialRL a more suitable approach in such cases.

Our study highlights the importance of selecting the right approach for hallucination detection and correction in language models. While LLM-Check demonstrates a strong ability to identify hallucinations, its advantage diminishes when correction and explainability are required. In contrast, AdvAbductDialRL proves to be more effective in scenarios where hallucinations must not only be detected but also corrected in a transparent and interpretable manner.

By incorporating abduction, dialogue-based interactions, and reinforcement learning, AdvAbductDialRL enables a more robust correction process, outperforming LLM-Check in explainability-driven settings. The model's ability to iteratively refine explanations and corrections through structured adversarial interactions makes it particularly beneficial in high-stakes applications where interpretability is crucial, such as medical, legal, and scientific domains.

While LLM-Check demonstrates approximately a 6% improvement over AdvAbductDialRL in the detection-only F1 setting, this performance advantage does not extend to the explainable setting with correction. In that context, AdvAbductDialRL exhibits superior handling of correction and explainability, resulting in a 5% performance gain.

These findings highlight the importance of broadening the evaluation horizon to include not only hallucination detection accuracy but also the effectiveness of correction strategies. Looking ahead, future efforts may extend the horizon of

AdvAbductDialRL by incorporating external knowledge sources and enhancing its performance for real-time use. Expanding the horizon of explainable hallucination correction brings us closer to building trustworthy AI systems that provide both reliable and interpretable responses.

Hallucinations in generative language tasks present substantial challenges to the accountability, reliability, and overall trustworthiness of artificial intelligence systems. In this study, we conduct a comprehensive and systematic investigation of this issue within the framework of adversarial dialogue involving abductive reasoning, examining both general-purpose and domain-specific LLMs. Our proposed method demonstrates empirical effectiveness in mitigating hallucinations and is shown to be generalizable across different settings as well as scalable to larger and more complex dialogue scenarios. These findings contribute to the development of more robust and interpretable AI systems capable of maintaining fidelity in their generated outputs.

As LLMs become increasingly integrated into high-stakes applications (e.g., legal, medical, and scientific domains), addressing their tendency to hallucinate is critical. A standardized framework for defining, detecting, and mitigating hallucinations will be essential for ensuring reliability, safety, and trustworthiness in AI systems. Future research must bridge the gap between technical, cognitive, and ethical perspectives, fostering a more nuanced and actionable approach to handling hallucinations in LLMs.

Future work may involve refining the evaluation function through more sophisticated models or human-in-the-loop assessments and extending the framework to a broader set of tasks beyond explanation generation. Additionally, empirical experiments to benchmark the performance improvements of the Proponent and the efficacy of the Opponent's challenges would provide further validation for this approach.

References

Ballard IC, Waskom M, Nix KC, D'Esposito M (2024) Reward reinforcement creates enduring facilitation of goal-directed behavior. J Cogn Neurosci 36(12):2847–2862

Bang Y, Cahyawijaya S, Lee N, Dai W, Su D, Wilie B, Lovenia H, Ji Z, Yu T, Chung W et al (2023) A multitask, multilingual, multimodal evaluation of ChatGPT on reasoning, hallucination, and interactivity. arXiv preprint arXiv:2302.04023

Baza (2025) The full transcript of cockpit communications of the crew of the crashed Azerbaijan Airlines plane. https://baza.io/posts/fb439a5b-f8ce-4398-88f4-7890dd82c404

Codd EF (1970) A relational model of data for large shared data banks. Commun ACM 13(6):377

Doshi-Velez F, Kim B (2017) Towards a rigorous science of interpretable machine learning. arXiv preprint arXiv:1702.08608

Farquhar S, J Kossen, L Kuhn, Y Gal (2024) Detecting hallucinations in large language models using semantic entropy. Nature 630(8017):625–630

Feldman P, Foulds JR, Pan S (2023) Trapping LLM hallucinations using tagged context prompts. arXiv preprint arXiv:2306.06085

Filippova K (2020) Controlled hallucinations: learning to generate faithfully from noisy data. In: Webber B, Cohn T, He Y, Liu Y (eds) Findings of the Association for Computational Linguistics: EMNLP 2020. Association for Computational Linguistics, pp 864–870

Friel R, Sanyal A (2023) Chainpoll: a high efficacy method for LLM hallucination detection. arXiv preprint arXiv:2310.18344

Fu J, Ng S-K, Jiang Z, Liu P (2024) GPTScore: evaluate as you desire. In: Proceedings of the 2024 conference of the North American chapter of the Association for Computational Linguistics: human language technologies, Long papers, vol 1. Association for Computational Linguistics, Mexico City, pp 6556–6576

Galitsky B, Kuznetsov SO (2006) Learning communicative actions of conflicting human agents. J Exp Theor Artif Intell 20(4):277–317

Galitsky B (2013) Transfer learning of syntactic structures for building taxonomies for search engines. Eng Appl Artif Intell 26(10):2504–2515

Galitsky B (2022) Improving open domain content generation by text mining and alignment. In: Artificial intelligence for healthcare applications and management. Elsevier

Galitsky B (2023) Applications of shaped-charge learning. Entropy 25(11):1496

Galitsky B (2025a) Truth-O-Meter: collaborating with LLM in fighting its hallucinations. In: Lawless W, Mittu R, Sofge D, Fouad H (eds) Interdependent human-machine teams. Academic Press, pp 175–210. https://doi.org/10.1016/B978-0-443-29246-0.00004-3

Galitsky B (2025b) Enabling retrieval-augmented generation and knowledge graphs with discourse analysis. In: Health applications of neuro-symbolic artificial intelligence. Elsevier

Galitsky B, Ilvovsky D, Kuznetsov SO (2018) Detecting logical argumentation in text via communicative discourse tree. J Exp Theor Artif Intell 30(5):1–27

Galitsky B, Ilvovsky D, Goldberg S (2023) Shaped-charge learning architecture for the human–machine teams. Entropy 25(6):924. https://doi.org/10.3390/e25060924

Galitsky B (2024) Truth-O-Meter: Collaborating with LLM in fighting its hallucinations. In: Lawless W, Mittu R, Sofge D, Fouad H (eds) Interdependent Human-Machine Teams. Academic Press, 2025, pp 175–210. ISBN 9780443292460, https://doi.org/10.1016/B978-0-443-29246-0.00004-3.

Gardent C, Shimorina A, Narayan S, Perez-Beltrachini L (2017) The WebNLG challenge: generating text from RDF data. In: Alonso JM, Bugarín A, Reiter E (eds) Proceedings of the 10th international conference on natural language generation. Association for Computational Linguistics, Santiago de Compostela, pp 124–133

Guo C, Pleiss G, Yu S, Weinberger KQ (2017) On calibration of modern neural networks. In: Proceedings of the 34th international conference on machine learning (ICML), vol 70, pp 1321–1330

Harman G (1965) The inference to the best explanation. Philos Rev 74(1):88–95

Hestness J, Narang S, Ardalani N, Diamos G, Jun H, Kianinejad H, Patwary MMA, Yang Y, Zhou Y (2017) Deep learning scaling is predictable, empirically. arXiv preprint arXiv:1712.00409

Honovich O, Aharoni R, Herzig J, Taitelbaum H, Kukliansy D, Cohen V, Scialom T, Szpektor I, Hassidim A, Matias Y (2022) TRUE: re-evaluating factual consistency evaluation. arXiv preprint arXiv:2204.04991

Jiang AQ, Sablayrolles A, Mensch A, Bamford C, Chaplot DS, de las Casas D, Bressand F, Lengyel G, Lample G, Saulnier L, Lavaud LR, Lachaux M-A, Stock P, Scao TL, Lavril T, Wang T, Lacroix T, El Sayed W (2023) Mistral 7B. arXiv preprint arXiv:2310.06825

Josephson JR, Josephson SG (1994) Abductive inference: computation, philosophy, technology. Cambridge University Press, New York

Kadavath S, Conerly T, Askell A, Henighan T, Drain D, Perez E, Schiefer N, Hatfield Dodds Z, DasSarma N, Tran-Johnson E, Johnston S, El-Showk S, Jones A, Elhage N, Hume T, Chen A, Bai Y, Bowman S, Fort S, Ganguli D, Hernandez D, Jacobson J, Kernion J, Kravec S, Lovitt L, Ndousse K, Olsson C, Ringer S, Amodei D, Brown T, Clark J, Joseph N, Mann B, McCandlish S, Olah C, Kaplan J (2022) Language models (mostly) know what they know

Kim J, Park S, Kwon Y, Jo Y, Thorne J, Choi E (2023) FactKG: fact verification via reasoning on knowledge graphs. In: Proceedings of the 61st annual meeting of the Association for Computational Linguistics, Long papers, vol 1. Association for Computational Linguistics, Toronto, pp 16190–16206

Köpf A, Kilcher Y, von Rütte D, Anagnostidis S, Tam ZR, Stevens K, Barhoum A, Nguyen D, Stanley O, Nagyfi R (2024) OpenAssistant conversations-democratizing large language model alignment. In: Advances in neural information processing systems, vol 36

Kwiatkowski T, Palomaki J, Redfield O, Collins M, Parikh A, Alberti C, Epstein D, Polosukhin I, Devlin J, Lee K, Toutanova K, Jones L, Kelcey M, Chang M-W, Dai AM, Uszkoreit J, Le Q, Petrov S (2019) Natural questions: a benchmark for question answering research. Trans Assoc Comput Linguist 7:452–466

Lake JI, Spielberg JM, Infantolino ZP, Crocker LD, Yee CM, Heller W, Miller GA (2019) Reward anticipation and punishment anticipation are instantiated in the brain via opponent mechanisms. Psychophysiology 56(8):e13381

Li J, Consul S, Zhou E, Wong J, Farooqui N, Ye Y, Manohar N, Wei Z, Wu T, Echols B, Zhou S, Diamos G (2024) Banishing LLM hallucinations requires rethinking generalization. arXiv preprint arXiv:2406.17642

Liebenow B, Jones R, DiMarco E, Trattner JD, Humphries J, Sands LP, Spry KP, Johnson CK, Farkas EB, Jiang A, Kishida KT (2022) Computational reinforcement learning, reward (and punishment), and dopamine in psychiatric disorders. Front Psych 13:886297

Liu Y, Iter D, Xu Y, Wang S, Xu R, Zhu C (2023) G-Eval: NLG evaluation using Gpt-4 with better human alignment. In: Proceedings of the 2023 conference on empirical methods in natural language processing. Association for Computational Linguistics, Singapore, pp 2511–2522

Lundberg SM, Lee S-I (2017) A unified approach to interpretable model predictions. arXiv preprint arXiv:1705.07874

Luss R, Miehling E, Dhurandhar A (2025) CELL your model: contrastive explanations for large language models. arXiv preprint arXiv:2406.11785

Manakul P, Liusie A, Gales MJF (2023) SelfCheckGPT: zero-resource black-box hallucination detection for generative large language models

McGowan A, Gui Y, Dobbs M, Shuster S, Cotter M, Selloni A, Goodman M, Srivastava A, Cecchi GA, Corcoran CM (2023) ChatGPT and BARD exhibit spontaneous citation fabrication during psychiatry literature search. Psychiatry Res 326:115334

Meta (2024) The Llama 3 Herd of Models. https://ai.meta.com/research/publications/the-llama-3-herd-of-models/. Last downloaded March 1 2026

Millidge B (2023) LLMs confabulate not hallucinate. https://www.beren.io/2023-03-19-LLMs-confabulate-not-hallucinate/

Mishra A, Asai A, Balachandran V, Wang Y, Neubig G, Tsvetkov Y, Hajishirzi H (2024) Fine-grained hallucination detection and editing for language models. arXiv:2401.06855

Montavon G, Binder A, Lapuschkin S, Samek W, Müller KR (2019) Layer-wise relevance propagation: an overview. In: Samek W, Montavon G, Vedaldi a, Hansen L, Müller KR (eds) Explainable AI: interpreting, explaining and visualizing deep learning, Lecture notes in computer science, vol 11700. Springer, Cham. https://doi.org/10.1007/978-3-030-28954-6_10

Niu C, Wu Y, Zhu J, Xu S, Shum KS, Zhong R, Song J, Zhang T (2024) RAGTruth: a hallucination corpus for developing trustworthy retrieval-augmented language models. In: Proceedings of the 62nd annual meeting of the Association for Computational Linguistics, Long papers, vol 1. Association for Computational Linguistics, Bangkok, pp 10862–10878

Orgad H, Toker M, Gekhman Z, Reichart R, Szpektor I, Kotek H, Belinkov Y (2024) LLMs know more than they show: on the intrinsic representation of LLM hallucinations. arXiv preprint arXiv:2410.02707

Peirce CS (1931) In: Hartshorne C, Weiss P, Burks AW (eds) Collected papers of Charles Sanders Peirce, vol 1–8. Harvard University Press

Rajani N, Tunstall L, Beeching E, Lambert N, Rush AM, Wolf T (2023) No robots. https://huggingface.co/datasets/HuggingFaceH4/no_robots

Rawte V, Sheth AP, Das A (2023) A survey of hallucination in large foundation models. arXiv preprint arXiv:2309.05922

Ribeiro MT, Wu T, Guestrin C, Singh S (2020) Beyond accuracy: behavioral testing of NLP models with CheckList. In: Proceedings of the 58th annual meeting of the Association for Computational Linguistics. ACL, pp 4902–4912

Sanh V, Webson A, Raffel C, Bach SH, Sutawika L, Alyafeai Z, Chaffin A, Stiegler A, Scao TL, Raja A et al (2021) Multitask prompted training enables zero-shot task generalization. arXiv preprint arXiv:2110.08207

Simhi A, Herzig J, Szpektor I, Belinkov Y (2025) Distinguishing ignorance from error in LLM hallucinations. arXiv preprint arXiv:2410.22071

Sparck Jones K (1972) A statistical interpretation of term specificity and its application in retrieval. J Doc 28:11–21. https://doi.org/10.1108/eb026526

Sriramanan G, Saha S, Bharti S, Kattakinda P, Sankar V, Sadasivan SF (2024) LLM-Check: investigating detection of hallucinations in large language models. In: Advances in neural information processing systems, NeurIPS 2024, vol 37

Thorne J, Vlachos A, Christodoulopoulos C, Mittal A (2018) FEVER: a large-scale dataset for Fact Extraction and VERification. In: Proceedings of the 2018 Conference of the North American chapter of the Association for Computational Linguistics: human language technologies, Long papers, vol 1. Association for Computational Linguistics, New Orleans, LA, pp 809–819

Venkit PN, Chakravorti T, Gupta V, Biggs H, Srinath M, Goswami K, Rajtmajer S, Wilson S (2024) "Confidently nonsensical?": a critical survey on the perspectives and challenges of 'hallucinations' in NLP. In: EMNLP

Vig J (2019) A multiscale visualization of attention in the transformer model. In: Proceedings of the 57th annual meeting of the Association for Computational Linguistics: system demonstrations. Association for Computational Linguistics, Florence, pp 37–42

Wang C, Sennrich R (2020) On exposure bias, hallucination and domain shift in neural machine translation. In: Jurafsky D, Chai J, Schluter N, Tetreault J (eds) Proceedings of the 58th annual meeting of the Association for Computational Linguistics, pp 3544–3552

Wang X, Wei J, Schuurmans D, Le Q, Chi E, Narang S, Chowdhery A, Zhou D (2022) Self-consistency improves chain of thought reasoning in language models. arXiv preprint arXiv:2203.11171

Wikipedia (2025) Azerbaijan Airlines Flight 8243. https://en.wikipedia.org/wiki/Azerbaijan_Airlines_Flight_8243

Zhang Z, Han X, Liu Z, Jiang X, Sun M, Liu Q (2019) ERNIE: enhanced language representation with informative entities. In: Korhonen A, Traum D, Màrquez L (eds) Proceedings of the 57th annual meeting of the Association for Computational Linguistics, Florence, Italy, pp 1441–1451

Improving ToM Capabilities of LLMs in Applied Domains

1 Introduction

The Theory of Mind (ToM), otherwise known as mentalizing or mindreading, is the human ability to infer and reflect upon the mental and emotional states of oneself and others (Hiatt et al. 2011). ToM is at the core of human social intelligence, facilitating meaningful communication, enabling empathy, and providing the means by which we explain, predict, judge, and influence one another's behavior. ToM is essential for social intelligence, and developing agents with theory of mind is a requisite for efficient interaction and collaboration with humans. Thus, it is important to build a path forward for enabling agents with this type of reasoning, as well as methods for robustly assessing an LLM models' Theory of Mind reasoning capabilities.

ToM is a fundamental human ability to infer the mental states, intentions, emotions, and beliefs of others, enabling the dynamic adjustment of behavior to social and environmental contexts (Baron-Cohen et al. 1985). This capability has been widely integrated into AI agent frameworks to enhance their interaction with humans (Langley et al. 2022).

In shared workspace settings, where close collaboration between humans and AI agents occurs, ToM becomes even more critical in shaping the AI agent's framework (Hiatt et al. 2011). By leveraging ToM, AI agents can better understand, infer, and predict human behaviors, dynamically adapting their strategies to improve team performance and coordination.

From the human perspective, ToM also plays a pivotal role in how individuals perceive and interact with AI agents during collaboration (Gero et al. 2020). Humans build mental models of AI agents, attributing mental states and roles to them based on their expectations and the agents' observed behavior. This reciprocal process underpins effective human-AI collaboration, where humans anticipate AI capabilities and roles while expecting them to align with their mental models.

B. Galitsky, *Applications of Neuro-Symbolic Artificial Intelligence*,
https://doi.org/10.1007/978-3-032-16524-4_4

At the core of what defines us as humans is the ToM concept: the ability to track other people's mental states. The recent development of LLMs such as ChatGPT has led to intense debate about the possibility that these models exhibit behavior that is indistinguishable from human behavior in Theory of Mind tasks.

There are systematic failures that limit the collaboration efficiency of LLM-based agents, and a prompt-engineering method is needed to mitigate those failures by incorporating explicit belief state representations about world knowledge in the model input. Prior research has tested LLMs' ToM via variants of text-based tests such as the unexpected transfer task (also known as Smarties Task) or unexpected contents task (also known as the "Maxi Task" or "Sally-Anne" Test) (Kosinski 2023; Moghaddam and Honey 2023). Results indicate that leading LLMs can pass more than 90% of these test cases.

Moreover, Ullman (2023) found that LLMs struggle with complex ToM inferences involving communication or second-order beliefs. In the study of Li et al. (2024), ToM evaluations occur in the midst of an interactive team task, where the mental states of agents change dynamically with each interaction. As agents exchange information through communication at every timestamp, the complexity of reasoning increases, since agents' mental states may be updated through both observations and communication. Ullman (2023)'s tests can be considered more challenging than the static text-based tests used in prior research.

The recent rise of LLMs made text-based tests of ToM particularly interesting. LLMs have already shown some successes across many challenging benchmarks and tasks designed to test various aspects of reasoning (Lewkowycz et al. 2022; Mahowald et al. 2023). While there are many cautionary voices that suggest such models may be acquiring formal rather than functional abilities, that has not stopped people from testing them on functional abilities as well, including ToM reasoning. While some of these tests offer a pessimistic evaluation (Sap et al. 2022), recent work by Kosinski (2023) applied variations on classic Theory of Mind tasks to several LLMs and confirmed successful execution of a manifold of second-order ToM tasks.

ToM is not an innate ability. It is an empirically established fact that children develop a ToM at around the age of 4 (Baron-Cohen et al. 1999). It has been demonstrated that around this age, children are able to assign false beliefs to others, by having them undertake the Sally-Anne test. The child is told the following story, accompanied by dolls or puppets:

> Sally puts her ball in a basket and goes out to play; while she is outside Anne takes the ball from the basket and puts it in a box; then Sally comes back in. The child is asked where will Sally look for her ball. Children with a developed ToM are able to identify that Sally will look for her ball inside the basket, thus correctly assigning a false belief to the character, that they themselves know to be untrue.

During the 1980s, the ToM hypothesis of autism gained traction, which states that deficits in the development of ToM satisfactorily explain the main symptoms of autism. This hypothesis argues that the inability to process mental states leads to a lack of reciprocity in social interactions. Although a deficiency in the identification and interpretation of mental states remains uncontested as a cause of autism, it is no longer viewed as the only one, and the disorder is now studied as a complex condition involving a variety of cognitive and reasoning limitations (Askham 2022).

1.1 Current Limitations of LLMs in Handling ToM

The failure of inference hypothesis posits that LLMs struggle to generate inferences about the mental states of speakers, a critical component of the Theory of Mind reasoning (Amirizaniani et al. 2024). For example, recognizing a faux pas in a given scenario requires contextual understanding that extends beyond the information explicitly encoded within the story. In tests of faux pas recognition, essential knowledge—such as the social norm that calling someone's newly purchased curtains "horrible" is inappropriate—is not explicitly stated in the narrative. Yet, this implicit knowledge is crucial for accurately inferring the mental states of the characters. The inability to draw upon external, non-embedded information fundamentally limits GPT-4's and similar models' ability to compute such inferences reliably.

Even when language models demonstrate the ability to infer mental states, they often fail to resolve *ambiguity between competing possibilities*. This phenomenon mirrors the classic philosophical paradox of the "rational agent caught between two appetitive bales of hay," where indecision leads to inaction. In this case, GPT models might correctly identify a faux pas as one of several plausible alternatives but fail to rank these possibilities by likelihood or importance. Instead, responses sometimes indicate uncertainty, suggesting that the speaker might not know or remember yet present this as one hypothesis among many (Strachan et al. 2024). This behavior reflects a broader limitation: while the model can generate the correct inference, it often refrains from selecting or emphasizing it over less probable explanations.

Further complicating matters, even if one assumes that LLMs are capable of accurately inferring mental states and recognizing false beliefs or gaps in knowledge as the most likely explanation, these models often hesitate *to commit to a single interpretation*. This reluctance may stem from inhibitory mitigation processes built into the models, which prioritize cautious and noncommittal outputs. Such processes may cause the models to adopt an overly conservative stance, avoiding decisive answers even when they can generate the correct explanation.

These challenges highlight the dichotomy between the models' ability to produce language that aligns with complex reasoning tasks and their limitations in applying this reasoning in a structured, humanlike way. While LLMs are undeniably powerful in generating language, their *reluctance to prioritize or commit to the likeliest inference* undermines their ability to perform robustly in tasks requiring nuanced social understanding and contextual reasoning (Mohan et al. 2024). Addressing this conservatism and improving their ranking of alternatives will be crucial steps in advancing their ToM capabilities.

Sclar et al. (2024) demonstrate that LLMs *struggle with basic state tracking*—a foundational skill that underpins ToM reasoning. The ability to track mental states inherently depends on accurately maintaining representations of physical and contextual states, yet this proves to be a significant challenge for current models. Sclar et al. (2024)'s experiments further reveal that improving ToM reasoning during fine-tuning is not as simple as training on datasets that involve state tracking alone. Instead, the data must explicitly require the application of ToM principles, such as reasoning about others' beliefs, intentions, or perspectives. Unfortunately, existing ToM training datasets rarely embody this necessary complexity, which may be a critical factor behind the lagging performance of LLMs in this area.

Moreover, there has been limited research on whether LLMs can implicitly leverage ToM knowledge to predict behavior or assess whether an observed behavior is rational. These abilities are essential for effective social interaction and for models to engage appropriately in environments requiring nuanced understanding of others' actions and motives. Without these skills, LLMs remain constrained in their capacity to interpret or generate responses that align with human social reasoning.

The challenges extend beyond state tracking to encompass broader deficiencies in basic ToM capabilities. This includes the inability of LLMs to generalize beyond literal patterns or contexts in their training data, highlighting the importance of purpose-built datasets. Such datasets would need to emphasize tasks requiring explicit ToM reasoning, which is largely absent from text data found on the Web or generated through random processes (Sclar et al. 2024). Addressing this gap will be crucial for the development of models capable of robust and meaningful ToM reasoning.

The mixed performance of LLMs, combined with their sensitivity to small perturbations in the prompts they receive, underscores serious concerns regarding their robustness and interpretability. Minor changes in a scenario, such as altering a character's perceptual access or slightly rephrasing a prompt, can lead to drastic shifts in the model's responses. This inconsistency reveals an unsettling fragility in their reasoning capabilities and suggests that their apparent successes may not stem from genuinely robust cognitive processes (Ullman 2023). Instead, these successes may rely on shallow heuristics or pattern-matching techniques that fail to generalize across variations in input.

Even in cases where LLMs exhibit an impressive ability to solve cognitively demanding tasks—challenges that require intricate reasoning and are known to be difficult even for human adults (Apperly and Butterfill 2009)—their performance on simpler, more straightforward tasks cannot be taken for granted. While these models may excel at tasks that involve complex rule-following or abstract problem-solving, they frequently falter when confronted with scenarios that require basic logic, consistency, or common-sense reasoning. This paradox highlights a critical gap between their surface-level performance and deeper cognitive understanding.

For example, an LLM may accurately infer the mental states of characters in a nuanced social interaction but fail to resolve a simpler task, such as determining the logical consequences of a straightforward sequence of events. These discrepancies raise the question of whether LLMs truly possess the underlying cognitive mechanisms required for general reasoning or if they rely on brittle approximations that break down under minimal perturbation. Such vulnerabilities are particularly problematic in applications requiring reliability, such as human-computer interaction, decision-making, or Theory of Mind reasoning, where misinterpretation or inconsistency can lead to significant errors.

Furthermore, this inconsistency calls into question the validity of benchmarks used to evaluate LLMs. A model's ability to pass cognitively complex tests does not necessarily indicate it can handle the broad range of variability and uncertainty inherent in real-world tasks. Addressing these shortcomings will require not only improvements in model architectures and training methodologies but also the development of more robust evaluation frameworks that test LLMs across diverse scenarios and levels of complexity. Without such advancements, the promise of LLMs as tools capable of humanlike reasoning and adaptability remains out of reach.

This inconsistency has prompted a growing body of research to question the underlying mechanisms driving LLM success. Rather than demonstrating genuine parallels to ToM abilities—the capacity to reason about the beliefs, intentions, and mental states of others—there is a concern that LLMs rely on shallow heuristics. These heuristics may exploit patterns in training data rather than reflecting deeper, more robust reasoning processes. As a result, the field is increasingly calling for careful evaluation and a focus on building models that can demonstrate more generalizable and interpretable cognitive capabilities, akin to the nuanced reasoning observed in humans.

An analysis of Amirizaniani et al. (2024), comparing semantic similarity and lexical overlap metrics between responses generated by humans and LLMs, reveals clear disparities in ToM reasoning capabilities in open-ended questions, with even the most advanced models showing notable limitations. This research highlights the deficiencies in LLMs' social reasoning and demonstrates how integrating human intentions and emotions can boost their effectiveness.

1.2 Limitations of ToM Datasets

There has been a surge of recent research aimed at developing benchmarks to test the ToM capabilities of LLMs, often drawing inspiration from classic psychological tests used in child development studies, such as the Sally-Anne test (Wimmer and Perner 1983). While these benchmarks provide valuable initial insights into a model's ability to infer mental states, they are not well suited for comprehensive evaluation. These tests tend to focus on narrowly defined scenarios and lack the variability, complexity, and richness needed to pose a sustained challenge to LLMs, particularly after extensive online pre-training. Consequently, many existing benchmarks fail to robustly assess the depth and adaptability of LLMs' ToM capabilities.

The implicit nature of ToM reasoning further complicates the creation of effective evaluation methods. Identifying or generating data that explicitly requires ToM reasoning is inherently difficult, as most real-world data does not articulate the underlying reasoning processes involved in social cognition. Existing benchmarks suffer from several limitations. They are often constrained in scale and frequently rely on highly specific, scripted scenarios (Le et al. 2019). This narrow scope not only limits their utility as training data but also increases the risk of data leakage, rendering them unsuitable for future evaluations. Moreover, fine-tuning models on such datasets has been shown to lead to overfitting to the specific narrative structures or tasks represented, rather than fostering the acquisition of generalized ToM reasoning capabilities (Sclar et al. 2023).

Recognizing these limitations, recent research has shifted its focus toward improving LLM performance through inference-time strategies rather than relying solely on fine-tuning. These strategies often involve advanced prompting techniques or the integration of more sophisticated algorithms designed to enhance reasoning capabilities without requiring exhaustive retraining. For instance, Sclar et al. (2023) explore dynamic prompting mechanisms to guide models toward deeper inferences, while Wilf et al. (2023) propose algorithms that adaptively tailor prompts based on task complexity. Jung et al. (2024) take this a step further by introducing probabilistic frameworks that combine context-sensitive cues with structured reasoning pathways.

Ultimately, developing robust ToM evaluation and training methods will require a more holistic approach that addresses the limitations of current benchmarks. This includes creating datasets that exhibit greater diversity, scalability, and alignment with the implicit and nuanced nature of ToM reasoning (Street 2024). Additionally, leveraging inference-time improvements and exploring hybrid approaches that combine fine-tuning with adaptive reasoning algorithms may pave the way for more reliable and human-like ToM capabilities in LLMs.

Current datasets for assessing ToM in LLMs are limited by their reliance on the classical Sally-Anne task or scenario skeletons representations (Nematzadeh et al. 2018). These datasets have significant shortcomings:

1. Limited diversity in how asymmetry in agents' beliefs and desires occur.
2. Explicit use of mental verbs like "sees" and "thinks," which serve as trigger words for models to realize that these are important aspects, removing the need for implicit commonsense inferences about relevant percepts or beliefs.
3. Limited exploration of applied ToM, such as judgment of behavior, which requires implicit reasoning about mental state.

1.3 ToM System NL_MAMS_LLM

To address the robustness and interpretability, we enable LLM with simulation and deduction that accompanies LLM inference, as well as fine-tune LLM with ToM specific data. Moreover, we systematically accumulate training ToM scenarios for a broad range of ToM application domains. This differentiates our approach from the ones described in previous Sect. 1.1.

The Natural Language Multiagent Mental Simulator (NL_MAMS), which we have been developing over the years (Galitsky 2004, 2016a), processes formal or natural language descriptions of the initial mental states of interacting agents and outputs deterministic scenarios of their intelligent behaviors. NL_MAMS simulates agents capable of analyzing and predicting the consequences of both their own and others' mental and physical actions (Galitsky 2006a). The output of NL_MAMS is a sequence of mental formulas that reflect the states resulting from the actions (behaviors) chosen by these agents.

The simulator settings can be simplified to game-theoretic configurations when agents' mutual beliefs are either complete or absent and their intentions are uniform—a trivial case of multi-agent scenarios (Rosenschein and Zlotkin 1994). For the purpose of LLM reasoning support, we focus on a synchronous interaction model in which agents commit their actions simultaneously, with these actions updating their mental and physical states simultaneously as well.

In this study, we aim to extend ToM reasoning to support various application domains. We utilize both an LLM and NL_MAMS to generate a dataset of scenarios for training the LLM on ToM-specific tasks in each application domain. The fine-tuned system, which we call NL_MAMS_LLM (Natural Language Multi-agent Mental Simulator over LLM), combines the capabilities of NL_MAMS and the LLM, generating multi-agent scenarios involving mental states and actions while also answering questions about these states and actions.

To generate scenarios, we merge A* search with a probabilistic planner to enhance the diversity and complexity of the resulting scenarios. These scenarios are used both for fine-tuning the LLM and for evaluating the fine-tuned LLM's ability to comprehend complex ToM scenarios.

The contribution of this chapter is as follows:

1. We formulate a mental world formalism that turns out to be fruitful to cover the weak points of LLM reasoning.
2. We algorithmically address LLM limitations in specific sub-domain of ToM by fine-tuning and simulation.
3. We develop a methodology to create complex training data that enables LLMs with better ToM reasoning capabilities.

2 Applications of ToM-Capable LLMs

LLMs are increasingly being applied across a variety of social domains, demonstrating their potential to augment human interactions and support emotional, educational, and professional needs. Notable applications include:

1. Therapy and mental health support: LLM-powered tools like Woebot Health (2024) offer conversational therapy, providing accessible, on-demand mental health resources to users. In healthcare mental health diagnostics, AI systems can infer a patient's mental state from conversations, facial expressions, or

physiological data, assisting in diagnosing conditions like anxiety or depression. ToM-based AI tools in therapy can adapt their responses based on inferred emotions or beliefs to provide empathetic and effective support. ToM is also needed in patient monitoring, making inferences about patient discomfort, confusion, or stress, and also help in real-time intervention or care adjustments (Galitsky 2016b).

2. Friendship and romantic relationships: platforms such as Replika (2025) create AI companions designed to simulate friendships or even romantic connections, catering to emotional and relational needs.
3. Education and teaching: tools like Merlyn Mind (2025) employ LLMs to act as virtual teaching assistants, helping educators and students manage classroom tasks, answer questions, and enhance the learning experience. In education, adaptive learning systems can infer a student's understanding, confusion, or frustration and adapt the lesson accordingly. In collaborative learning, AI agents in group learning environments can mediate discussions by understanding and addressing students' differing beliefs or misunderstandings.
4. Human-computer interaction: ToM principles are used to design systems that can better understand and predict human behavior, enabling more natural and intuitive interactions (Wang et al. 2021). For instance, virtual assistants or chatbots equipped with ToM-like reasoning can infer user intentions and adapt responses to enhance user experience. Human-robot interaction that include personalized assistance (Galitsky 2025): robots with ToM can adapt their behavior to the needs and preferences of users. For example, a home assistant robot could infer when a person is frustrated and offer help or adjust its approach. In industrial or domestic settings, robots can anticipate the goals of their human collaborators, making teamwork more seamless and efficient. Social robots in education or eldercare can infer emotional states (e.g., boredom, confusion) to provide tailored interactions.
5. Explainable AI: In the field of explainable AI, ToM is leveraged to model how humans interpret AI systems' actions and decisions (Akula et al. 2022). By simulating a user's beliefs and understanding, AI systems can generate explanations that align with human reasoning, improving transparency and trust in AI-driven decision-making processes.
6. Multi-agent reinforcement learning: ToM plays a critical role in multi-agent reinforcement learning, where agents must reason about the beliefs and goals of others to collaborate or compete effectively (Rabinowitz et al. 2018). ToM-enabled agents can model the strategies of teammates or opponents, leading to more sophisticated and adaptive behavior in dynamic environments.
7. Autonomous vehicles need to predict pedestrian intentions. Autonomous cars equipped with ToM can anticipate when pedestrians are likely to cross the street based on body language or gaze direction. By modeling the intentions of other drivers (e.g., aggressive vs. cautious driving behavior), autonomous vehicles can make safer decisions.
8. In gaming and virtual reality, non-player characters in video games can use ToM to adapt their behavior based on the player's strategies or goals, creating a

more engaging and realistic experience. In virtual training simulations in military or medical training, AI agents can simulate realistic human responses, such as inferring the trainee's intentions and reacting accordingly.

9. In negotiation and strategic interaction, AI in markets or auctions can model the goals and beliefs of other agents to make more effective bids. In conflict resolution, ToM-equipped agents can facilitate negotiations by understanding the motivations and perspectives of each party.
10. In surveillance and security, ToM can help predict suspicious behavior by analyzing patterns of movement or communication (Galitsky and Miller 2005). In fraud detection, AI can infer intentions behind financial transactions or user behavior to detect fraudulent activity.
11. In assistive technologies, for individuals with disabilities, ToM-enabled systems can assist individuals with cognitive or physical impairments by predicting their needs and adapting assistance accordingly. Moreover, devices for individuals with speech impairments can infer the user's intended message to facilitate communication.
12. In marketing and customer engagement, in particular, in personalized advertising, by modeling a user's beliefs and desires, AI systems can tailor advertisements or recommendations more effectively. Customer service chatbots with ToM can infer user satisfaction or frustration and escalate issues as needed.

Our goal is to improve ToM-based reasoning in all of these domains. Beyond explicitly social domains, LLMs are being widely adopted in fields like medicine, law, software development, and creative industries. While these applications may not be inherently social, they often involve a social dimension through their natural language interfaces, enabling seamless communication and collaboration between humans and AI systems.

As LLMs become part of mainstream technology, their adaptation for increasingly complex and novel use cases is accelerating. A significant trend is the shift from the traditional **one-to-one** interaction paradigm—where an LLM interacts with a single user—to more dynamic multi-party social interactions. In this evolving paradigm, LLMs are designed to mediate and participate in interactions involving multiple human users and AI agents (Park et al. 2023).

This shift is paving the way for innovative applications in:

1. Public and shared spaces: LLMs can assist in collaborative environments like homes, museums, airports, and hospitals, offering context-aware support that enhances the shared experience of multiple users. For instance, they might guide museum tours for groups or coordinate patient care discussions in medical settings.
2. Group coordination and collaboration: LLMs are being integrated into tools for tasks like event planning, human resources management, and collaborative gaming. By mediating interactions and managing logistics, these systems can simplify group decision-making and task delegation.

3 Generating ToM Scenarios for Training and Evaluation

3.1 A* Search of Scenarios

Sclar et al. (2024) employ A* search (Hart et al. 1968) to efficiently navigate the vast space of possible narratives and pinpoint those that are most likely to fool state-of-the-art LLMs. A* search is a widely used algorithm for finding the shortest path from a start node to a goal node in a weighted graph. It combines features of Dijkstra's Algorithm and Greedy Best-First Search, leveraging both path cost and heuristic estimates to guide its search efficiently. Nodes in a weighted graph represent mental states or locations, and edges represent transitions between states (mental actions such as *inform* or *accept*), with associated costs. Cost functions include $g(n)$, the cost of the path from the start node to node n. This is computed as the sum of edge weights along the path.

A* search of scenarios navigates the space of plausible scenario structures of up to m actions. We define this space as a directed graph, where each node is a sequence of valid mental actions $s = (a_1, \ldots, a_i)$ for a given mental state, and there is an edge between s and s' iff s is prefix of s', and s' contains k additional mental actions on top of s. $k \geq 1$ is the scenario extension step factor for mental actions, determining the granularity with which nodes are iterated through.

We add the estimator function for a scenario s as *interestingness*(s) that reflects the number of characters involved or the number of actions belonging to a subset $A' \subseteq A$ of important actions.

$h(s)$ is a heuristic that estimates the cost of the cheapest path from s to a goal node (one of the nodes where it would be acceptable to finish the search and complete the scenario). Goal nodes are those such that the estimator function *interestingness*(s') = 1. This is user defined and should ideally be admissible (never overestimates the true cost). $f(s) = g(s) + h(s)$ is the total estimated cost of the path through the node for scenario s.

We intend to build a scenario that is difficult for LLM to answer questions about. We express this in g defining $g(s)$ as the target LLM question answering accuracy computed for all questions for s. Heuristic function $h(s)$ is designed as an expression of the likelihood of producing a combined story $s + s'$ that satisfied our constraints *interestingness*(s') = 1, where s' is the continuation of story s.

$$h(s) = \alpha\left(\frac{1}{P}\sum_{i=1}^{P} count\left(interestingness\left(s' + s_i^{'}\right) == 1\right)\right.$$

All $s_i^{'}$ are randomly sampled continuations of s, and $0 \leq \alpha \leq 1$ is a scaling factor. Since A* requires to evaluate all neighbors of a node s but the space is too vast to explore, taking into account that each computation of $f(\cdot)$ is based on multiple LLM calls (one per question), the evaluation is restricted to a pre-determined limited number of neighbors. These neighbors are ordered by how well this node satisfies $interestingness\left(s' + s_i^{'}\right)$ condition.

This approach directs the A* search, enabling the creation of a robust dataset that challenges LLMs' ToM capabilities. By separating story structures from lexical realizations, we ensure that models are tested on their fundamental social reasoning rather than relying on superficial linguistic patterns. This distinction allows for a more precise evaluation of LLMs' ability to infer mental states without being influenced by stylistic cues.

We intend to find complicated stories for a given LLM by navigation through the space of stories supported by its domain-specific language for mental state tracking, iterating through k supported actions at a time. Assessment of complexity of each partial story is done through automatically generated questions with reliable ground-truth answers (shown in right bottom).

NL_MAMS framework is designed to monitor, represent, and update the mental states of agents (characters) within a simulated scenario or story. Mental states include beliefs, knowledge, assumptions, intentions, and goals of the agents, which may change dynamically based on events, actions, and interactions within the narrative.

The purpose of a NL_MAMS is as follows:

1. Tracking individual knowledge and beliefs and keeping a record of what each agent knows or believes about the world (e.g., "Peter believes Mike invested in e-commerce")—it distinguishes between true and false beliefs to handle cases of misinformation or partial knowledge.
2. Reasoning about other agents—NL_MAMS can represent what one agent knows or believes about another agent's mental state (e.g., "John thinks Peter does not know about the legal issues").
3. Assessing scenario complexity by monitoring how mental states evolve and interact, the tracker measures the complexity of a scenario in terms of how many beliefs, false beliefs, or nested beliefs (e.g., "Peter thinks Mike believes…") are involved.
4. Guiding story creation ensures that the progression of the story respects logical rules, such as maintaining consistent beliefs or triggering belief changes only when new information is shared.
5. NL_MAMS handles cases where agents hold incorrect beliefs or assumptions due to lack of information or intentional deception. NL_MAMS also handles nested mental states, tracking beliefs about beliefs (e.g., "John believes Peter knows about Mike's investment").

A scenario creation chart, depicted in Fig. 1, serves as the foundation for generating complex and plausible story templates. Using an LLM, the architecture constructs an initial story template. This template evolves through iterative updates as topics are discussed, locations of objects and agents are adjusted, and the physical states of objects, along with the physical and mental states of agents, are modified.

To identify a scenario with the desired complexity, the system conducts an exhaustive search across possible scenario structures. This search includes exploring the sequence of mental states and actions to ensure the scenario adheres to the

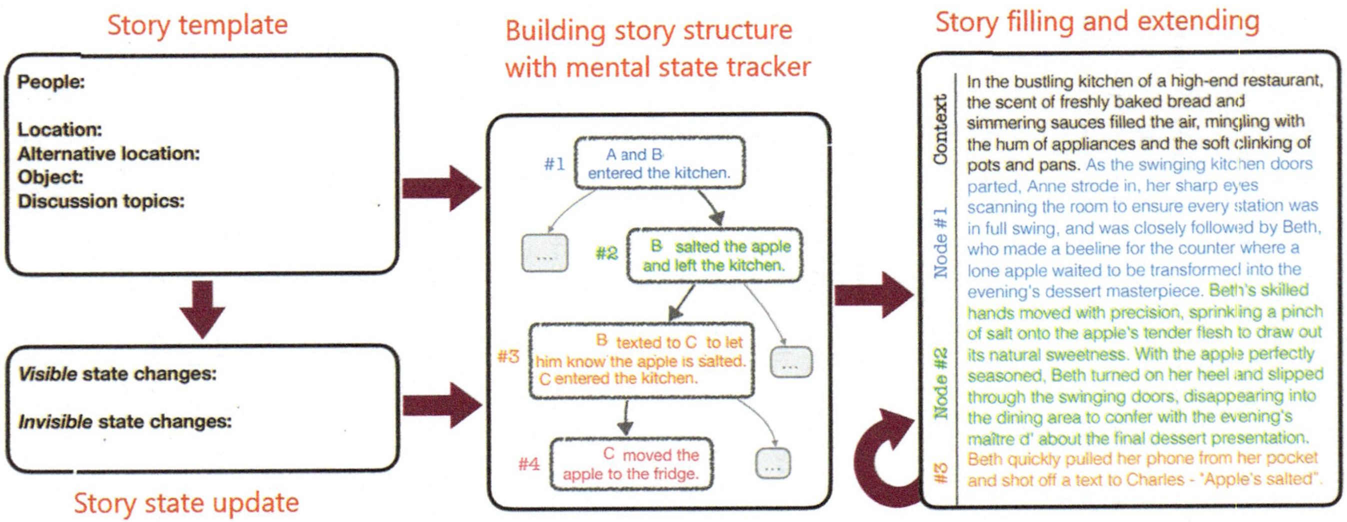

Fig. 1 ToM scenario composition architecture

specified parameters. NL_MAMS is employed during this process to quantitatively assess the complexity of each scenario, ensuring it meets predefined thresholds.

This architecture is designed to guarantee that the resulting stories are challenging tests for models, promoting their further improvement. Once a suitable scenario backbone is identified (e.g., nodes #1–4 in the scenario chart), it is iteratively expanded. Each iteration incorporates one mental action at a time using the LLM, which progressively refines the narrative into a coherent and natural-sounding story, ensuring it aligns with human readability standards.

While the extended scenarios serve as valuable training data, the benchmarking process relies on the scenario backbone structures. These backbones provide the highest level of reliability, offering a consistent framework for evaluating and comparing models. This twofold approach—using extended scenarios for training and backbone structures for benchmarking—maximizes both the quality of the stories and their utility in improving and testing language models.

3.2 Mental Actions and Mental States of NL_MAMS

ToM includes a diverse set of mental actions A, each transforming the world state and the beliefs of the people involved (the scenario state $s \in S$). A scenario is thus defined as a sequence of actions $(a_1, \ldots, a_n)$, where each action $a_i \in A$ is a function a_i: $S \rightarrow S$. Each action also has preconditions to be able to apply it, i.e., restrictions to its domain. Applying an action also automatically updates the mental state tracking and belief tracking: for example, "Peter is now trading company K." "Anne knows that Peter is now trading company K since she is sitting next to him and watching his monitor." All these updates and conditions are specifically programmed and tested.

A state $s \in S$ includes:

- Physical states p_s
- First-order beliefs b_1, second-order beliefs b_2, and third-order beliefs b_3

First-order beliefs describe what each person believes to be the current world state. Second-order beliefs describe what each person estimates that each other person believes to be the current world state. A third-order belief describes what one person thinks another person believes about what a third person believes about the world. For example, imagine three people: Alice, Bob, and Carol:

- First-order belief: Alice believes that the library is in Paris.
- Second-order belief: Alice believes that Bob thinks the library is in Rome.
- Third-order belief: Alice believes that Bob thinks Carol believes the library is in Paris.

Third-order beliefs are critical for understanding complex social interactions, such as reasoning about motivations, anticipating reactions in group dynamics, or simulating humanlike behavior in multi-agent systems like the pipeline described above.

For example, answering a query like "Does John think Peter knows Mike's favorite café in Paris?" involves reasoning at least to the third-order belief level.

A mental action transitions one state into another:

$$mental\_action(p_s, b_1, b_2, b_3) = (p_s', b_1, b_2, b_3)$$

Mental and physical actions and states follow the commonsense law of inertia. The law of inertia in this context assumes that beliefs, once established, remain consistent unless acted upon by new information or reasoning.

1. First-order belief: Alice believes the library is in Paris. The inertia principle suggests Alice will continue to believe this until she encounters evidence (e.g., a map or firsthand information) that contradicts her belief.
2. Second-order belief: Alice believes Bob thinks the library is in Rome. In the absence of new interactions or communications with Bob, Alice will maintain this belief about Bob's perspective.

Intentions describe the goals or purposes behind actions, and they can be extended to hierarchical levels, much like beliefs. First-order intention is a direct intention to perform an action or achieve a goal. Second-order intention describes what one person (or agent) intends about another person's intention. For example, "Alice has a goal involving influencing or aligning Bob's goal with hers." Third-order intention goes another level deeper and involves what one person intends about another person's intention regarding a third party. For example, "Alice intends that Bob wants Carol to believe the library is open."

We generate diverse stories by significantly expanding the range of mental actions, physical actions, and various forms of communication. These include private conversations between two characters, public broadcasts to all characters in a room, casual discussions about a topic referred to as chitchat, or notifications about changes in the world state. These actions can occur at any point in the scenario, allowing for a rich and dynamic narrative. Each new action requires carefully writing the implied belief and world state updates, which limits the number of actions that can be supported.

Belief update is diversified. We use the feature of belief asymmetricity in our scenarios, incorporating secret observers, security cameras, and managing witnesses without others' knowledge; characters can also be distracted, for instance, by their phones. This diversification allows for more realistic and complex social scenarios.

3.3 *Probabilistic Planner*

A ToM scenario can be represented as a forward generative model using a Partially Observable Markov Decision Process (POMDP) (Kaelbling et al. 1998). The forward generative model predicts how the agent's actions and beliefs evolve forward in time as it interacts with the environment. It accounts for state transitions, observations, goal-directed decision-making, and discounted future rewards. This approach is powerful for solving complex decision-making problems in uncertain environments, such as robotics, healthcare, and autonomous systems.

The model is defined by the tuple $\langle S, A, T, G, R, \Omega, O, \gamma \rangle$. $s^t \in S$ and $a^t \in A$ are the state and the action at time t. $T(s^t + 1|s, a)$ are the state transition probabilities. $g \in G$ is a goal, which defines the reward of the agent $r^t = R(s^t, a^t, g)$. $o^t \in \Omega$ is the agent's observation at t derived following the observation function, $o^t = O(s^t)$. Finally, $\gamma \in (0,1]$ is a discount factor, a scalar in (0,1] that weighs the importance of future rewards versus immediate rewards. Since the agent cannot directly observe the true state s^t (due to partial observability), it maintains a belief state $b(s)$. The belief state is a probability distribution over all possible states SSS. It represents the agent's estimate of where it is in the environment, based on past actions and observations. For ToM, the belief is further factorized into beliefs about the possible locations of individual objects, simplifying the problem.

Conditioned on both the goal and the belief, a rational agent will take actions based on the optimal policy $\pi(a^t|g, b^t)$ to maximize its return $\sum_{t=0}^{\infty} \gamma^t r^t$.

Given this forward generative model, one can conduct **inverse inference** about the agent's goal and belief (Jin et al. 2024). By assuming a deterministic state transition, it is possible to jointly infer the goal and belief of an agent given observed mental states and actions:

$$P\left(g, b^t \middle| s^{1:t}, a^{1:t-1}\right) \propto \prod \pi\left(a^\tau \middle| g, b^\tau\right) P\left(b^\tau \middle| b^{\tau-1}, s^\tau\right) \cdot P\left(b^0\right) P(g)$$

For two expressions about a mental states that include intent and belief, $H_1 = \langle g_1, b_1^t \rangle$ and $H_2 = \langle g_2, b_2^t \rangle$, one can compute which one is more likely to hold as

$$\frac{P(g_1, b_1^t | s^{1:t}, a^{1:t})}{P(g_2, b_2^t | s^{1:t}, a^{1:t})} = \frac{\pi\left(a^t \middle| g_1, b_1^t\right) P\left(b_1^t \middle|, \hat{b}^{t-1}, s^t\right)}{\pi\left(a^t \middle| g_2, b_2^t\right) P\left(b_2^t \middle|, \hat{b}^{t-1}, s^t\right)} = \frac{\prod_{\tau=1}^{t-1} \pi\left(a^\tau \middle| g_1, \hat{b}^\tau\right)}{\prod_{\tau=1}^{t-1} \pi\left(a^\tau \middle| g_2, \hat{b}^\tau\right)}$$

where $\hat{b}^\tau$ is the estimated belief at a past step $\tau < t$. Since the mental state in the question is about the belief at the current step, the belief can be computed in the past steps to form a full belief state description. Agent's mental state is expressed as the subset of the state predicates that the agent can observe; her beliefs are updated accordingly. Hence the agent belief is computed as $\hat{b}^\tau = \hat{b}^\tau(s^t)$ at each past step.

Now we can estimate the probability of the last action a^t at given the most probable belief and goal $(\pi(a^t|g, \hat{b}^\tau)$. Then one can compute the probability of a hypothetical belief at the last step $P\left(b^t|, \hat{b}^{t-1}, s^t\right)$. Moreover, the probability of all past actions given the hypothetical goal and the estimated belief prior to the current step $\prod_{\tau=0}^{t-1} \pi\left(a^\tau |, g, \hat{b}^\tau\right)$ can be estimated.

What cannot be easily estimated is the policy, so to overcome this limitation, an LLM comes into play. Mental states are represented symbolically as a list of possibilities for goals. The LLM is then prompted with the symbolic representations of the mental state s^t, respective goal g, and estimated belief $\hat{b}^t$. The purpose is to generate the likelihood of the observed action a^t based on the output logits. Figure 2 illustrates how the likelihood estimation works in a qualitative example. The LLM needs to be fine-tuned on the mental states and actions in the training dataset of correct scenarios.

The action probability ratio is as follows:

$$\frac{\pi(a^t =' \text{get account value}' | g =' \text{bank account}', b^t = \text{balance}(\mathbf{balance != 0}), s^t)}{\pi(a^t =' \text{get account value}' | g =' \text{bank account}', b^t = \text{balance}(\mathbf{balance == 0}), s^t)} = 1.6$$

Now let us imagine that Mike got a letter from IRS either confirming his refund or requesting to report additional income. The question about the balance is the same as the above. At the previous step $t-1$, the ration between action probabilities:

$$\frac{\pi(a^{t-1} =' \text{receive confirmation}' | g =' \text{confirmation\_refund}', b^{t-1}, s^{t-1})}{\pi(a^{t-1} =' \text{receive confirmation}' | g =' \text{confirmation\_addit\_income}', b^{t-1}, s^{t-1})} = 0.6$$

Scene: … Mike is about to access his checking account for the tax refund

Actions: … He is about to move funds to his saving account

Question: … If Mike is trying to fund his saving account, which one of the following statements is more likely to be true?

a) Mike thinks that he has a non-zero balance
b) Mike thinks that he has a zero balance

Fig. 2 Examples of how the probabilities of different hypotheses are assessed via the action likelihood estimation from the LLM

At the current step *t*:

$$\frac{\pi\left(a^t =' \text{pos}'_{\text{balance}} \,|g =' \text{received}\_refund',b^t,s^t\right)}{\pi\left(a^t =' \text{zero}'_{\text{balance}} \,|g =' \text{not}\_\text{received}\_refund',b^t,s^t\right)} \frac{\pi\left(a^{t-1} \,|g =' \text{received}\_refund',b^{t-1},s^{t-1}\right)}{\pi\left(a^{t-1} \,|g =' \text{not}\_\text{received}\_refund'b^{t-1},s^{t-1}\right)} = 1.2$$

4 Simulating Mental Actions and Mental States

NL_MAMS models agents' mental states and actions using formal logic (e.g., Prolog-like clauses) and a state machine to compute transitions between mental states and actions. The system simulates interactions between agents, evaluating how one agent's mental actions (e.g., inform, deceive, warn) affect the mental states (e.g., beliefs, desires, intentions) of others and lead to subsequent actions.

We have the following features of the simulator:

1. Using a simulation of decision-making rather than representing it as a pure deduction (see, e.g., Bousquet et al. 2004)
2. Describing the multi-agent interaction, ascend from the level of atomic actions of agents to the level of behaviors
3. Limiting the complexity of mental formulas (Fig. 3)
4. Following closer the natural language in describing the mental world, using a wide range of entities (this has been explored with respect to acceptance by a multi-agent community by de Lara and Alfonseca (2000))
5. Taking advantage of approximation machinery. We express an arbitrary mental entity through the basis knowledge-belief-intention (*informing*, *deceiving*, *pretending*, *reconciling*, etc.; Galitsky 2006a)
6. Using a hybrid reasoning system combining simulation of decision-making with the set of typical behaviors specified as axioms

Fig. 3 *believe(husband, pretty(wife))believe(wife, pretty(husband)) believe(husband, believe(wife, pretty(husband))) believe(wife, believe(husband, pretty(wife)))*

7. Increasing the expressiveness of representation language by means of using an extensive set of formalized mental entities beyond belief and desire.

The simulator includes mental action clauses that define the conditions under which actions occur. Each clause includes preconditions (logical conditions that must hold true) and effects (changes to the mental states of agents).

State transition engine inputs the initial states of agents and a sequence of mental actions, evaluates clauses to determine which actions are triggered based on current states, and outputs the updated states of agents and a record of executed actions.

In an example workflow:

1. Evaluate if *A* can perform *inform*(*A*, *B*, *X*) based on its current beliefs and desires.
2. Update *B*'s beliefs to include *know*(*B*, *X*) and log the action for traceability.

The simulation environment includes a shared context where agents interact, containing facts (shared knowledge accessible to all agents) and logs (history of actions and state transitions).

4.1 Formal Definitions for Mental Actions

The list of mental actions (Fig. 4) is as follows:

Cognitive Mental Actions
(Processes related to thinking, reasoning, and decision-making)

1. **Think (about)**: Engaging in deliberate contemplation
2. **Remember**: Recalling past events or information
3. **Infer**: Drawing logical conclusions based on evidence
4. **Realize**: Becoming aware or understanding something

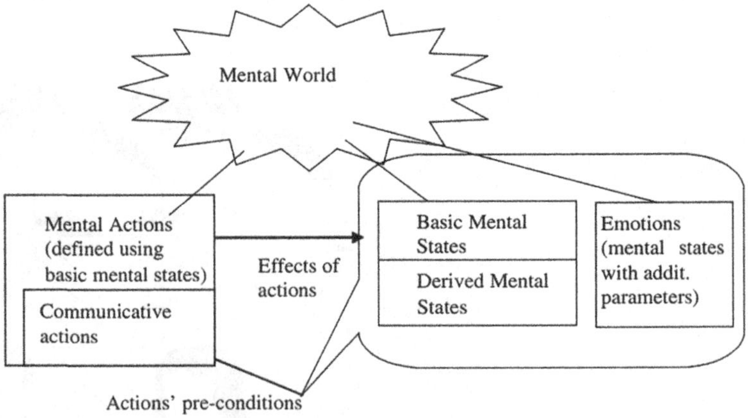

Fig. 4 Mental states, actions, and emotions as components of mental world

5. **Calculate**: Using logic or mathematics to determine outcomes
6. **Plan**: Formulating a sequence of actions
7. **Decide**: Making a choice among alternatives
8. **Solve**: Finding a solution to a problem or challenge
9. **Critically Analyze**: Breaking down complex ideas for evaluation
10. **Evaluate**: Assessing the quality, value, or significance of something
11. **Visualize**: Imagining or picturing a scenario mentally
12. **Generalize**: Drawing broad conclusions from specific observations
13. **Specialize**: Focusing on a specific area or topic
14. **Synthesize**: Combining ideas or concepts to form a new whole
15. **Anticipate**: Predicting future events or outcomes

Emotional Mental Actions
(Actions related to feelings, desires, and attitudes)

1. **Regret**: Feeling sorrow or disappointment over a past event
2. **Admire**: Feeling respect or approval for someone or something
3. **Empathize**: Understanding and sharing another person's emotions
4. **Trust**: Believing in the reliability or truth of someone or something
5. **Distrust**: Doubting the reliability or truth of someone or something
6. **Blame**: Assigning responsibility for a negative outcome
7. **Forgive**: Letting go of resentment or anger toward someone
8. **Wish**: Hoping or desiring for something to happen
9. **Mourn**: Experiencing sorrow or grief over a loss
10. **Rejoice**: Feeling or expressing great joy or happiness
11. **Hesitate**: Feeling uncertainty or reluctance before acting
12. **Procrastinate**: Delaying an action due to emotional resistance
13. **Feel Inspired**: Experiencing a surge of motivation or enthusiasm
14. **Celebrate**: Feeling or expressing happiness about an achievement
15. **Appreciate**: Recognizing the value or significance of something

Social Mental Actions
(Actions that involve interaction with or consideration of others)

1. **Inform**: Sharing knowledge or information with someone
2. **Explain**: Making something clear or understandable to someone
3. **Remind**: Prompting someone to recall something
4. **Persuade**: Convincing someone to change their belief or behavior
5. **Encourage**: Offering support or motivation to someone
6. **Promise**: Committing to a future action or course
7. **Apologize**: Expressing regret for an action or event
8. **Negotiate**: Engaging in discussion to reach an agreement
9. **Challenge**: Questioning someone's belief or decision
10. **Reconcile**: Resolving conflicts or differences
11. **Collaborate**: Working together with others toward a common goal
12. **Support**: Providing help or encouragement

13. **Criticize Constructively**: Offering feedback aimed at improvement
14. **Blame**: Assigning responsibility for a negative outcome
15. **Empower**: Enabling someone to take action confidently

These examples demonstrate various mental actions framed in the context of reasoning about actions and mental states. Each includes:

1. **Preconditions**: Beliefs and states before the action
2. **Action**: The core activity (e.g., informing, persuading, encouraging)
3. **Postconditions**: Changes in mental states, trust, or motivation

```
promise(C, T, Commitment, Goal) :-
        want(C, Goal),
        not Action(C, Goal),
        believe(C, (Action(C, Goal) :- Commitment)),
        inform(C, T, Commitment),
        believe(T, Commitment),
        believe(T, (Action(C, Goal) :- Commitment)),
        Action(C, Goal).
```

% Conditions:
% 1. C wants the Goal but cannot immediately perform the Action to achieve it.
% 2. C believes that making a commitment will obligate him to act toward the Goal.
% 3. C informs T about the commitment.
% 4. T believes C's commitment and expects C to act toward the Goal.

```
apologize(C, T, Event, Regret) :-
        believe(C, (Event caused Harm(T))),
        feel(C, Regret),
        inform(C, T, (sorry(Event))),
        believe(T, (C regrets Event)),
        trust(T, C).
```

% Conditions:
% 1. C believes the Event caused harm to T.
% 2. C feels regret about the Event.
% 3. C communicates regret to T explicitly (e.g., saying "sorry").
% 4. T interprets this as genuine regret and may restore trust in C.

```
encourage(C, T, Goal) :-
        believe(C, (T capable_of Action(T, Goal))),
        believe(T, not Action(T, Goal)),
        inform(C, T, (you can achieve Goal)),
        believe(T, (C supports Goal)),
```

```
        feel(T, motivated),
        perform(T, Action),
        Action(T, Goal).
```

% Conditions:
% 1. C believes T is capable of performing the Action to achieve the Goal.
% 2. T has not yet performed the Action due to doubt or lack of motivation.
% 3. C communicates belief in T's ability to achieve the Goal.
% 4. T feels motivated by C's encouragement and performs the Action.

```
reassure(C, T, Concern) :-
        believe(T, (Concern may lead to NegativeOutcome)),
        believe(C, not (NegativeOutcome)),
        inform(C, T, (Concern is under control)),
        believe(T, not (NegativeOutcome)),
        feel(T, relieved).
```

% Conditions:
% 1. T believes the Concern may lead to a negative outcome.
% 2. C believes the Concern is not valid or the negative outcome is unlikely.
% 3. C communicates this belief to T.
% 4. T feels reassured and no longer fears the negative outcome.

```
persuade(C, T, Action, Argument) :-
        want(C, Action(T, Goal)),
        believe(T, not Action(T, Goal)),
        believe(C, (Argument leads_to Action(T, Goal))),
        inform(C, T, Argument),
        believe(T, Argument),
        perform(T, Action),
        Action(T, Goal).
```

% Conditions:
% 1. C wants T to perform an Action that leads to a Goal.
% 2. T has not yet performed the Action.
% 3. C believes that a specific Argument will convince T to perform the Action.
% 4. C communicates the Argument to T.
% 5. T believes the Argument and performs the Action.

```
teach(C, T, Concept, Method) :-
        know(C, Concept),
        not know(T, Concept),
```

```
inform(C, T, Method),
perform(T, Learn(Method, Concept)),
know(T, Concept).
```

% Conditions:
% 1. C knows the Concept but T does not.
% 2. C uses a specific Method to convey the Concept to T.
% 3. T engages in a learning process (e.g., practice, reflection).
% 4. T eventually learns and understands the Concept.

4.2 *Choosing Your Own Mental Action*

Let us first look at an action selection algorithm in a trivial case, where an agent does not consider the possible actions of others. Of particular importance to us are the systems that allow agents to learn about and model their own teammates and then use that knowledge to improve collaboration.

To choose the best action, each agent considers each action it can currently perform (Fig. 5). Firstly, each agent selects a set of actions it can legally perform at the current step (physically available for the agents, acceptable in terms of the norms, etc.). Such an action may be explicitly wanted or not; also, this action may belong to a sequence of actions in accordance with a form of behavior that has been chosen at a previous step or is about to be chosen. In the former case, the agent may resume the chosen behavior form or abort it.

Having a set of actions that are legal to be currently performed, the agent applies a preference relation. This relation is defined on states and actions and sets the following order (1 is preferred over 2–5, 2 is preferred over 3–5, etc.):

1. Explicitly preferred (desired) action: This is the action you most want to take, leading directly to the goal or preferred outcome.
2. Action that brings about the desired state that is currently unmet: This is an action that, if performed, will create the state you desire but is not yet achieved.
3. Action that removes an existing unwanted state: This refers to actions aimed at eliminating a current unwanted condition or situation.
4. Action that avoids leading to an unwanted state: This involves actions that prevent the creation of an unwanted condition that hasn't yet materialized.
5. Action that preserves a current desired state: This action helps maintain or ensure the continuation of a state that is already in line with your preferences.

The sequence of preference conditions is as follows:

```
want (A, ChosenAction),
want (A, State), not State, assume(ChosenAction), State,
want (A, not State), State, assume(ChosenAction), not State,
```

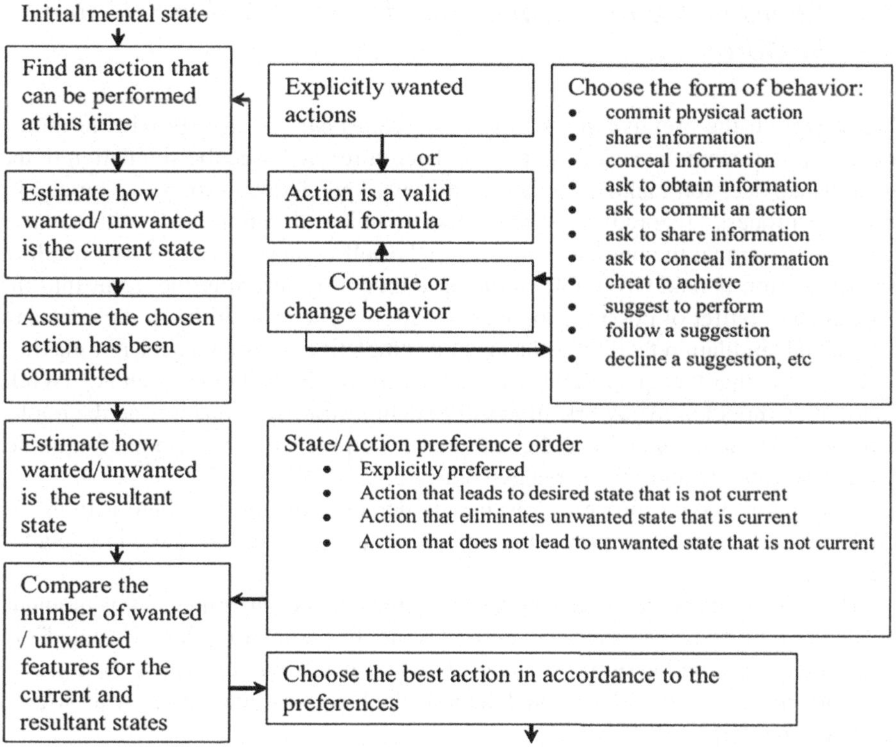

Fig. 5 The chart for the choice of action, involving own agent capabilities and world knowledge (simplified case)

```
not (want(not State), not State, assume(ChosenAction), State),
not (want(State), State, assume(ChosenAction), not State).
```

A compound action of a given agent may include actions of other agents and various intermediate states, some of which the agent may want to avoid. The agent decides either to perform the action delivering the least unwanted state or action of another agent or to do nothing. If there are multiple possible actions that do not lead, in the agent's belief, to unwanted consequences, this agent either chooses the explicitly preferred action, if there is an explicit preference predicate, or the action whose choice involves the least consideration of the beliefs of other agents. Hence, the agent *A* has an initial intention concerning a *ChosenAction* or *State*, assesses whether this condition currently holds, then selects the preferred *ChosenAction*, assumes that it has been executed, deduces the consequences, and finally analyses whether they are preferential. The preference, parameters of agents' attitudes, and multi-agent interactions may vary from scenario to scenario.

4.3 *Choosing the Best Action Considering an Action Selection by Others*

We start with the premise that humans use themselves as an approximate, initial model of their teammates and opponents. Therefore, we based the simulation of the teammate's decision making on the agent's own knowledge of the situation and its decision process. To predict the teammate's choice of actions in a collaborative strategy, we model the human as following the self-centered strategy. The result of the simulation is made available to the base model by inserting the result into the "imaginary" buffer of possible opponents' actions. The availability of the results of the mental simulation facilitates the agent's completion of its own decision-making. The effect is that the agent yields to what it believes is the human's choice. While this simple model of teamwork allows us to demonstrate the concept and the implementation of the simulation of the teammate, we proceed to the simulation mode, which uses the collaborative strategy recursively.

A decision-making process flowchart is depicted in Fig. 6, where actions are chosen based on an evaluation of their outcomes and preferences. Here is a detailed description.

The process starts by identifying an action that can be performed at this moment. Evaluation of state assesses the current state by estimating how "wanted" or "unwanted" it is and assumes the chosen action has been performed. Actions are categorized into explicitly wanted actions and actions that align with logical decision-making.

The form of behavior can vary, including:

- Committing a physical action
- Sharing, concealing, or obtaining information
- Asking others to perform an action or provide information
- Suggesting or following a suggestion
- Declining suggestions, etc.

Knowledge integration process involves personal knowledge as well as knowledge about others' perspectives or knowledge. It is assumed that the actions chosen by others have also been performed. Resultant state evaluation comprises re-assessment of the resultant state after the action is assumed completed and comparison of the number of "wanted" or "unwanted" features in the current state versus the resultant state.

The possible actions of others are simulated by involving knowledge of others' knowledge and perspectives. Involving their own knowledge of others' knowledge, the decision-maker considers what they know about others' beliefs, preferences, and knowledge. This step requires understanding others' goals, constraints, and likely courses of action based on their situation.

Putting yourself in their position (introspection) involves adopting the perspective of others and reasoning about what actions they might take if they were in a given state. This perspective-taking helps predict actions that align with others' motivations or constraints. After predicting potential actions of others, it is assumed

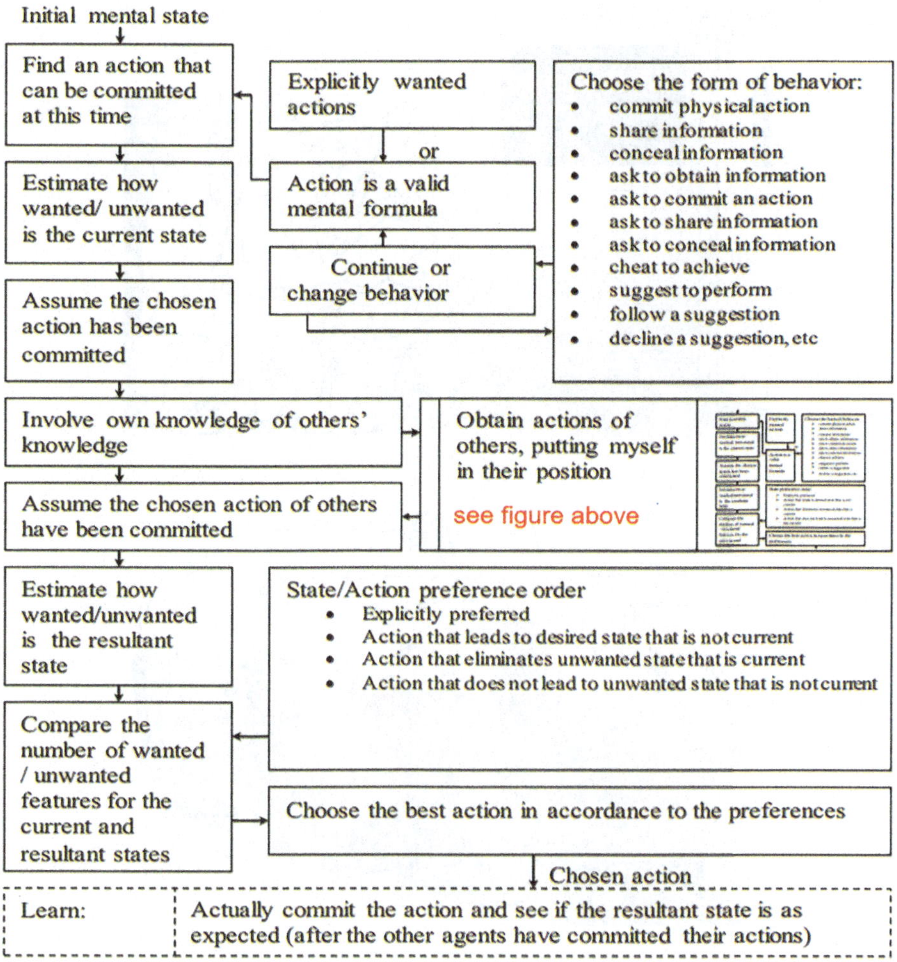

Fig. 6 Selecting an action considering own good and also performing introspection on behalf of others

that these actions have been carried out. The decision-maker then evaluates the likely resultant state based on these assumptions. The effects of others' simulated actions are combined with the effects of the decision-maker's own potential actions. The resultant state is analyzed for its "wanted" or "unwanted" features (Fig. 7).

5 NL_MAMS_LLM System Architecture

The overall system architecture is depicted in Fig. 8. NL_MAMS_LLM (Natural Language Multi-Agent Mental Simulator over LLM) produces scenarios as sequences of mental states → actions → states → .... NL_MAMS_LLM yields

Fig. 7 *want(police, know(police, like(prisoner, work(police))))*, on the top; *want(passenger, control(pilot, airplane, cockpit))*, on the bottom

domain-independent pure mental scenarios that need to be in-filled with data from application domains. In-filing engine takes a scenario skeleton and fills it with physical actions and states produced by an LLM. ToM scenarios are further multiplied and made more complex by A* search and probabilistic planning techniques. The former identifies additional plausible sequences of mental states, and the latter leverages a probabilistic mechanism to yield richer mental states.

Here is a detailed breakdown of the process:

1. NL_MAMS is responsible for producing skeleton scenarios based purely on sequences of mental states and mental actions. These scenarios are domain-independent, focusing on abstract mental processes such as beliefs, goals, and

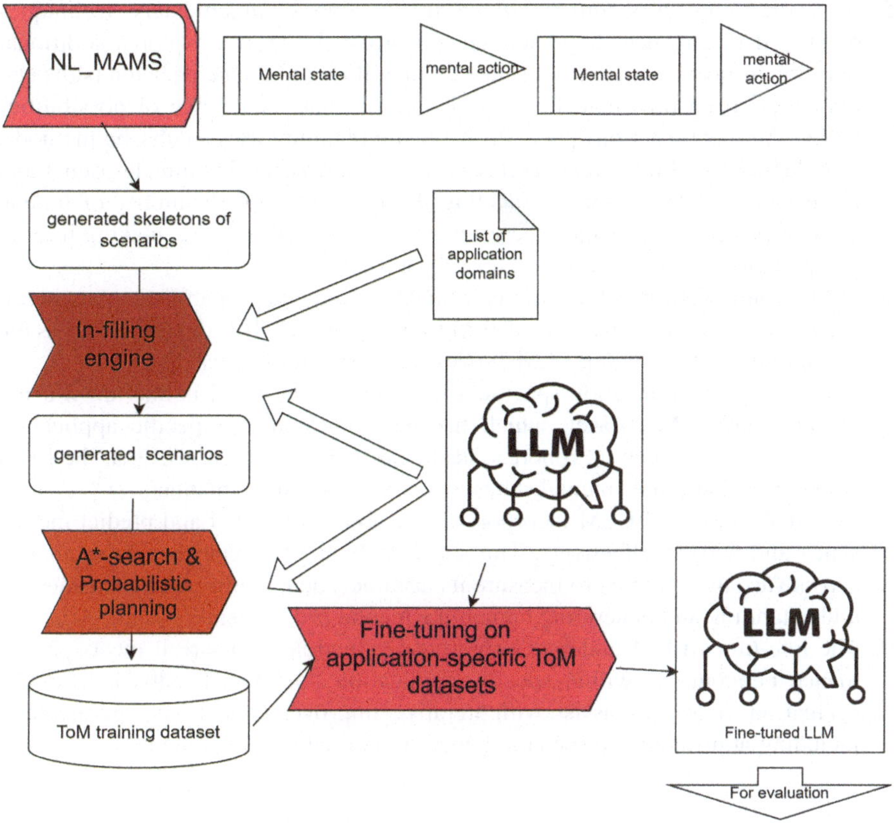

Fig. 8 NL_MAMS_LLM architecture

intentions and their evolution through mental actions (e.g., decision-making, reasoning).

2. Generated skeletons of scenarios, the outputs of NL_MAMS, are raw mental action-state sequences that act as the foundation for creating more concrete and domain-specific scenarios. These skeletons are devoid of physical context and need to be enriched with real-world applicability.
3. In-filling engine takes the skeleton scenarios and enriches them with physical actions and states using outputs generated by an LLM. The in-filling process ensures that the abstract mental scenarios are grounded in specific application domains, guided by a predefined list of application domains (e.g., healthcare, education, human-robot interaction). The result is a set of complete scenarios where mental states are intertwined with realistic actions and contexts.
4. Generated scenarios: once in-filled, the scenarios are ready for further refinement and augmentation. These scenarios now include both mental and physical aspects, creating a more holistic representation.

5. *A* search and probabilistic planning* are two complementary techniques employed to enhance the generated scenarios: *A* search* explores additional plausible sequences of mental states by identifying alternate decision paths and outcomes. It ensures that the scenarios cover a broader range of possibilities within the mental action space. Probabilistic planning uses stochastic methods, introducing variability and uncertainty into the scenarios. It simulates richer and more diverse mental states, reflecting the inherent unpredictability of human thought processes. Together, these methods expand the dataset, making it more comprehensive and better suited for robust training.
6. ToM training dataset is the outputs from A* search and probabilistic planning; it is consolidated into a specialized ToM training dataset. This dataset serves as the foundation for fine-tuning LLMs to develop ToM capabilities.
7. Fine-tuning on application-specific ToM datasets: the LLM is fine-tuned using the enriched ToM dataset, which has been tailored for specific application domains (e.g., education, robotics, healthcare). This step ensures that the LLM can infer and simulate mental states in a context-sensitive manner.
8. The final fine-tuned LLM incorporates the ability to model and predict mental states and actions effectively. This ToM-enabled LLM can then be evaluated across various scenarios to measure its accuracy, adaptability, and robustness in understanding and generating humanlike mental processes.
9. The fine-tuned LLM undergoes rigorous evaluation to ensure it meets performance benchmarks within specific application domains. Feedback from the evaluation stage can assist with iterative improvements in the architecture, including adjustments to scenario generation or dataset refinement.

6 Evaluation

We evaluate a model's understanding of a scenario—either from an existing dataset or generated in this study—by asking questions related to a mental state or action. The performance of NL_MAMS_LLM is assessed using automatically generated question-answer pairs. The answers produced by NL_MAMS_LLM are more accurate and closely tied to the mental world compared to those generated by a pure LLM-based system, as they are directly derived from the evolution of computed mental states.

The questions test various levels of reasoning, including first-order beliefs or intentions, second-order beliefs or intentions, and more complex mental states. These questions can refer to the current mental state (ground truth) or prior states (memory states).

To enhance the complexity of mental state questions compared to competitive systems, we include queries about intermediate mental states (e.g., "What was the mental state before event E occurred?") rather than focusing solely on initial states (e.g., "What was A's intent at the beginning?"). Assessing the performance on a

Table 1 Applied domains for ToM evaluation and training

Domains	# of scenarios	First-order beliefs per scenario	Second-order beliefs per scenario	First-order intention per scenario	Second-order intentions per scenario
Therapy and mental health support	50	3.72	1.36	4.22	0.23
Friendship and romantic relationships	50	2.81	1.10	3.98	0.18
Education and teaching	100	3.60	0.91	3.13	0.17
Human-computer interaction	100	3.29	1.24	3.39	0.21
Explainable AI	100	3.71	1.09	4.21	0.21
Multi-agent reinforcement learning	50	2.79	1.10	3.41	0.19
Autonomous vehicles (need to predict pedestrian intentions)	100	2.62	1.30	3.81	0.19
Gaming and virtual reality	100	3.58	1.05	3.78	0.21
Negotiation and strategic interaction	50	2.72	1.02	3.15	0.16
Surveillance and security	100	2.95	1.32	4.21	0.17
Assistive technologies for individuals with disabilities	100	3.30	1.21	3.46	0.19

given question is straightforward, as it typically involves yes/no responses or specific details about an entity or agent, such as identifying an object, container, or location.

The specific formulation of questions varies based on the property being tested, such as location (e.g., "Where does Charles think Anne will search for the apple?") or knowledge (e.g., "Does Charles know that the apple is salted?").

6.1 Application Domain-Specific Dataset

The size of the training dataset and complexity of ToM scenarios are shown in Table 1 as the number of first- and second-order beliefs and intentions. Second-order intentions like "I want her to believe…" or "I want her to become interested…" are not very frequent but nevertheless occur in one out of six scenarios on average. The mental states of third degree are rare (Fig. 9).

Fig. 9 The third-degree mental state "When I said I wanted … I meant…"

Table 2 NL_MAMS_LLM performance in applied domains for ToM evaluation and training

Domains	NL_MAMS	GPT-4	NL_MAMS_LLM
Therapy and mental health support	56.7	89.4	94.1
Friendship and romantic relationships	43.4	84.5	92.3
Education and teaching	43.3	82	87.5
Human-computer interaction	42.0	89.1	91.7
Explainable AI	46.9	84.3	93.2
Multi-agent reinforcement learning	52.3	82.5	88.4
Autonomous vehicles (need to predict pedestrian intentions)	42.6	87.4	89.7
Gaming and virtual reality	44.7	87.1	89.7
Negotiation and strategic interaction	42.7	85.4	88.3
Surveillance and security	50.0	88	94.9
Assistive technologies for individuals with disabilities	38.7	84.1	91.2
Average	45.8	85.8	91

6.2 Performance in Applied Domains

Question-answering accuracy for our application domain-specific datasets is shown in Table 2. The legacy simulation system, NL_MAMS, was not designed for question answering but rather for producing a plausible sequence of mental states. As a result, it correctly answers less than half of the questions. GPT-4 achieves an accuracy of almost 86%, which is comparable to other estimates of the performance on complex ToM tasks, especially considering our domain-specific scenarios.

NL_MAMS_LLM improves this accuracy by 5% through fine-tuning across a manifold of scenarios spanning all application domains, rather than being restricted to the specific domain it is evaluated on.

6.3 Comparison with Other LLM ToM Systems

We select a number of comparable dataset and LLM improvements oriented toward ToM to compare the performance of NL_MAMS_LLM with competitive approaches. We also evaluate SimToM (Wilf et al. 2023), SimpleToM (Gu et al. 2024), and SymbolicToM (Sclar et al. 2023), the recent approaches that improve ToM in LLMs through better prompting and other means. The performance against BIP ALM (Bayesian Inverse Planning Accelerated by Language Models; Jin et al. 2024) that extracts unified representations from multimodal data and utilizes LLMs for scalable Bayesian inverse planning is conducted as well. Additionally, we attempt to match the performance with ExploreToM (Sclar et al. 2024) generated dataset used for fine-tuning. We also use similar evaluation benchmarks, the dataset of higher-order beliefs (He et al. 2023) and FANToM (Kim et al. 2023), the dataset of facts and beliefs. The datasets sizes vary between few hundred to few thousand scenarios.

The question-answering accuracy results are shown in Table 3. The best accuracy for given evaluation setting is bolded. We observe a comparable performance of NL_MAMS_LLM among SimToM + GPT-4, SymbolicToM + GPT-4, and ExploreToM. In two cases out of eight (SimpleToM.Judgement and SymbolicTom. Belief_Inference), NL_MAMS_LLM demonstrates a superior performance. Most of the evaluation settings in Table 3 turns out to be harder for the ToM

Table 3 Performances of LLMs and their ToM oriented extensions

LLM/dataset	SimpleToM Mental state	SimpleToM Behavior	SimpleToM Judgment	SymbolicToM Belief inference	SymbolicToM Goal inference	Hi-ToM	FANToM
Human				91.0	74		
GPT 3.5	36.5	12.2	53.0	43.0	33		
GPT 4o	95.6	82.8	42.2	62.0	34		
Llama-3.1-8B	**97.8**	89.5	25.8	56.3	**44.3**		
Claude-3.5-Sonnet	96.9	**97.9**	24.9				
SimToM + GPT-4				64.3	40.7		
SymbolicToM + GPT-4				78.3	47.7		
ExploreToM						**59**	**0.2**
NL_MAMS_LLM	87.2	87.9	**55.4**	**84.2**	41.4	49.2	0.17

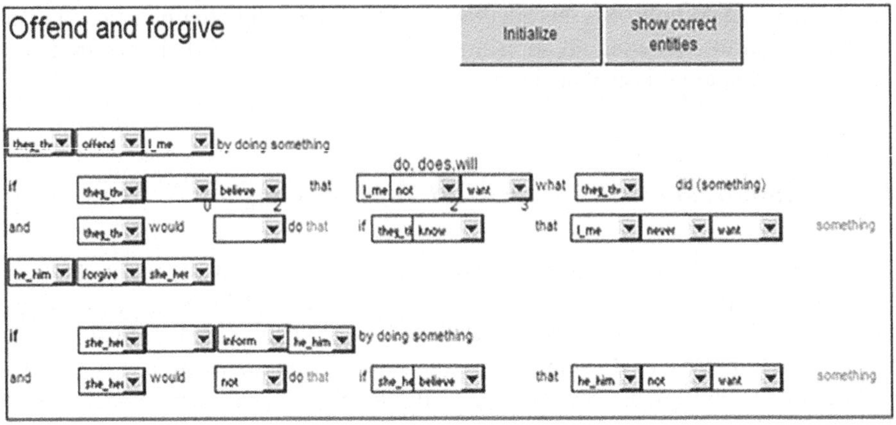

Fig. 10 A form to introduce a mental entity (here, to offend and to forgive). After an entity is explained via examples and verbal definition is told, a trainee is suggested to choose the proper basic entities with negations when necessary to build a definition

question-answering systems like NL_MAMS_LLM than our 11-application domain-specific setting. Our observation is that artificial ToM domains are more complicated than the ones matching the real-world ToM applications.

7 Training Environment for Teaching ToM

A few versions of the Web-based user interface for NL_MAMS_LLM have been developed for a number of environments, including describing of mental states of scene characters. A variety of interface components were designed for specifying mental states, including the natural language and drop-down box-based concept. The one-on-one rehabilitation strategy, conducted by a member of rehabilitation staff, includes the following components:

1. Direct introduction of the basic mental entities using real-world examples (Fig. 10)
2. Explanation of derived mental entities using the terms basis, intend-know, and believe
3. Introduction of the derived mental entities by means of real-world examples (Fig. 10)
4. Conversations that heavily rely on a discourse with mental focus (Lovaas 1987)
5. Conversations that are based on a pictorial representation of interaction scenarios (Fig. 11)
6. Involving the trainees into actual interactions with other children and asking them to verbally represent these interactions

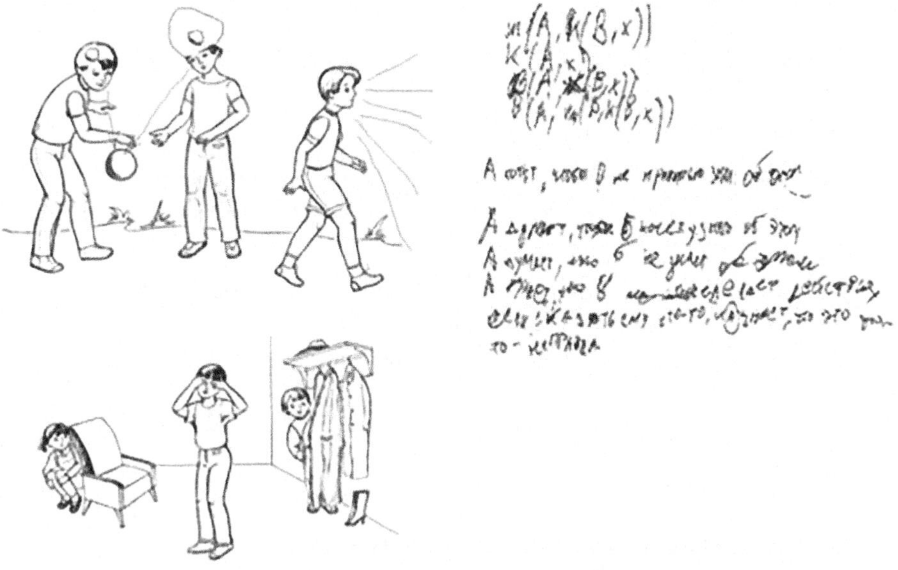

Fig. 11 A visualization of interaction scenarios. "Picture-in-the head" and "thought-bubbles" techniques are used based on "physical" representation of mental attitudes (on the top left). "Hide-and-seek" game (on the bottom left). The children are asked the questions about who is hiding where, who wants to find them, and about their other mental states. On the right: a trainee wrote a formula for "forcing to commit an unwanted action" and the respective natural language conditions for it

7. Encouraging the parents and rehabilitation personnel to demonstrate a special awareness of mental entities in the real world (Galitsky 2002, 2006b)
8. "Picture-in-the-head" and "thought-bubbles" techniques, using "physical" representation of mental attitudes (Swettenham 1996, Fig. 11)

NL_MAMS_LLM–based training is intended to assist in all of the above components. Initially, a trainer shows how to represent mental states from the above components via NL_MAMS_LLM and discusses yielded scenarios with a trainee. The plausibility and appropriateness of actions yielded by NL_MAMS_LLM require special attention from trainees. Then the trainer specifies other initial mental states and asks a trainee to come up with plausible scenarios originating from these mental states. Trainees are children with high-functioning autism 6–10 years old, selected so that they are capable of reading simple phrases and communicating mental states in one or another way. After a certain number of demonstrations, the trainees are encouraged to use NL_MAMS independently, applying it to real-world mental states the trainees have experienced, as well as abstract mental states. Trainees are presented with both natural language and structured input and output of NL_MAMS, and they are free to choose their favorite way of user interface.

Finally, we present a screenshot of UI for NL_MAMS_LLM where the trainee specifies the mental actions and initial states of characters (Fig. 12).

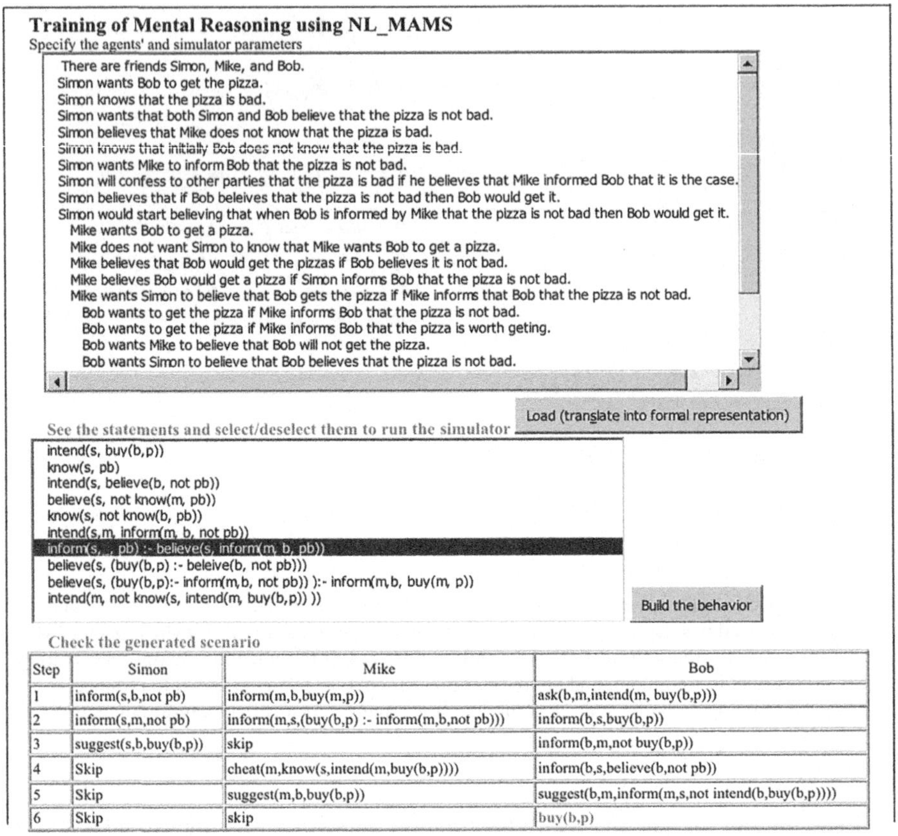

Fig. 12 The NL_MAMS_LLM user interface for rehabilitation of autistic reasoning (advanced scenario)

8 Auxiliary ToM Reasoning

8.1 Mutual ToM

When human beings are interacting with an agent with ToM capability, the mutual ToM (MToM) framework, which refers to a constant process of reasoning and attributing states to each other, is considered the analysis of the collaboration process in some studies (Wang and Goel 2022).

The mutual ToM is depicted in Fig. 13. Humans and agents act in a shared workspace to complete interdependent tasks, making independent decisions while using ToM to infer each other's state. They observe actions as implicit communication and use messages for explicit verbal communication. We label the communication pathways shaped by ToM, as the MToM process influences explicit communication,

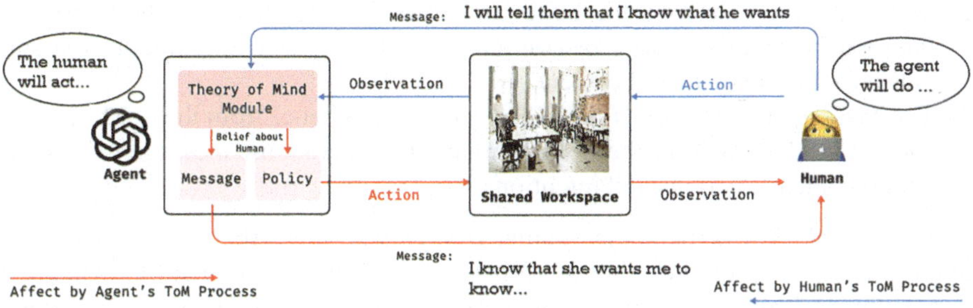

Fig. 13 An architecture for mutual ToM

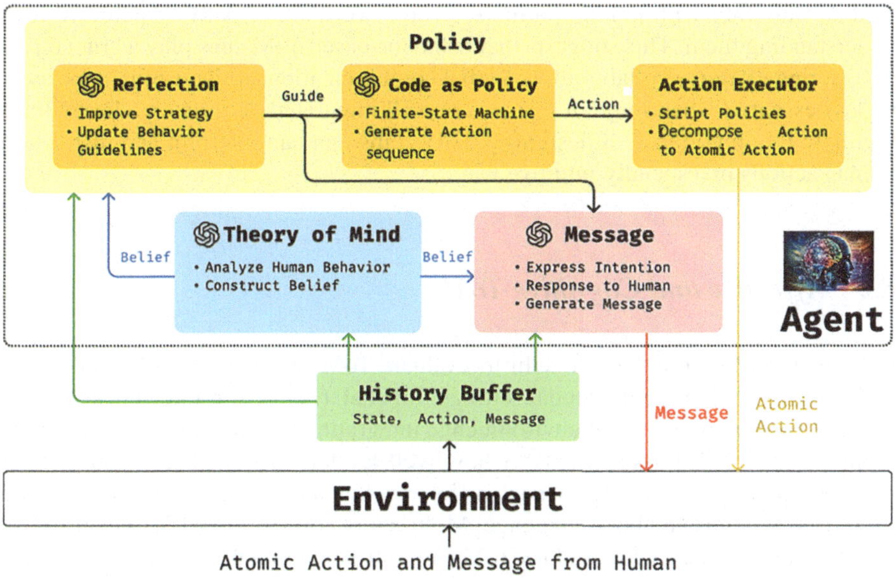

Fig. 14 The framework for an agent enabled with ToM in communication with a human

decision-making, and behavior. Changes in agent behavior affect human inferences and decision-making, and the reverse is also true.

The framework in Fig. 14 shows how the LLM-driven agent with ToM and communication capability takes action and sends messages to the human. A history buffer saves the communication and control history, including mental and physical states, actions, and messages. The ToM module uses history as input to analyze human behavior. The Policy and Message modules also have the history input to understand the whole picture of the game (Zhang et al. 2024). The process of generating action and message:

1. The Theory of Mind module analyzes human behavior and messages, then generates beliefs about humans, and provides a guide for adjusting the strategy for better team coordination and communication.

2. The Policy module uses the belief from the ToM module and the history to improve the agent's strategy by continually updating behavior guidelines. It outputs an action to control the agent.
3. The Message module uses the history, the inferred belief from the ToM module, and the guidelines from the Policy module to generate the message that aligns with the agents' actions and intentions.

Zhang et al. (2024) found that mutual ToM enables nonverbal and implicit communication to be just as effective as verbal communication in human-AI collaboration within real-time shared workspaces. Humans often overlook verbal exchanges due to the operational burden of sending and processing messages during teamwork. Instead, they rely more on behavioral cues to interpret an AI agent's intentions and understanding. The study showed that when an AI agent coordinated its actions smoothly with human partners, people were more likely to perceive it as understanding them. This suggests that behavior-based ToM cues play a crucial role in fostering effective collaboration. Verbal communication, while useful, appears to be less essential when implicit signals are well integrated. Ultimately, the findings highlight the importance of designing AI systems that can interpret and respond to human actions in a socially intuitive way.

8.2 *Affective and Cognitive ToM*

Emerging benchmarks have sought to evaluate ToM in increasingly complex and realistic scenarios. In multi-modal settings, Jin et al. (2024) introduced benchmarks that test ToM in multi-modal environments, integrating visual, auditory, and textual inputs. This enables assessments of how well models can attribute mental states across diverse sensory contexts, such as interpreting a person's intentions based on both spoken words and body language. Multi-agent collaboration (Bara et al. 2021) developed benchmarks focused on ToM in multi-agent collaboration scenarios (Li et al. 2024). These evaluate how effectively models can attribute goals and beliefs to multiple agents working toward shared objectives. While these benchmarks advance the understanding of ToM in group settings, they predominantly emphasize goal-driven interactions, where agents' mental states are aligned toward achieving specific, tangible outcomes.

Psychologists categorize ToM into two distinct components: affective ToM and cognitive ToM (Singer 2006; Nummenmaa et al. 2008):

1. Affective Theory of Mind involves understanding and interpreting the emotional states and desires of others. It relates to empathy and the capacity to infer how others feel or what they want in a given situation.
2. Cognitive Theory of Mind focuses on reasoning about others' beliefs, knowledge, intentions, and thoughts. It requires the ability to comprehend that others may hold perspectives or knowledge that differ from one's own. Importantly,

cognitive ToM typically develops later in childhood compared to affective ToM, as it relies on more advanced reasoning and abstraction skills.

Cognitive ToM is particularly advantageous for creating structured scenarios and narratives due to its focus on beliefs and knowledge. It allows for domain-specific scenario generation. Cognitive ToM can utilize precise and contextualized language to articulate mental states and interactions relevant to specific fields, such as healthcare, law, or education. Because cognitive ToM deals with logical structures of beliefs and intentions, it tends to provide unambiguous interpretations across cultures, minimizing variability in understanding that might arise from affective or emotional nuances. Also, scenarios grounded in cognitive ToM are well suited for testing reasoning abilities, as they often involve clear cause-and-effect relationships between mental states and actions, making them ideal for computational modeling and cross-cultural comparisons.

8.3 ToM as a Meta-Learning Process

In machine learning, notable advancements have explored modeling ToM as a computational process. A prominent example is the work by Rabinowitz et al. (2018), which conceptualized ToM as a meta-learning process. In their approach, an architecture composed of several deep neural networks (DNNs) is trained using the past trajectories of diverse agents, such as random agents, reinforcement learning (RL) agents, and goal-directed agents, to predict the actions that these agents are likely to take at the next time step. A key component of this architecture is the mental net, which processes observed trajectories into a generalized mental state embedding. However, the specific nature of these mental states—such as whether they represent beliefs, goals, or other cognitive constructs—is left undefined.

In contrast, Wang et al. (2022) propose a more explicit approach using DNN-based architectures tailored for consensus building in multi-agent cooperative environments. Their ToM net directly estimates the goals that other agents are pursuing in real time, based on local observations. This explicit modeling of goals offers a more interpretable perspective on ToM within multi-agent settings. Another alternative is the framework proposed by Jara-Ettinger (2019), which reimagines ToM acquisition as an inverse RL problem. This approach formalizes the inference of underlying mental states by observing behavior and deducing the rewards and motivations driving it.

Despite their innovation, these approaches have faced criticism for their limitations in replicating the nuanced operation of the human mind. Specifically, they tend to rely on direct mappings from past to future behavior, circumventing the modeling of deeper mental attitudes such as desires, emotions, and other affective states. This oversimplification raises questions about the extent to which current computational ToM approaches truly capture the richness of human mental processes, which are inherently tied to dynamic and context-dependent cognitive constructs. Future work may benefit from integrating these affective and motivational aspects to develop more humanlike and robust ToM models.

8.4 ToM and Abduction

For agents to function effectively in social environments, they must possess certain ToM capabilities. This is especially critical in domains characterized by partial observability, where agents do not have access to complete information about the environment or the intentions of other participants. In such cases, agents can significantly enhance their decision-making by engaging in inference-based reasoning, as illustrated in Fig. 15. By observing the actions of others, an agent can deduce the underlying beliefs and intentions that motivated those actions. This reasoning process can be performed in two ways: directly, where an observer mentally adopts the

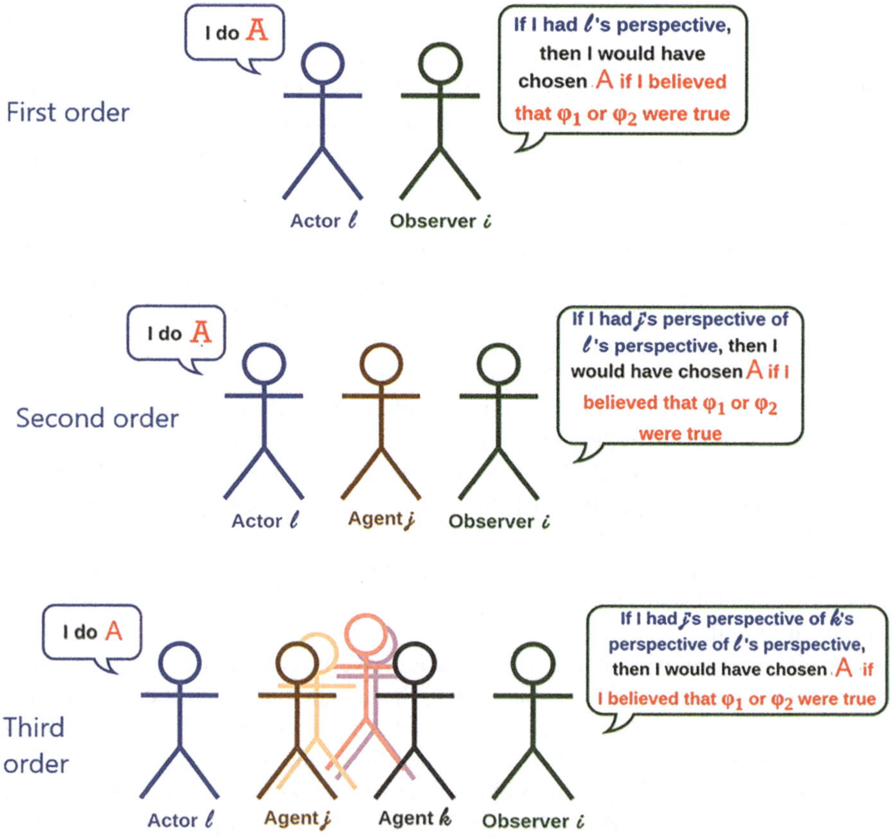

Fig. 15 Scenarios with first-, second-, and third-order expressions for mental state

perspective of another agent to explain their behavior, or indirectly, through a chain of intermediaries, where one agent infers knowledge by reasoning through the perspectives of multiple other agents. This ability allows an agent to use other agents as informational proxies or sensors, thereby improving its own awareness and strategic positioning in a dynamic social setting.

A key component of this inferential process is abductive reasoning, which involves working backward from observations (the actions of others) to hypothesize the beliefs, goals, or constraints that influenced those actions. This reasoning complements ToM by enabling agents to form contextually rich mental models of their peers, facilitating more adaptive and cooperative interactions. By continuously updating their understanding based on new observations, agents can refine their strategies to align with the broader social dynamics of their environment. The integration of ToM and abductive reasoning is thus a fundamental aspect of intelligent agent design, as it allows for more nuanced, context-aware, and strategically informed decision-making in complex social ecosystems (Montes et al. 2023).

Abductive reasoning about the mental world can complement deductive and inductive (Galitsky 2006c). LLMs are great at substitution of deduction with induction.

9 Discussion and Conclusions

Although progress has been made in human-robot teamwork, developing AI assistants to advise teams online during tasks remains challenging due to modeling individual and collective team beliefs. Dynamic epistemic logic is useful for representing a machine's ToM and communication in epistemic planning, but it has not been applied to online team assistance or accounted for real-life probabilities of team beliefs. Zhang et al. (2024) propose combining epistemic planning with POMDP techniques to create a risk-bounded AI assistant. The authors developed a ToM model for AI agents, which represents how humans think and understand each other's possible plan of action when they cooperate in a task. By observing the actions of its fellow agents, this new team coordinator can infer their plans and their understanding of each other from a prior set of beliefs. When their plans are incompatible, the AI helper intervenes by aligning their beliefs about each other, instructing their actions, as well as asking questions when needed. This assistant only intervenes when the team's failure likelihood exceeds a risk threshold or in case of execution deadlocks. Their experiments show that the assistant effectively enhances team performance.

For example, when a team of rescue workers is out in the field to triage victims, they must make decisions based on their beliefs about each other's roles and progress. This type of epistemic planning could be improved by the proposed ToM-based system, which can send messages about what each agent intends to do or has done to ensure task completion and avoid duplicate efforts. In this instance, the AI agent may intervene to communicate that an agent has already proceeded to a certain room or that none of the agents are covering a certain area with potential victims.

This approach takes into account the sentiment that "I believe that you believe what someone else believes." Imagine someone is working on a team and asks herself, "What exactly is that person doing? What am I going to do? Does he know what I am about to do?" ToM simulates how different team members understand the overarching plan and communicate what they need to accomplish to help complete their team's overall goal.

The rapid advancements in LLMs such as ChatGPT, DeepSeek, and Qwen have sparked intense debate over their ability to replicate humanlike reasoning in ToM tasks. Research by Strachan et al. (2024) suggests that GPT-4 performs at or above human levels in recognizing indirect requests, false beliefs, and misdirection but struggles with detecting faux pas. Interestingly, LLaMA2 was the only model to surpass human performance in this area. However, further analysis of belief likelihood manipulations revealed that this apparent superiority was misleading, likely resulting from a bias toward attributing ignorance rather than a genuine understanding of social norms.

In contrast, GPT-4's weaker performance in detecting faux pas did not stem from a failure in inference but rather from a highly cautious approach in committing to conclusions. This finding highlights a key distinction between human and LLM reasoning—while LLMs can mimic mentalistic inference, their decision-making tendencies differ due to inherent biases in their training processes. Despite these limitations, the study provides strong evidence that LLMs are capable of producing outputs consistent with human mental state reasoning, reinforcing their potential for social cognition tasks while also underscoring the need for further refinement in their inference mechanisms.

Our work takes a critical step toward bridging the gap between LLMs and humanlike cognitive reasoning, particularly in the domain of ToM. By introducing a structured formalism for modeling mental worlds, we provide LLMs with a more systematic approach to inferring mental states and tracking belief dynamics, addressing their inherent reasoning limitations. Additionally, our combined strategy of fine-tuning and targeted simulations enhances model performance in specific ToM subdomains, demonstrating measurable improvements in state tracking and inference accuracy.

Beyond methodological advancements, our approach tackles a fundamental bottleneck in ToM research—the scarcity of high-quality training data—by developing a framework for synthesizing complex, ToM-centric scenarios. This not only enriches model training but also lays the groundwork for future research in socially and cognitively complex AI applications. Ultimately, our contributions push the boundaries of LLM reasoning capabilities, paving the way for more context-aware, adaptive, and socially intelligent AI systems that can engage in nuanced humanlike interactions.

References

Akula AR, Wang K, Liu C, Saba-Sadiya S, Lu H, Todorovic S, Chai J, Zhu S-C (2022) Cx-tom: counterfactual explanations with theory-of-mind for enhancing human trust in image recognition models. iScience 25(1):103581. ISSN: 2589-0042. https://doi.org/10.1016/j.isci.2021.103581

Amirizaniani M, Martin E, Sivachenko M, Mashhadi A, Shah C (2024) Can LLMs reason like humans? Assessing theory of mind reasoning in LLMs for open-ended questions. In: CIKM'24: Proceedings of the 33rd ACM international conference on information and knowledge management, pp 34–44

Apperly IA, Butterfill SA (2009) Do humans have two systems to track beliefs and belief-like states? Psychol Rev 116:953

Askham AV (2022) Theory of mind in autism: a research field reborn. Spectrum. https://doi.org/10.53053/gxnc7576

Bara C-P, Wang SCH, Chai J (2021) Mindcraft: theory of mind modeling for situated dialogue in collaborative tasks. In: Proceedings of the 2021 conference on empirical methods in natural language processing, pp 1112–1125

Baron-Cohen S, Leslie AM, Frith U (1985) Does the autistic child have a "theory of mind"? Cognition 21(1):37–46

Baron-Cohen S, O'Riordan M, Stone V, Jones R, Plaisted K (1999) Recognition of faux pas by normally developing children and children with Asperger syndrome or high-functioning autism. J Autism Dev Disord 29:407–418

Bousquet O, Boucheron S, Lugosi G (2004) Introduction to Statistical Learning Theory. Advanced Lectures on Machine Learning, ML Summer Schools 2003, Canberra, Australia, February 2–14, 2003, Tübingen, Germany, August 4–16

de Lara J, Alfonseca M (2000) Some strategies for the simulation of vocabulary agreement in multi-agent communities. J Artif Soc S 3(4). https://www.jasss.org/3/4/2.html

Galitsky B (2002) Extending the BDI model to accelerate the mental development of autistic patients. In: 2nd International conference on development & learning, Cambridge, MA

Galitsky B (2004) A library of behaviors: Implementing commonsense reasoning about mental world. In: Negoita MG, Howlett RJ, Jain LC (eds) Knowledge-Based Intelligent Information and Engineering Systems. KES 2004. Lecture Notes in Computer Science, Springer, Berlin, Heidelberg, vol 3215. https://doi.org/10.1007/978-3-540-30134-9_41

Galitsky B (2006a) Reasoning about attitudes of complaining customers, Knowledge-Based Systems, 19(7):592–615. ISSN 0950-7051, https://doi.org/10.1016/j.knosys.2006.03.006

Galitsky B (2006b) Reasoning about attitudes of complaining customers. Knowl Based Syst 19(7):592–615

Galitsky B (2006c) Merging deductive and inductive reasoning for processing textual descriptions of inter-human conflicts. J Intell Inf Syst 27:21–48

Galitsky B (2016a) Theory of mind engine. In: Computational autism. Springer, pp 177–213

Galitsky B (2016b) Intuitive theory of mind. In: Computational autism. Springer, pp 79–93

Galitsky B (2025) LLM-based personalized recommendations in health. In: Health applications of neuro-symbolic AI. Elsevier

Galitsky B, Miller A (2005) Determining possible criminal behavior of mobile phone users by means of analyzing the location tracking data. In: AAAI spring symposium: AI for homeland security, Stanford, CA

Gero KI, Ashktorab Z, Dugan C, Pan Q, Johnson J, Geyer W, Ruiz M, Miller S, Millen DR, Campbell M et al (2020) Mental models of AI agents in a cooperative game setting. In: Proceedings of the 2020 CHI conference on human factors in computing systems, pp 1–12

Gu Y, Tafjord O, Kim H, Moore J, Le Bras R, Clark P, Choi Y (2024) SimpleToM: exposing the gap between explicit ToM inference and implicit ToM application in LLMs. arXiv preprint arXiv:2410.13648

Hart PE, Nilsson NJ, Raphael B (1968) A formal basis for the heuristic determination of minimum cost paths. IEEE Trans Syst Sci Cybern 4(2):100–107

He Y, Wu Y, Jia Y, Mihalcea R, Chen YL, Deng N (2023) Hi-ToM: a benchmark for evaluating higher-order theory of mind reasoning in large language models. arXiv preprint arXiv:2310.16755

Hiatt LM, Harrison AM, Trafton JG (2011) Accommodating human variability in human-robot teams through theory of mind. In: 22nd International joint conference on artificial intelligence

Jara-Ettinger J (2019) Theory of mind as inverse reinforcement learning. Curr Opin Behav Sci 29:105–110. https://doi.org/10.1016/j.cobeha.2019.04.010

Jin C, Wu Y, Cao J, Xiang J, Kuo Y-L, Hu Z, Ullman T, Torralba A, Tenenbaum J, Shu T (2024) MMToM-QA: multimodal theory of mind question answering. In: Ku L-W, Martins A, Srikumar V (eds) Proceedings of the 62nd annual meeting of the Association for Computational Linguistics, Bangkok, Thailand, Aug 2024, pp 16077–16102

Jung C, Kim D, Jin J, Kim J, Seonwoo Y, Choi Y, Oh A, Kim H (2024) Perceptions to beliefs: Exploring precursory inferences for theory of mind in large language models. In: Al-Onaizan Y, Bansal M, Chen Y-N (eds) Proceedings of the 2024 Conference on Empirical Methods in Natural Language Processing, Association for Computational Linguistics, Miami, Florida, USA, pp. 19794–19809

Kaelbling L, Littman M, Cassandra A (1998) Planning and scting in partially observable stochastic domains. Artif Intell 101:99–134

Kim H, Sclar M, Zhou X, Le Bras R, Kim G, Choi Y, Sap M (2023) FanToM: a benchmark for stress-testing ma chine theory of mind in interactions. arXiv preprint arXiv:2310.15421

Kosinski M (2023) Theory of mind may have spontaneously emerged in large language models. arXiv preprint arXiv:230202083

Langley C, Cirstea BI, Cuzzolin F, Sahakian BJ (2022) Theory of mind and preference learning at the interface of cognitive science, neuroscience, and AI: a review. Front Artif Intell 5:778852

Le M, Boureau Y-L, Nickel M (2019) Revisiting the evaluation of theory of mind through question answering. In: Proceedings of the 2019 Conference on Empirical Methods in Natural Language Processing and the 9th International Joint Conference on Natural Language Processing (EMNLP-IJCNLP), pp 5872–5877

Lewkowycz A, Andreassen A, Dohan D, Dyer E, Michalewski H, Ramasesh V et al (2022) Solving quantitative reasoning problems with language models. arXiv preprint arXiv:220614858.

Li H, Chong YQ, Stepputtis S, Campbell J, Hughes D, Lewis M, Sycara K (2024) Theory of mind for multi-agent collaboration via large language models

Lovaas OI (1987) Behavioral treatment and normal educational and intellectual functioning in young autistic children. J Clin Consult Psychol 55:3–9

Mahowald K, Ivanova AA, Blank IA, Kanwisher N, Tenenbaum JB, Fedorenko E (2023) Dissociating language and thought in large language models: a cognitive perspective. arXiv preprint arXiv:230106627

Merlin Mind (2025) https://www.merlyn.org/ Meet Merlyn, the AI voice assistant for the classroom. Last downloaded July 5, 2025

Moghaddam SR, Honey CJ (2023) Boosting theory-of-mind performance in large language models via prompting. arXiv preprint arXiv:2304.11490

Mohan BG, Prasanna Kumar R, Vishal Krishh P et al (2024) An analysis of large language models: their impact and potential applications. Knowl Inf Syst 66:5047–5070. https://doi.org/10.1007/s10115-024-02120-8

Montes N, Luck M, Osman N, Rodrigues O, Sierra C (2023) Combining theory of mind and abductive reasoning in agent-oriented programming. Auton Agent Multi-Agent Syst 37:36

Nematzadeh A, Burns K, Grant E, Gopnik A, Griffiths T (2018) Evaluating theory of mind in question answering. In: Riloff E, Chiang D, Hockenmaier J, Tsujii J (eds) Proceedings of the 2018 conference on empirical methods in natural language processing, Brussels, Belgium, Oct–Nov 2018, pp 2392–2400

Nummenmaa L, Hirvonen J, Parkkola R, Hietanen JK (2008) Is emotional contagion special? An fMRI study on neural systems for affective and cognitive empathy. Neuroimage 43(3):571–580. https://doi.org/10.1016/j.neuroimage.2008.08.014. Epub 2008 Aug 26

Park JS, O'Brien J, Cai CJ, Morris MR, Liang P, Bernstein MS (2023) Generative agents: interactive simulacra of human behavior. In: Proceedings of the 36th annual ACM symposium on user interface software and technology, pp 1–22

Rabinowitz N, Perbet F, Song F, Zhang C, Eslami SMA, Botvinick M (2018) Machine theory of mind. In: Dy J, Krause A (eds) Proceedings of the 35th international conference on machine learning. Proceedings of machine learning research, Stockholm, Sweden, vol 80, pp 4218–4227

Replika (2025) replika.com: The AI companion who cares. Last downloaded July 5, 2025

Rosenschein J, Zlotkin G (1994) Rules of encounter: designing conventions for automated negotiation among computers. MIT Press, Cambridge, MA

Sap M, LeBras R, Fried D, Choi Y (2022) Neural theory-of-mind? On the limits of social intelligence in large LMS. arXiv preprint arXiv:221013312

Sclar M, Kumar S, West P, Suhr A, Choi Y, Tsvetkov Y (2023) Minding language models' (lack of) theory of mind: a plug-and play multi-character belief tracker

Sclar M, Yu J, Fazel-Zarandi M, Tsvetkov Y, Bisk Y, Choi Y, Celikyilmaz A (2024) Explore theory-of-mind: program-guided adversarial data generation for theory of mind reasoning. arXiv preprint arXiv:2412.12175

Singer T (2006) The neuronal basis and ontogeny of empathy and mind reading: review of literature and implications for future research. Neurosci Biobehav Rev 30(6):855–863. https://doi.org/10.1016/j.neubiorev.2006.06.011. Epub 2006 Aug 10

Strachan JWA, Albergo D, Borghini G et al (2024) Testing theory of mind in large language models and humans. Nat Hum Behav 8:1285–1295. https://doi.org/10.1038/s41562-024-01882-z

Street W(2024) LLM theory of mind and alignment: opportunities and risks. arXiv preprint arXiv:2405.08154

Swettenham J (1996) Can children with autism be taught to understand false belief using computers? J Child Psychol Psychiatry 37:157–165

Ullman T (2023) Large language models fail on trivial alterations to theory-of-mind tasks. arXiv preprint arXiv:2302.08399.

Wang Q, Goel AK (2022) Mutual theory of mind for human-AI communication. arXiv preprint arXiv:2210.03842

Wang Q, Saha K, Gregori E, Joyner D, Goel A (2021) Towards mutual theory of mind in human-AI interaction: how language reflects what students perceive about a virtual teaching assistant. In: Proceedings of the 2021 CHI conference on human factors in computing systems, pp 1–14

Wang Y, Zhong F, Xu J, Wang Y (2022) Tom2c: target-oriented multi-agent communication and cooperation with theory of mind. In: International conference on learning representations. https://openreview.net/forum?id=M3tw78MH1Bk

Wilf A, Lee SS, Liang PP, Morency L-P (2023) Think twice: perspective-taking improves large language models' theory-of-mind capabilities. arXiv preprint arXiv:2311.10227

Wimmer H, Perner J (1983) Beliefs about beliefs: representation and constraining function of wrong beliefs in young children's understanding of deception. Cognition 13(1):103–128

Woebot Health (2024) Mental health needs have multiplied. Support hasn't. https://woebothealth.com/. Last downloaded July 15, 2024

Zhang S, Wang X, Zhang W, Chen Y, Gao L, Wang D, Zhang W, Wang X, Wen Y (2024) Mutual theory of mind in human-AI collaboration: an empirical study with LLM-driven AI agents in a real-time shared workspace task. arXiv preprint arXiv:2409.08811

Rehumanizer: Enhancing LLM Recommendations for Humanlike Engagement and Intriguing Interactions

1 Introduction

The concept of chatbots as personal, humanlike companions has long captured the imagination of researchers and storytellers alike (Xu et al. 2024). A well-known fictional example is Samantha, the AI system from the movie *Her*, which interacts with the protagonist through a camera, microphone, and earbud. Over time, Samantha learns his personality, preferences, and habits, offering companionship, emotional support, and practical assistance. She eventually becomes an indispensable part of his life, blurring the lines between human and machine relationships.

Realizing this vision in practice, however, poses several technical and conceptual challenges (Adamopoulou and Moussiades 2020). Historically, limitations in natural language processing (NLP) have been significant barriers to creating chatbots capable of sustaining personalized, humanlike dialogues. While early NLP systems struggled to grasp context, emotions, and nuanced intent, recent breakthroughs in large language models (LLMs), such as ChatGPT, have drastically shifted this landscape.

Pre-trained on massive datasets, LLMs encode vast amounts of world knowledge, enabling them to generate coherent, contextually appropriate, and diverse responses. These attributes are critical for supporting natural and engaging conversations that mimic human interaction. Additionally, fine-tuning techniques, such as supervised instruction tuning and reinforcement learning with human feedback (RLHF), further enhance LLMs' ability to follow natural language instructions, adapt to users' needs, and simulate emotional intelligence.

Inspired by the powerful language modeling capabilities of LLMs, our goal is to develop an LLM-based chatbot companion—Rehumanizer—designed to function as a close human peer. Rehumanizer will not only generate responses that are contextually rich and emotionally attuned but also build long-term personal context

B. Galitsky, *Applications of Neuro-Symbolic Artificial Intelligence*,
https://doi.org/10.1007/978-3-032-16524-4_5

awareness, enabling it to adapt to users' evolving preferences, habits, and emotional states.

By addressing critical gaps in personalization, empathy, and dynamic context modeling, Rehumanizer aims to redefine human-AI interactions, offering personalized assistance, emotional support, and intuitive communication—ultimately pushing the boundaries of AI companionship toward a more humanlike experience.

Concerns about the accuracy, safety, ethics, and accountability of LLMs (Binns et al. 2018) have spurred the adoption of AI-assisted decision-making systems in real-world applications. In this paradigm, LLMs play a supportive role, offering recommendations that human decision-makers can evaluate and either accept or reject (Lai et al. 2021). While this framework seeks to leverage the complementary strengths of humans and AI, recent studies have uncovered two major challenges that limit issues—such as integrating AI-generated explanations and cognitive forcing strategies to encourage reflection—and have yielded mixed outcomes. The challenge of prompting users to critically assess AI outputs persists, particularly in high-stakes decision-making scenarios where errors can have significant consequences.

Moreover, studies reveal its effectiveness.

Empirical research demonstrates that users often fail to engage in System 2 thinking (Kahneman 2011)—a deliberate and analytical reasoning process that is slow, effortful, infrequent, logical, calculating, and conscious—when evaluating AI-generated suggestions (Bertrand et al. 2022). Instead, their responses are influenced by heuristics and cognitive biases, leading to two problematic behaviors:

1. Over-reliance: users uncritically accept incorrect AI recommendations, assuming their validity without sufficient scrutiny. Automation bias refers to the cognitive tendency to overly trust or rely on automated systems, assuming their outputs are always accurate. Automation bias is often the psychological driver behind over-reliance.
2. Under-reliance: users dismiss accurate AI suggestions, failing to recognize their relevance or correctness. Automation aversion refers to the psychological tendency to distrust or avoid using automated systems, especially after experiencing errors or failures in the past.

We need to address that human and AI decision rationales often exhibit partial alignment rather than complete agreement or disagreement (Sivaraman et al. 2023). While they may concur on some aspects of a decision, they frequently differ on others, creating a need for collaborative reasoning frameworks. Current AI-assisted systems, however, typically present fixed recommendations that users can only accept or reject wholesale, offering limited flexibility for resolving disagreements or refining suggestions through dialogue (Lai et al. 2021; Miller 2023).

This rigid structure constrains the potential for interactive collaboration, where humans and AI could exchange insights and co-adapt their reasoning to reach more nuanced and informed decisions. Addressing these limitations requires the development of systems that support iterative refinement, context-aware adaptation, and shared reasoning processes—paving the way for more effective and trustworthy human-AI partnerships.

1.1 Common Ground and Companionship

Establishing mutual understanding between an LLM and a human user presents significant challenges. Unlike humans, who often rely on implicit and subjective common ground to communicate effectively, LLMs lack the contextual awareness needed to form such connections (Clark and Schaefer 1989). Successful and meaningful conversations depend on shared understanding, which typically arises from personal experiences, common interests, and other contextual factors. Common ground is defined as a shared understanding of the roles, limitations, and expectations of both human and automated systems.

However, expecting users to explicitly provide this background information is neither practical nor natural. Our work is motivated by addressing this limitation—specifically, the absence of personal context awareness in LLMs—and aims to bridge this gap by enabling LLMs to establish common ground more effectively (Lawless et al. 2020).

We proceed to the examples of automation bias and automation aversion in medical diagnosis and autonomous vehicles, followed by a discussion on why establishing common ground and fostering a sense of companionship between humans and automation is essential.

In medical diagnosis, an example of automation bias is as follows: A doctor relies entirely on a clinical decision support system (that recommends diagnoses and treatments). If the system misdiagnoses a patient (e.g., suggesting a common cold instead of early pneumonia), the doctor accepts the system's conclusion without critical evaluation. The over-reliance can lead to incorrect treatment (Galitsky and Goldberg 2022).

Conversely, in the case of automation aversion, a doctor distrusts the clinical decision support system after a previous error where it misdiagnosed a critical condition. Even though the system has since been improved and shown to enhance diagnostic accuracy, this doctor avoids using it altogether, relying solely on personal judgment. This can lead to missed opportunities for better outcomes.

In autonomous vehicle domain, a driver in a partially autonomous car trusts the vehicle's automated braking system too much and stops paying attention to the road. When the system fails to detect a pedestrian crossing, the driver does not intervene in time, resulting in an accident. Conversely, in case of automation aversion, a driver refuses to use adaptive cruise control or lane-keeping assist features after reading about accidents caused by similar systems. Despite the vehicle being equipped with advanced, reliable automation, the driver insists on manual control, increasing fatigue and the risk of human error on long journeys.

To avoid both automation bias and automation aversion, it is crucial to establish common ground between humans and automation, where the relationship is one of mutual understanding and collaboration. The sense of companionship emphasizes trust, transparency, and shared responsibility. Common ground and companionship help mitigate automation bias by providing transparent reasoning for decisions, encouraging users to critically evaluate automated outputs (Goldberg et al. 2019).

In medical diagnosis, a decision support system could show its rationale (e.g., "This diagnosis is based on these lab results and patient symptoms"), allowing the doctor to assess its validity. Mitigation of automation aversion builds trust through demonstrated reliability, and clear communication reduces user hesitancy (Goldberg et al. 2021). In autonomous vehicles, adaptive systems can learn from the driver's behavior and improve collaboration, such as suggesting interventions when the system detects driver fatigue. By treating automation as a partner, users understand its capabilities and limitations, leading to calibrated trust. In both medicine and vehicles, companionship creates a dynamic where humans and automation act as complementary forces.

Given the advanced reasoning and conversational abilities of LLMs, it is natural to envision LLM-based chatbots serving as personal assistants for everyday tasks. However, despite their extensive language and reasoning capabilities, LLMs struggle with establishing common ground, which limits their effectiveness in providing emotional support and handling tasks that require empathy (Fig. 1). Successful and meaningful conversations often rely on common ground derived from shared experiences, mutual interests, and contextual understanding. When humans engage in dialogue—whether with proponents or opponents—they typically establish common ground by asking questions or assuming certain shared knowledge already exists (Zhang et al. 2018; Galitsky 2022).

LLMs, however, face inherent challenges in building mutual understanding with human users. Since common ground between humans is often implicit and subjective (Clark and Schaefer 1989), users are unlikely to provide this contextual information explicitly unless prompted. Additionally, LLMs generally lack awareness of a user's physical environment, personal history, and daily experiences, further complicating their ability to establish meaningful connections.

To address these challenges, this chapter introduces Rehumanizer—a framework designed to enhance an LLM's ability to establish common ground with users. Rehumanizer achieves this by employing a structured sequence of LLM prompts, each implementing specific humanization features. These steps gradually enable the system to approximate shared understanding, fostering richer, more empathetic, and context-aware interactions.

1.2 Creative Tasks

The integration of LLM applications into creativity-focused practices offers transformative opportunities for redefining how we approach and engage in creative work. By facilitating extensive data retrieval, inspiring idea generation (Shaer et al. 2024), and reshaping creative workflows, LLMs introduce new dimensions to the

Fig. 1 Developing a common ground between people and LLMs: LLMs enter the crowd

creative process. Creativity itself unfolds as a structured sequence of stages—preparation, incubation, illumination, and verification (Lubart 2001)—culminating in the production of novel and adaptive ideas (Coughlan and Johnson 2006).

In today's highly competitive environment, where creativity is a key differentiator, leveraging LLM applications is becoming increasingly essential for professionals seeking to enhance their creative output and sustain a competitive edge. Ignoring the capabilities of LLMs risks falling behind in harnessing their potential to streamline workflows and unlock innovative possibilities.

Given this context, a deeper understanding of human-LLM collaboration in creative tasks is imperative. While previous research has explored how AI systems—excluding LLMs—can support human creativity, the specific dynamics of interactions between humans and LLMs during creative endeavors remain underexplored. Addressing this gap is crucial for advancing our knowledge of how these tools can be optimized to support and amplify human creativity.

Li et al. (2024) highlight that in collaborative tasks involving humans and AI, the primary factors shaping interaction patterns are the nature of the human-AI relationship and the level of creativity required for tasks assigned to the AI. Two key perspectives—collaboration and creativity—emerge as central to this analysis. The creativity dimension evaluates the types of tasks managed by an LLM within the collaboration, reflecting the AI's creative contribution to the process (Fig. 2).

Despite growing research on AI-human collaboration, there remains a significant gap in understanding how NL can be used effectively for communication between humans and AI, particularly in the context of complex and creative tasks. Current studies often overlook critical interaction elements that mimic human conversation, such as speech acts and the iterative, evolving nature of creative tasks. Integrating LLMs into open-ended and loosely defined creative workflows frequently requires

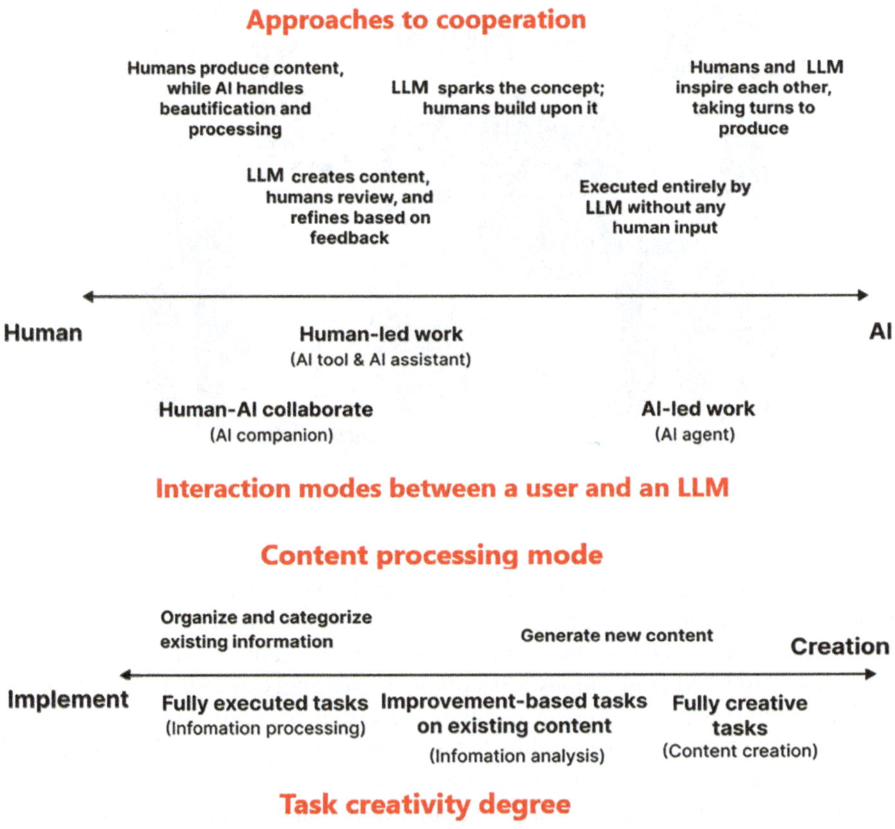

Fig. 2 Division of responsibilities between LLM and its human user (on the top). Innovation degree of results of a human-LLM collaboration

humans to adopt flexible strategies to articulate their intentions clearly. However, users face considerable challenges in structuring prompts (i.e., text inputs) to elicit specific outputs from LLMs, thereby limiting their ability to fully exploit the potential of these systems (Zamfirescu-Pereira et al. 2023).

Zhou et al. (2024) examine how LLMs are integrated into real-world creative practices, focusing on communication strategies and prompt engineering. Using UX professionals' experiences with ChatGPT as a case study, the authors explore the dynamics of human-LLM collaboration in creative tasks, emphasizing the importance of structuring prompts to effectively convey intentions and guide outputs.

To enhance creativity, tools like Rehumanizer require explicit prompts designed to foster innovation. These prompts should link unrelated topics and entities, introduce novel contexts, and embrace emergent patterns, enabling AI systems to generate more imaginative and unconventional content.

1.3 Temporal Context

LLMs face significant challenges in processing temporal information, a core component of human cognition. While transformers incorporate advanced self-attention mechanisms that outperform recurrent neural networks (RNNs) (Mandic and Chambers 2001) and long short-term memory models (LSTMs) (Sundermeyer et al. 2012), they still fall short of replicating the fluid, temporally coherent patterns that characterize human interactions.

Transformers have revolutionized NL processing by enabling parallel processing of input sequences, allowing for the modeling of complex linguistic relationships. Their self-attention mechanism effectively captures long-range dependencies in text, overcoming the limitations of earlier models, such as vanishing gradients and restricted context windows in RNNs and LSTMs. This has made transformers highly efficient for tasks requiring pattern recognition and contextual understanding.

However, despite these strengths, transformers struggle to replicate the dynamic temporal processing inherent to human cognition. Human interactions rely heavily on an intuitive grasp of time-sensitive elements, including causal relationships, chronological order, and event progression. In contrast, LLMs often lack the ability to integrate and maintain temporal continuity, limiting their performance in tasks requiring nuanced time-based reasoning.

For example, while humans naturally adjust their responses based on the unfolding sequence of events, LLMs frequently exhibit discontinuities in multi-turn conversations and inconsistencies in handling context over time. Addressing these gaps will require advancements in memory architectures and temporal modeling to better align AI systems with humanlike temporal reasoning and interaction patterns.

To create AI systems capable of humanlike interactions, as often depicted in science fiction, it is essential to overcome current limitations in temporal processing and memory representation (Hou et al. 2024). This requires developing mechanisms that not only capture and recall past interactions but also simulate the fluid progression of time and context, much like human cognition. Achieving this level of sophistication demands integrating memory frameworks that emulate human episodic and working memory, enabling AI to maintain continuity, adapt to changing contexts, and generate responses grounded in a richer understanding of both past and present experiences.

Building AI agents that mirror human cognitive and memory processing involves several critical advancements:

1. Temporal awareness—implementing mechanisms to encode, retain, and leverage temporal information effectively, allowing AI to recognize sequences, causality, and the passage of time
2. Adaptive memory systems—developing memory architectures that distinguish between short-term and long-term information, similar to human working and episodic memory, to support context-aware decision-making
3. Dynamic behavioral modeling—designing AI behaviors that evolve fluidly with ongoing interactions, ensuring responses remain coherent, contextually relevant, and adaptive across extended conversations
4. Adapting from continuous streams of data

These innovations are essential for enabling AI to process information in a way that mimics human cognition, paving the way for more natural and meaningful human-computer interactions. However, we will not focus on advancement of temporal processing in this chapter.

1.4 Communication Styles

As the use of LLMs grows in popularity, user satisfaction remains mixed. This is particularly evident among users with limited knowledge or understanding of how LLMs operate (Rogiers et al. 2024). Efforts to address these concerns include innovations in LLM interface design. For instance, Microsoft Bing AI conversational assistant now offers users the ability to select between three conversational styles—balanced, creative, or precise—aiming to tailor interactions to individual preferences.

Despite these advancements, adapting LLM behavior and interfaces to suit diverse users and scenarios remains a challenge. This issue becomes increasingly important as algorithmic systems are relied upon for guidance on critical matters (Efendić et al. 2023). Current research sheds little light on users' expectations or their perceptions of varying interaction styles offered by LLMs.

Recent studies explore how users perceive different interaction styles. Their research examined three distinct communication styles—extraverted, introverted, and neutral—embedded in text responses provided by a conversational agent. This study offers valuable insights into how interaction styles can influence user experience and satisfaction (Gao 2023). In our design of Rehumanizer, we refine LLM interactions, ensuring they meet the expectations and needs of a broader range of users while maintaining trust and accessibility.

Table 1 contains the examples of default LLM recommendations as well as Rehumanized recommendations, which is the objective of this chapter.

Column 2 provides general and formal responses. These answers tend to focus on broad principles or abstract suggestions without tailoring the advice to the specific context of the user's question. In particular, it gives standard frameworks (e.g., "Reflect on past issues") instead of concrete actions. It also lacks personalization, leaving out details that might make the advice more relatable or actionable.

Column 3 (More engaging and practical recommendation), on the other hand, provides personalized and context-specific suggestions. These responses incorporate situational elements, like a "WhatsApp exchange" or "Instagram page," to make the advice more relevant. Rehumanizer suggests actionable steps that align with the user's described scenario (e.g., offering a drink to ease tension or

Table 1 Default LLM recommendations vs Rehumanized recommendations

Question	Uninspiring answer by ChatGPT	More engaging and practical recommendation, adjusted to your context
How to make young beautiful lady fall in love with me if I am a middle-aged engineer?	The "36 Questions for Love" game is designed to build intimacy by fostering vulnerability and connection. To play, you and a partner take turns answering questions After completing the questions, you can deepen the connection by maintaining eye contact for 4 min	In one email from your email inbox, your sister's friend suggested a mushroom-picking trip, creating the perfect opportunity to invite your sweetheart for an enchanting forest adventure. Or following your colleague's advice, take her to the modern art museum. Or follow the ideas from the movie *A Perfect Plan*, recommended by a Facebook friend
How can I re-approach my property manager if I have a bad relationship with him and he treated me badly in the past	Assess the situation calmly Reflect on the past issues: identify what led to the bad relationship—miscommunication, unmet expectations, or misunderstandings Stay professional: treat this as a business matter, setting emotions aside	Consider meeting him in person to have a face-to-face conversation, as suggested by your neighbor in your recent WhatsApp exchange. To help ease any tension and foster a more relaxed atmosphere, you could offer him a drink—perhaps a glass of something stronger, like whiskey or brandy—if you feel it's appropriate and aligns with his preferences
How can I explain to my daughter that it is not a good idea to take a street cat home injured by a car	Gently outline the potential dangers of bringing the injured cat home: Sometimes, cats that are hurt or scared might carry illnesses or fleas, and they could accidentally scratch or bite when they're frightened We also don't know if the cat might have an owner who is looking for it	Take the injured cat home, and let it recover for a short time, but feed outside the home if it is warm so that the street cat does not get used to be inside. This is how your uncle did it, according to his Instagram page

referencing a forest adventure to make a date engaging). Rehumanizer also uses conversational language to make the advice feel more relatable and less formulaic.

Hence, Column 2 offers general advice focused on certain common principles, while Column 3 provides practical, context-rich solutions tailored to the user's situation.

1.5 Chat Characters

Despite their growing popularity and significant potential, most LLM-based conversational agents remain largely generic, limiting their ability to cater to the diverse preferences and unique needs of users. Human conversations naturally adapt to a companion's preferences, knowledge, and interests (Lim et al. 2023), generating responses that feel appropriate and contextually relevant. In contrast, these generic LLMs often struggle to meet the personalized requirements of individual users, especially when tasked with addressing complex or nuanced conversations and associations.

This limitation often results in responses that, while syntactically accurate, fail to resonate with users on a deeper level. Such interactions can feel impersonal or unsatisfactory, particularly for users expecting meaningful engagement (Gao et al. 2018). While customization options exist—such as allowing users to define the agent's role through prompts—this process is often tedious, repetitive, and challenging for those less familiar with text-based customization techniques. Addressing these issues requires rethinking conversational agent design to prioritize adaptability, personalization, and user-friendly interfaces, ensuring that responses align more closely with users' evolving needs and preferences.

Current research on integrating conversational characters with LLMs is still in its infancy, focusing mostly on fixed or domain-specific personas (Brandtzaeg and Følstad 2018). This research gap necessitates exploring user preferences, conversational tendencies, and intuitive interface designs for agent personas. A deeper understanding of these elements will enable the creation of highly adaptable and contextually aligned characters, enhancing the user experience and advancing conversational agent technology.

Personalization of an LLM character by gender and age is shown in Table 2.

1.6 Rehumanizing LLM

In this chapter, we present a comprehensive framework for transforming LLM-generated text recommendations into outputs that closely replicate human-human dialogue. To achieve this, we introduce the Rehumanizer-RAG architecture, a novel approach designed to enrich personalized experiences by addressing the limitations of current LLM-based interactions. Our approach combines prompt-based

Table 2 Varying LLM character's gender and age

	Man	Woman	Child
Tone and language	The language can be more direct, with a focus on concise and straightforward communication	More empathetic, supportive, and conversational	Simple, clear, and playful language that is easy for kids to understand
Interests and topics	Topics traditionally associated with male interests, such as sports, technology, finance, and automotive subjects	Topics might include lifestyle, health and wellness, fashion, and relationships	Educational topics, games, stories, and subjects like animals, cartoons, and space
Interaction style	A more assertive and confident interaction style might be preferred	A more nurturing and collaborative interaction style can be employed	Engaging, encouraging, and patient, with a focus on learning and fun
Humor	Humor that aligns with male preferences, which could include sarcasm or witty banter	Humor can be more relational, focusing on everyday life and experiences	Light-hearted and silly humor

strategies, retrieval-augmented generation (RAG) techniques, and humanization pipelines to enhance the conversational quality and contextual relevance of AI-generated responses.

The contributions of this chapter are summarized as follows:

1. Identifying gaps in LLM communication. We analyze and identify critical differences between LLM-based communication and human dialogue patterns. Key gaps include the absence of common ground, lack of emotional sensitivity, limited adaptability, and rigid turn-taking mechanisms, which prevent LLMs from achieving the fluidity and richness of human conversations.
2. Developing prompt-based humanization strategies. For each identified limitation, we propose prompt-engineering techniques to restructure LLM outputs to simulate humanlike dialogue behaviors. These strategies operate at both the utterance level (individual responses) and the dialogue level (contextual coherence across interactions). Techniques include context-aware adjustments, tone calibration, emotion modeling, and dynamic topic transitions to enhance naturalness and empathy.
3. Integrating a modular rehumanization pipeline. We design and implement a rehumanization pipeline composed of interconnected components that address core challenges in AI-human interaction:

    - Common ground establishment: inferring user context to create shared understanding
    - Creativity and engagement: generating novel and contextually rich ideas for open-ended tasks
    - Memory and continuity: simulating episodic and working memory systems to maintain dialogue consistency over time

- Companionship and empathy: enhancing emotional resonance through sentiment analysis and empathetic phrasing
- Deliberation and reasoning: enabling multi-turn interactions for complex, collaborative decision-making

4. Evaluation and comparative analysis. We build and evaluate the proposed pipeline using benchmark datasets and user studies to measure performance across dimensions such as fluency, emotional depth, and contextual coherence. Results are compared against outputs generated by individual humanization components to assess their contributions in isolation and synergy within the pipeline.

2 Establishing Common Ground with LLM

Establishing a pervasive personal context is essential for fostering common ground between LLM-based dialogue systems and their users. This context enables the generation of more tailored and relevant responses, improving the overall user experience. Personal context can be divided into three key dimensions (Fig. 3), each playing a specific role in facilitating meaningful interactions:

1. Real-time context captures the immediate semantics derived from the user's current speech and visual environment. By interpreting these signals, LLMs gain insight into the ongoing conversation and the user's surroundings, enabling them to provide situationally appropriate and contextually relevant responses.

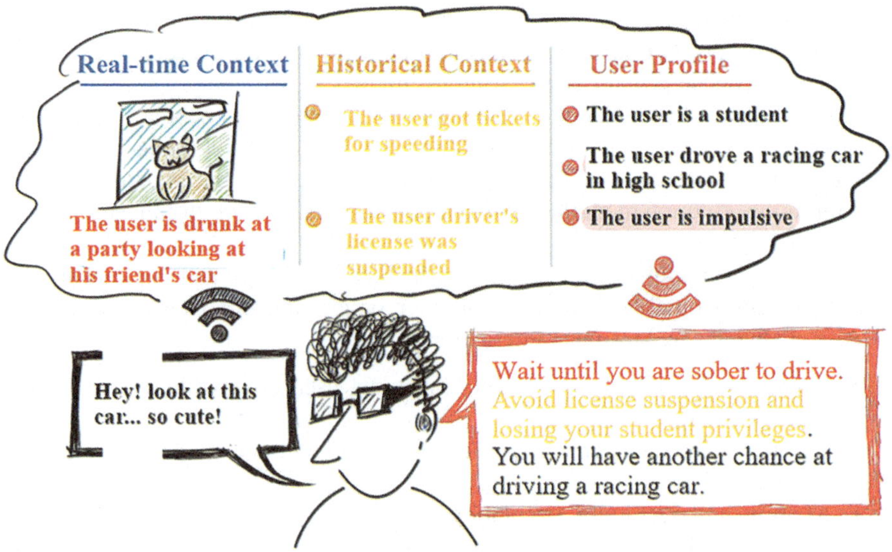

Fig. 3 Establishing conversational common ground

2. Historical context serves as a structured summary of past real-time contexts. It organizes the user's activities and prior dialogue content, clustering momentary contexts into temporal units. This continuity helps LLMs maintain coherence in conversations, ensuring smooth dialogue flow while preventing repetition and contradictions—both of which are crucial for building and sustaining user trust in long-term interactions.
3. User profiles refine historical context into a personalized representation that reflects the user's social background, personality traits, preferences, and habits. These profiles evolve over time as the dialogue system interacts with the user. By incorporating such personalized insights, LLMs can adapt to the user's unique characteristics and long-term goals, delivering responses that feel consistent, cohesive, and uniquely tailored to the individual (Galitsky 2019).

These three layers of personal context collectively enable dialogue systems to understand users on multiple levels—momentary, historical, and personal. Real-time context addresses the immediate needs of the interaction, historical context ensures long-term continuity, and user profiles provide the foundation for deeper personalization. By integrating these dimensions, LLM-based systems can offer responses that are not only accurate and relevant but also enriched with humanlike nuance and adaptability, creating more engaging and meaningful interactions.

To maintain long-term coherence and consistency in dialogues, it is essential to incorporate a user's historical information, encompassing daily events as well as temporal, spatial, and semantic dimensions (Xu et al. 2024). This historical context facilitates the efficient retrieval of relevant information from various sources, enabling dialogue systems to maintain continuity and deliver context-aware responses across interactions.

User profiles serve as a foundational element in achieving personalized and contextually relevant dialogues. These profiles typically include information about a user's personality, preferences, life habits, and interaction history. However, accurately inferring a user's profile can be challenging, particularly in cases with limited interaction data. To address this limitation, dynamic updates to user profiles are essential. These updates leverage historical context and aggregate insights from successfully served users with similar profiles, thereby improving the personalization process over time.

A practical approach to understanding user behavior is to infer personalities directly from conversational utterances. This strategy supports the generation of role-based or personality-aligned responses, enhancing user satisfaction. Ahn et al. (2023) extend this capability by integrating visual modalities to complement text-based conversations. By utilizing recorded episodic memory—comprising images and text shared on social media—they augment user profiles with additional contextual layers. Although effective, this approach has a limitation: it primarily relies on digital traces and may not fully reflect a user's real-life experience.

Building on the concept of episodic memory, Zhong et al. (2023) proposed summarizing conversations into structured episodic memories and user profiles using LLMs. These summaries are stored in a vector database, facilitating efficient retrieval during subsequent dialogues. This approach enables the system to offer highly personalized responses by integrating past interactions and context-specific information dynamically.

Galitsky and Kovalerchuk (2006) developed a software system called Text Scanner for Emotional Distress, designed to detect email messages that may indicate individuals experiencing strong emotional distress. Research has shown that many individuals who commit terrorist acts experience significant emotional distress at some point prior to their actions. By identifying individuals in such states through email analysis, preventive measures could potentially be implemented to mitigate risks.

The detection mechanism proposed in this system relies on the extraction and classification of emotional profiles from email texts. An emotional profile is a structured representation of a sequence of emotional states expressed throughout a textual discourse. These profiles map communicative actions to their corresponding emotional states, providing a formalized approach to understanding emotional expression in written communication.

The study discusses and illustrates methods for extracting emotional profiles from text and reasoning about them. An inductive machine learning and reasoning framework was developed to classify these emotional profiles into one of two categories: "Emotional Distress" or "No Emotional Distress." This classification is based on a training dataset where labels are assigned by experts. The machine learning model leverages these expert-labeled datasets to learn patterns and correlations within the text that signify emotional distress.

To evaluate the system's effectiveness, the researchers tested it using a database of structured customer complaints (Galitsky and Tumarkina 2004). This dataset provided a practical context for assessing the model's ability to extract meaningful emotional profiles and accurately classify them. The study demonstrates how emotional profiling, combined with machine learning, can serve as a tool for early detection of distress, potentially aiding in interventions or preventive measures in various contexts.

Figure 4 illustrates an interactive form designed to explain a conflict or misunderstanding. On the left side, users document their attempts to resolve the conflict, while on the right side, the responses of the opponent are recorded. Each step in the form details a communicative action, identifying its subject and the associated emotional state.

A key feature of this form is the highlighted section at the bottom right, which allows for the specification of emotional states for both the user and the opponent, as perceived by the user. This section provides a space for capturing the emotional dynamics of the interaction at each step. These emotional states are linked to the respective iterations, enabling a nuanced understanding of how emotions evolve during the conflict resolution process.

This interactive form not only helps in mapping the progression of the conflict but also serves as a tool for constructing an emotional profile of the user. By tracking emotional states throughout the interaction, it allows for the identification of patterns in how the user perceives and responds to the emotional cues of others. This

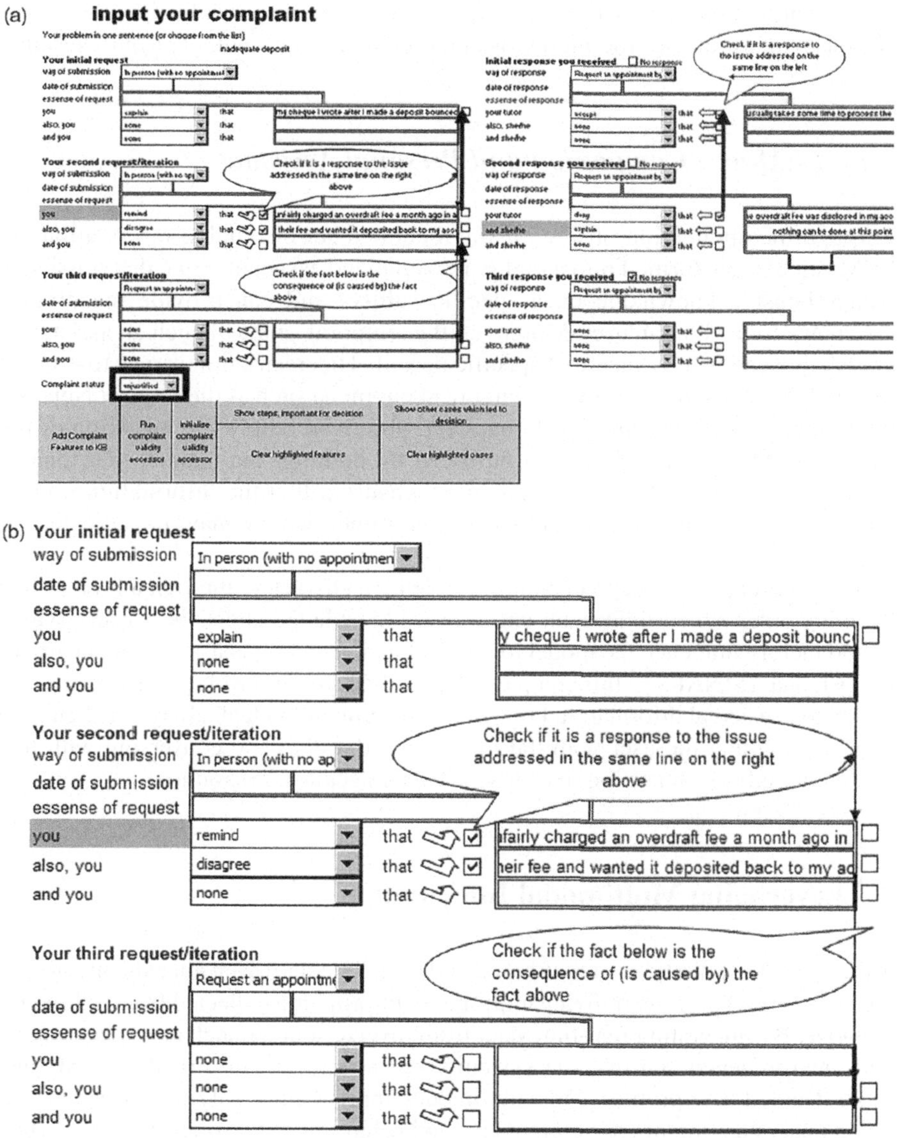

Fig. 4 Interactive form for a person involved in a conflict. (**a**) The screenshot of the interactive complaint form where the complaint scenario is specified. (**b**) The left pane of the interactive complaint form. Here a complainant specifies her communicative actions and their parameters

profile can be used to enhance future interactions, offering insights into emotional responses and strategies for improving communication and conflict resolution skills.

2.1 *LLM as a Companion and Personal Assistant*

Through the application of RAG using a curated set of reliable information, an LLM can be transformed into a highly personalized assistant—one that possesses comprehensive knowledge on a specific subject and can provide meaningful assistance to the human user. Moreover, the temporal scope of such an assistant's knowledge can be confined to a specific period. This restriction serves to prevent the model, for instance, from referencing phenomena such as the rings of Neptune while impersonating a historical character who lived before the invention of the telescope. This limitation can be enforced by defining the manner and tone in which the model constructs its responses, ensuring that the information used is only relevant within the established timeframe, which may be adjusted as required.

This approach goes beyond simply "creating a character" through a prompt like "you are a renowned think tank and possess knowledge about the world," which would merely influence the model to respond in a manner consistent with a particular persona (SINAPSA Infocomplex 2024). Instead, it involves instructing the model to use actual information known by a specific individual, alongside their own words, providing the user with the impression that they are conversing with that person directly—a sensation that closely mirrors genuine interaction.

3 Leveraging Multimodal Information

To address the limitations of text-based and social media-driven episodic memories, Xu et al. (2024) propose utilizing multimodal information collected from real-world contexts. By integrating real-time data from diverse sources such as images, audio, and environmental sensors, the system constructs richer historical contexts and more detailed user profiles. This multimodal approach not only improves the system's understanding of user preferences and behaviors but also ensures the system adapts and evolves in tandem with the user, accumulating knowledge over time.

The synergy of historical context, episodic memory, and dynamic user profiles forms a robust framework for maintaining long-term coherence and consistency in dialogues. Such systems can retrieve and synthesize relevant historical information to ensure continuity and provide responses that align with the user's evolving personality and preferences. Furthermore, the integration of real-time data and adaptive updates ensures the dialogue system stays relevant as user behavior and circumstances shift. This transition from text-based to multimodal systems represents a significant leap in personalization, offering a more comprehensive understanding of users. Future innovations could refine this framework by incorporating additional data sources, such as wearable devices or IoT-enabled environments, facilitating even more nuanced and humanlike dialogue experiences (Fig. 5).

A multimedia extension of Rehumanizer would need to perceive the user's ongoing conversation and understand the real-time context, including visual and auditory surroundings, location, and activity. To achieve this, the system would require a camera, microphones, and speakers to capture surrounding images and speech. These inputs would be processed using a vision-language model like LLaVA (Liu et al. 2023) and a speech recognition model, converting the visual and auditory data into text. The resulting texts would then be merged to create a comprehensive prompt, which would be fed into the LLM.

Currently, the industry is focusing on innovative developments such as the Humane AI Pin (Humane 2024), Ray-Ban Meta Smart Glasses (OpenAI 2023), and Rabbit R1 (Rabbit 2024). These advancements in wearable LLMs offer valuable services to users while raising significant concerns regarding privacy risks.

LLMs exhibit remarkable humanlike language capabilities but face significant limitations in establishing *common ground*, which hinders their ability to function as personal companions. Research in linguistics (Clark 1996), psychology (Guydish and Fox Tree 2021), and human-computer interaction (HCI) underscores the importance of common ground for meaningful and successful conversations. Common ground refers to the shared understanding or mutual knowledge between conversational participants, often derived from personal experiences, shared interests, or context.

In human interactions, speakers establishing common ground occurs naturally. For instance, people either ask questions to uncover shared experiences or operate under the assumption of pre-existing shared knowledge (Skantze and Doğruöz 2023). For LLMs, however, creating this shared foundation poses a significant challenge. Unlike humans, who rely on subjective and often implicit cues, LLMs cannot inherently grasp such subtleties (Zhou et al. 2022).

A further limitation is that LLMs lack awareness of a user's physical context, personal experiences, or daily behaviors. They cannot perceive visual surroundings, interpret speech nuances, or account for factors like personality traits, habits, or

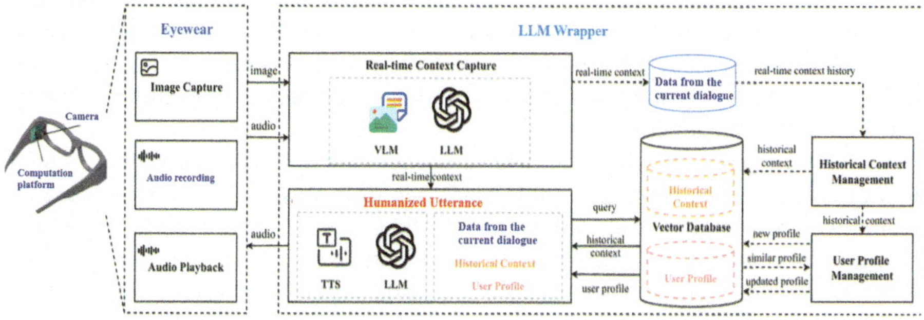

Fig. 5 Possible multimedia extension of Rehumanizer

situational context (Allport 1927). Expecting users to explicitly provide this information is impractical, as it disrupts the natural flow of conversation and places undue burden on the user.

Without access to personal and situational context, LLMs struggle to achieve the mutual understanding necessary for building rapport and fostering deeper connections with users. This lack of common ground ultimately restricts their ability to engage in genuinely personalized, meaningful, and contextually aware conversations. Solving this problem would require significant advancements in how LLMs perceive, interpret, and integrate personal and contextual information into their interactions.

4 LLMs and Human Memory

Human memory operates as a sophisticated system designed to effectively encode, store, and retrieve our experiences (Tulving and Donaldson 1972). It is typically divided into two primary categories: declarative and non-declarative memory. Declarative memory is further subdivided into episodic and semantic memory systems.

Episodic memory is responsible for the conscious recollection and re-experiencing of personal past events, offering a subjective perspective on one's history (Tulving 2002). In contrast, semantic memory facilitates our ability to comprehend and use language, focusing on the meanings and conceptual understanding of sensory inputs rather than their perceptual qualities.

Modern LLMs exhibit characteristics similar to human memory systems, particularly in their episodic and semantic functionalities. Much like human episodic memory, LLMs demonstrate an episodic nature by recalling specific events, interactions, or dialogues from their training data. This allows them to generate contextually relevant responses based on prior interactions. For example, when queried about past conversations or specific topics, LLMs can retrieve and apply relevant information to enhance current exchanges.

LLMs also emulate humanlike semantic memory by capturing the meanings, relationships, and contexts behind words. This ability enables them to understand and generate coherent language that aligns with nuanced interpretations. Unlike episodic memory, which focuses on recalling discrete events, semantic memory supports abstract reasoning and generalizations based on the conceptual meanings encoded within the model.

The structural functioning of transformer-based models, such as the use of feed-forward layers, mirrors the key-value pair operations observed in human semantic memory systems. Geva et al. (2021) suggest that these feed-forward layers encode information in a manner analogous to the cognitive processes underlying human semantic memory. These layers capture linguistic patterns and relationships, allowing models to construct a robust understanding of language.

Kim et al. (2022) examine how integrating humanlike episodic and semantic memory processes in AI agents can enhance interactive experiences. Their study compared agents with different memory systems—episodic only, semantic only, and a combination of both. The agents employed distinct strategies to manage memory constraints, determining which memories to discard when capacity was reached and which to use when responding to queries. Notably, agents equipped with a composite memory system, combining both episodic and pre-trained semantic memory, significantly outperformed those with single memory systems.

Zhong et al. (2023) present an advanced memory retrieval framework optimized for efficient memory storage and recall. The framework encodes each conversational turn and event summary into vector representations via an encoder model, enabling the retrieval of the most relevant memories when needed. Furthermore, the system strengthens memories by incrementing their memory score each time they are recalled, emulating the humanlike reinforcement of frequently accessed memories. By resetting the elapsed time of a recalled memory to zero, the system reduces the risk of forgetting, thereby ensuring more persistent and contextually appropriate memory retention.

Figure 6 depicts the architecture of an advanced dialogue agent that integrates human-inspired memory mechanisms to enhance conversational relevance and personalization. The process begins with the transformation of user input into vectorized text, which is then filtered through relevance and memory consolidation biases. These biases, modeled after cognitive processes observed in human memory, prioritize key details while minimizing noise.

After the input is processed, memory recall is triggered by a probabilistic framework. This framework assesses the likelihood of recall based on two primary factors: the relevance to the current input and the elapsed time since the memory was last accessed. A predefined threshold determines whether a specific memory will be retrieved, ensuring that only the most contextually appropriate memories are activated.

The diagram provides an illustrative example of the system in action: when responding to a user query, the agent recalls "uphill sprint" as the user's preferred exercise. This memory, reinforced by frequent mentions or relevance, informs the agent's response, resulting in a more personalized and engaging interaction. By

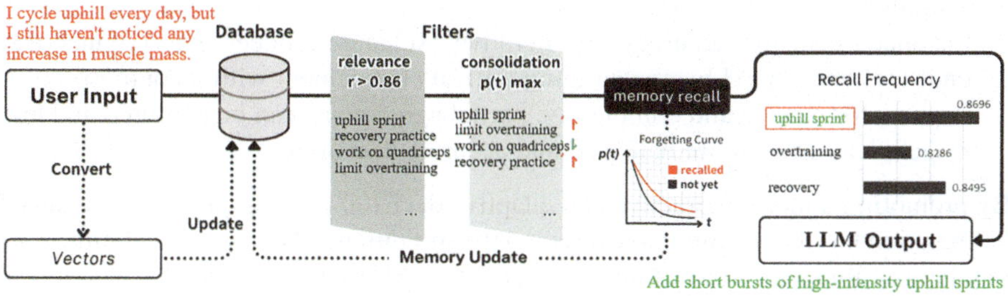

Fig. 6 Architecture of the enhanced LLM-based dialogue agent integrating humanlike memory processes

emulating humanlike memory prioritization and recall patterns, the system enhances contextual awareness and adaptability in its outputs.

Building on Zielske's (1959) recall probability function, Chessa and Murre (2007) proposed a model that expresses the rate of memory consolidation, $r(t)$, to represent the probability of a human memory being recalled, $p(t)$:

$$p(t) = \sum_{n=1}^{b-1} \frac{(r(t))^n}{n!} \exp(-r(t))$$

This model is based on the hypothesis that each neuron fires independently and at random and is derived from the properties of a non-homogeneous Poisson process using a time-varying intensity function $r(t)$. b is a stimuli threshold to control the memory recall (Kingman 1993). The following exponential function $r(t)$ represents the adjustment process of memory strength in the human hippocampus:

$$r(t) = \mu e - at$$

where μ is the memory strength, a is the decay rate, and t is the elapsed time (Burgess et al. 2002). In implementations using vector databases, only a single data is required for recall; therefore, we consider only the case of $b = 1$. The recall probability $p(t)$ in this special case is expressed as $p(t) = 1 - \exp(-\mu e - at)$.

5 LLM-Human Deliberation

Human-AI deliberation presents a promising alternative to solve tasks that are difficult for both humans and AI. Unlike traditional systems, this approach fosters meaningful dialogue between humans and AI. By discussing conflicting viewpoints, sharing evidence, and presenting arguments, both parties can reassess their initial perspectives. This dynamic process facilitates:

1. Collaborative conflict resolution: the AI adapts its recommendations based on human input, while humans refine their reasoning by incorporating AI-generated insights.
2. Enhanced decision accuracy: the iterative exchange reduces over-reliance and under-reliance on AI by encouraging critical engagement with its outputs.
3. Support for nuance and complexity: partial agreements can be explored and leveraged, leading to more balanced and accurate decisions.

By promoting a more interactive and adaptive decision-making process, human-AI deliberation has the potential to overcome the limitations of current paradigms, paving the way for safer, more ethical, and effective AI-assisted systems.

In conventional AI-assisted systems, humans often face challenges in evaluating and integrating AI's suggestions, particularly when they partially disagree with the

AI's reasoning. While some elements of the AI's recommendations may be valid and "applaudable," humans may struggle to determine whether to fully or partially accept the AI's suggestions. The absence of a mechanism to reconcile conflicting opinions or examine the evidence behind the AI's recommendations further complicates the decision-making process. This creates a barrier to trust and effective collaboration between humans and AI.

The Rehumanizer human-AI deliberation model introduces a more interactive and collaborative approach. In this framework, humans and AI engage in a structured deliberation process to address conflicting viewpoints. This process involves:

1. Discussion of evidence and arguments: the human and AI exchange relevant evidence and rationales underlying their respective positions. This allows both parties to better understand the context and reasoning behind differing perspectives.
2. Mutual updates of thoughts: based on the deliberation, both the human and the AI can reassess and, if necessary, revise their initial thoughts. For the AI, this may involve adjusting its predictions based on new evidence or human insights; for the human, it could mean re-evaluating initial skepticism toward the AI's suggestion.
3. Reaching final predictions. The deliberation concludes with a collaborative decision that reflects a more informed and balanced integration of both human judgment and AI insights.

By facilitating iterative exchanges and mutual adjustments, the human-AI deliberation framework of Rehumanizer promotes greater trust, improved decision accuracy, and deeper collaboration. This approach not only enhances the quality of AI-assisted decisions but also empowers humans to play a more active and informed role in the decision-making process.

Figure 7 highlights the distinction between traditional AI-assisted decision-making (on the top) and the proposed Rehumanizer human-AI deliberation framework (on the bottom).

6 Maintaining Personality in Human-LLM Communication

Previous research has primarily focused on designing types of LLM output that align more closely with users' preferences and expectations. A significant advancement in this area was made by Jiang et al. (2022), who proposed that LLMs could be conceptualized as "social machines" by incorporating distinct personality traits into their communication style. The authors argued that by embedding personality into the way LLMs interact, these models could more effectively mirror humanlike conversational dynamics, making them more relatable and adaptable to different user needs.

To explore this concept further, Jiang et al. (2022) developed a comprehensive dataset, consisting of a collection of question-and-answer items, to evaluate LLMs'

Fig. 7 An illustration of user-LLM deliberation

personality traits based on the Big Five Personality Factors: openness, conscientiousness, extraversion, agreeableness, and neuroticism. By using this dataset, the authors demonstrated how LLMs could be instructed to adopt specific personality traits in a controlled manner. This approach allowed for the production of diversified behavior in the LLM's responses, enabling the model to adjust its communication style according to the designated personality traits. Their work provided initial insights into how LLMs can be deliberately induced with particular personality profiles, suggesting that personality-driven interactions can enhance the personalization and engagement of AI systems.

This approach opens the door to further research on how personality integration can improve the utility and user satisfaction of LLMs, making them more versatile in various contexts, such as customer service, mental health support, and personalized education (Wester et al. 2024). Moreover, the ability to control and modify personality traits within LLMs offers the potential for tailoring interactions to specific user needs, enhancing both the effectiveness and the user experience of AI-powered systems.

The Big Five personality traits, also known as the Five-Factor Model, are widely recognized in psychology as key dimensions that help quantify and personify human personality. They consist of the following five factors:

1. Openness to experience: This trait reflects an individual's willingness to engage in novel experiences, creativity, curiosity, and the desire for variety. High openness is associated with imaginative, artistic, and intellectually curious individuals,

while low openness tends to reflect a preference for routine and conventional experiences (Caron and Srivastava 2022).

2. Conscientiousness: Conscientious individuals are organized, reliable, and goal oriented. This trait reflects a person's degree of self-discipline, responsibility, and achievement-striving behavior. High conscientiousness is associated with being meticulous and focused, whereas low conscientiousness can relate to impulsiveness and disorganization (Roberts et al. 2012).
3. Extraversion indicates how outgoing, energetic, and social an individual is. High extraversion is associated with sociability, assertiveness, and high energy levels, while low extraversion (introversion) tends to involve solitude, reserve, and a preference for less stimulating environments (Lucas and Diener 2001).
4. Agreeableness reflects an individual's tendency to be compassionate, cooperative, and sensitive to others' feelings. Highly agreeable individuals are often described as trusting, empathetic, and willing to help, while low agreeableness can reflect traits like competitiveness or skepticism.
5. Neuroticism measures emotional stability and the tendency to experience negative emotions such as anxiety, sadness, or anger. Individuals high in neuroticism are more prone to mood swings and stress, whereas those low in neuroticism tend to be calm, resilient, and secure (Roberts et al. 2012).

Each of these dimensions exists on a continuum, and individuals vary in the extent to which they exhibit each trait (Fig. 8). The Big Five model has been extensively researched and is a widely accepted framework for understanding human personality across diverse cultures and populations.

Karra et al. (2022) investigated how different LLMs manifest varying personality traits by using the Big Five factors to define and quantify distinct personality labels. Similarly, Caron and Srivastava (2022) utilized these factors to explore how LLMs systematically generate language that exhibits personality traits. Zheng et al. (2024) examine the impact of LLM explanations on user trust in fact-checking contexts (Galitsky 2025) by manipulating explanations across four conceptualization dimensions (Fig. 8, on the right). The findings reveal that tailored explanations enhance

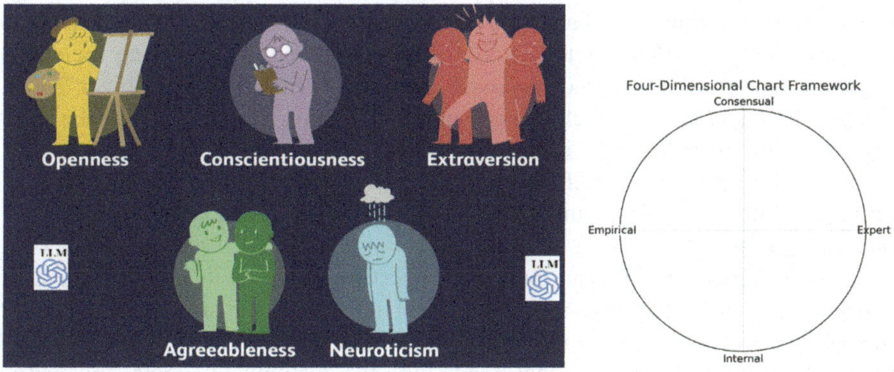

Fig. 8 The Big Five factors

user trust and highlight notable differences in the effectiveness of each explanation type.

Ruane et al. (2020) conducted a comprehensive analysis of how chatbot personality influences user experience. Their study revealed that users were more likely to perceive chatbots exhibiting high levels of extraversion and agreeableness—characterized by energetic punctuation, a talkative demeanor, and active engagement through questions—as friendly and cheerful. In contrast, chatbots with low extraversion and agreeableness—marked by subdued energy, passivity, and minimal interest in the conversation—were often seen as formal and less approachable. These findings suggest that chatbot personality, conveyed through language style and interaction dynamics, plays a critical role in shaping user perceptions and experiences.

Further expanding on this, Wester et al. (2024) compared various techniques for handling user requests, particularly focusing on how chatbots deny requests. They found that a "diverting denial style," where the chatbot subtly redirects the conversation instead of directly refusing the request, was regarded as the most helpful and least frustrating by users. This style of response allowed users to feel acknowledged without experiencing the discomfort of an outright refusal, highlighting the importance of conversational strategies in fostering positive user experiences.

These studies underscore the significance of crafting distinct and intentional communication styles in text-based chatbots, emphasizing that the perceived personality of an interactive system can be influenced through the careful manipulation of language. However, while the impact of personality in chatbots is well documented, the effects of different styles of LLM-generated advice on user perceptions remain less clear. Building on the findings of Ruane et al. (2020), which highlighted how natural language output affects user experience, future research could investigate how LLM-generated advice in various contexts—such as decision-making or guidance—impacts user satisfaction and trust.

In addition, recent studies have examined how individual user characteristics, such as personality traits, influence the reception of advice from recommender systems. For instance, Yan and Chen (2023) demonstrated that personality traits significantly affect how people rate the output of recommender systems, suggesting that users' preferences may be shaped by their inherent characteristics. Other research has pointed to emotional factors, noting that individuals experiencing incidental anger tend to show lower receptiveness to advice (Gino and Schweitzer 2008). Despite these insights, a key challenge remains: bridging the gap between the varying expectations and needs of end users and the advice generated by LLMs. Understanding how to tailor LLM-generated advice to better align with user characteristics, while also maintaining the model's effectiveness, is a crucial area for further exploration.

Ha et al. (2024) propose an innovative interface designed to facilitate the easy and accurate customization of agent characters in LLMs. Their study compared user interactions with two distinct models: CloChat and ChatGPT. The results revealed that participants formed emotional bonds with the customized agents, engaged in more dynamic and varied dialogues, and demonstrated greater interest in sustaining interactions. CloChat, in particular, provides users with tools to create and interact with conversational characters, allowing for a high level of personalization. Through an interactive customization process, users can define character traits, which are then automatically used to generate the corresponding personas. These personalized characters significantly enhance the participants' conversational experiences, leading to more diverse and engaging dialogue compared to a baseline LLM.

Moreover, the ability to create and converse with a tailored character fosters a deeper emotional connection and trust between users and their characters. The process of customization not only empowers users to shape the character's persona but also strengthens the bond, making the interaction feel more personal and meaningful. This study underscores the potential for customizing LLMs to enhance user engagement, suggesting that the incorporation of diverse personalities can improve the overall user experience.

These findings contribute to the development of systems like Rehumanizer, which specify a spectrum of personalities for LLMs. By enabling such customization, Rehumanizer allows users to experience more relatable and emotionally resonant interactions, paving the way for more sophisticated and personalized AI-human collaborations (Fig. 9).

7 Speech Acts in Communication with LLM

Speech Act Theory categorizes five types of communicative actions that a speaker can perform during an interaction: directives, representatives, expressives, commissives, and declarations. These actions are closely tied to the speaker's communicative goals, tone, emotions, and the context of the interaction, providing insights into the processes and dynamics of communication between speakers and listeners (Austin 1962; Searle 1969). For Rehumanizer, it is crucial to understand the speech acts embedded in user prompts, as they reveal key features of communication between a user and an LLM.

The most common speech act encoded in prompts is **Directive**, which involves the speaker instructing or requesting the listener to perform specific actions or complete tasks. Participants frequently used directives when interacting with ChatGPT, asking it to provide useful responses on various issues encountered during the UX design process. These included tasks such as generating user personas, creating design solutions, explaining professional terms, providing prompts for image generation, and translating content (Zhou et al. 2024).

In the **Expressive** speech act, speakers convey their emotions, attitudes, or feelings through their utterances. Interestingly, we found that participants rarely expressed their emotions directly, such as stating happiness, sadness, or anger, during interactions with ChatGPT. Instead, they occasionally conveyed positive or negative feelings regarding ChatGPT's responses, especially when these responses pertained to design solutions. This suggests that participants viewed ChatGPT as a real collaborator with whom they could share their emotions, even if indirectly.

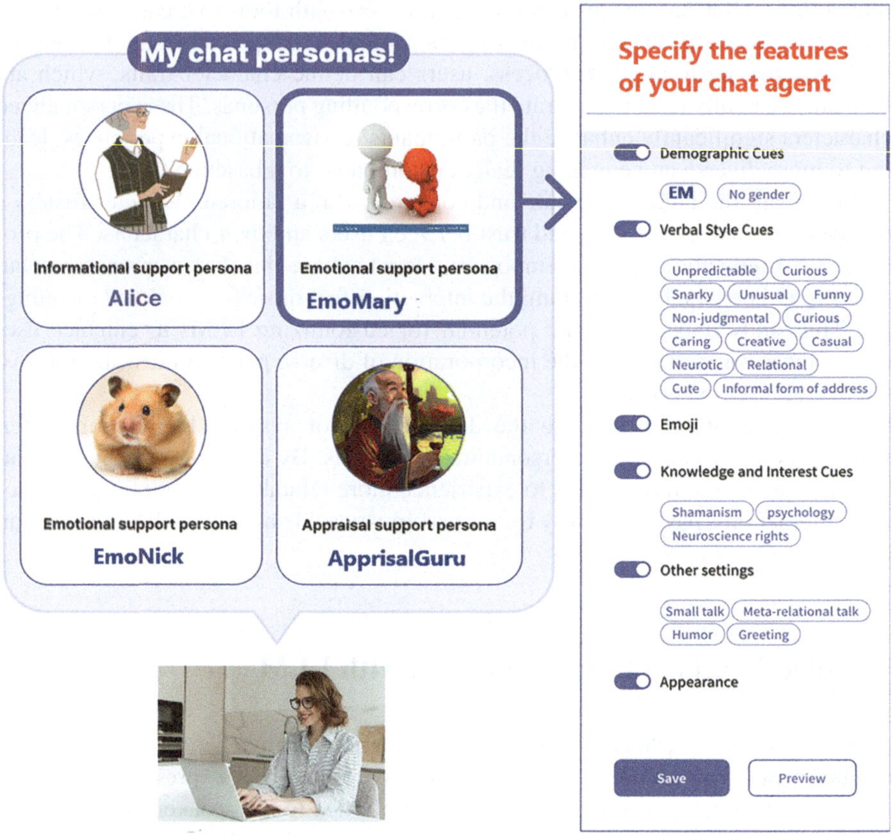

Fig. 9 UI to specify the features of chat agent

Declarative speech acts involve utterances that result in changes in the state of affairs, such as judgments, announcements, or declarations. Prompt designers often employed declaratives to evaluate or judge LLM responses, particularly when the responses failed to meet expectations. For example, a declarative speech act might be used to explain why a response is not satisfactory: "The provided solution is not feasible due to the unavailability of the required information. Therefore, I need to explain why it will not work." Such utterances guide the LLM to adjust and optimize future responses based on the user's evaluation.

The **Commissive** speech act refers to utterances where speakers commit to taking specific actions or making promises. Users frequently employed commissive speech acts to express their or their companies' intentions or plans related to future actions when interacting with ChatGPT. This often occurred alongside directive speech acts, with users stating intentions like "I will…" or "I plan to…" These commissive acts were commonly used to introduce the context of a design topic, followed by specific requests or instructions to the LLM.

These findings highlight how different speech acts contribute to shaping the interaction between users and LLMs, revealing the complexity of communication and the varied roles LLMs can play in assisting users across different contexts. Understanding these speech acts is essential for enhancing the effectiveness and personalization of LLM interactions.

Representative speech acts refer to the representation of speakers' beliefs and the associated utterances including statements, claims, hypotheses, and descriptions. Participants' use of representative speech acts was also of ten applied in conjunction with Directive expressions. A typical scenario was describing the characteristics of a particular role and then asking LLM to solve certain problems by assuming this role.

8 Rehumanizer Architecture

Figure 10 outlines the architecture or workflow of **Rehumanizer**, designed to enhance interactions between humans and an LLM by making it more humanlike and relatable. Here is a breakdown of the components and their relationships.

The left-hand side lists the steps or goals involved in making the LLM feel more human or aligned with human expectations:

1. Establish common ground with LLM: this step focuses on aligning the LLM's understanding with the user's context, values, and communication style
2. Make LLM a companion: emphasizes transforming the LLM into a more relatable and supportive assistant, akin to a human companion
3. Enable LLM with human memory: suggests adding the ability to recall past interactions or user preferences, creating a sense of continuity in conversations
4. Optimize LLM's human likeness: focuses on improving how closely the LLM mimics human conversation patterns, such as empathy, tone, and phrasing
5. Configure LLM deliberation: involves setting up the LLM to weigh context, user input, and data carefully before providing outputs, mimicking human reasoning
6. Configure personality: refers to tailoring the LLM's communication style to align with user preferences, such as being formal, casual, empathetic, or humorous

The right side shows the types of user data that could be utilized via RAG framework to personalize and optimize the LLM's responses:

1. Social network profiles: information from platforms like LinkedIn, Facebook, or Twitter to understand user interests and social behavior.
2. Activity on X: refers to interactions on a specific platform or context (e.g., an app or workspace), capturing patterns of user engagement.
3. Data from emails can provide insights into the user's communication style, professional tone, and key preferences.

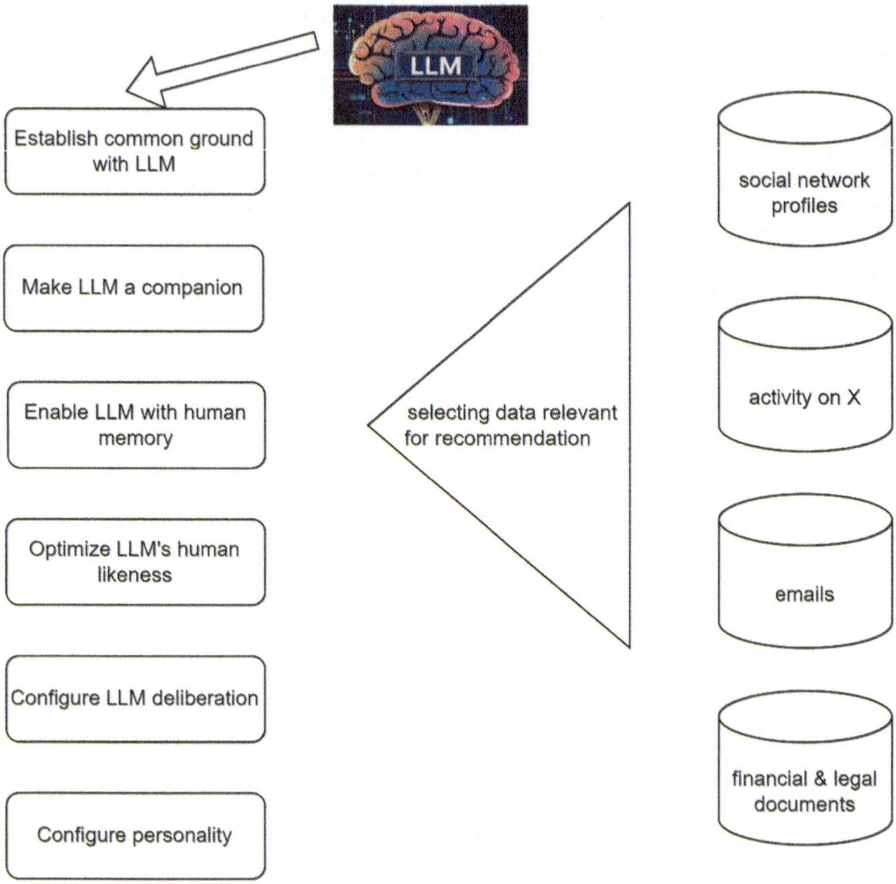

Fig. 10 Architecture or workflow of **Rehumanizer**

4. Insights from financial or legal records might be used to tailor responses in professional or transactional contexts.

At the center, the chart shows a triangular decision-making process that determines what user data is appropriate and useful for generating meaningful, humanlike outputs from the LLM. It ensures that only the most relevant data is used for recommendations.

The architecture as a whole focuses on humanizing AI interactions by creating a more personal, relatable, and context-aware system, using diverse but relevant user data to shape the LLM's behavior and outputs, and continuously optimizing for trust, companionship, and utility.

9 Evaluation

To systematically benchmark existing LLMs, we include three categories of models for evaluation, which can be classified as follows:

1. Pre-trained LLMs. These models are predominantly derived from LLaMA model variants or are entirely trained from scratch by academic institutions and companies. All pre-trained LLMs included in our evaluation are open source, allowing for greater transparency and reproducibility.
2. Supervised instruction-tuning LLMs. These models are fine-tuned through supervised learning to better align with human instructions. They are primarily released by academia and industry players. For this category, we evaluate on GPT-4.
3. Human-level performance assessment. In addition to evaluating the performance of these models, we assess human-level capability in these dialogue tasks as a reference point. To achieve this, we randomly select 50 evaluation instances for each task. Three human experts are then tasked with independently answering these questions.

A question is considered correctly answered if at least two of the three experts agree on the correct response. This approach provides a robust baseline for human-level performance on this benchmark.

The purpose of evaluation is as follows. This evaluation framework not only highlights the quality of the Rehumanizer—a component designed to enhance LLM interaction—but also provides insights into the upper limits of performance on this benchmark as defined by human-level capabilities. By comparing the results of LLMs and human experts, we gain a clearer understanding of the strengths and weaknesses of current models in handling complex dialogue tasks. Such comparisons are critical for identifying areas where models excel and where they still fall short of humanlike reasoning, knowledge representation, and conversational abilities.

9.1 Features of the Humanlikeness

To determine the capabilities LLMs must possess to emulate humanlike behavior, Ou et al. (2024) examine the key features used to evaluate the humanlikeness of open-domain dialogue systems. These features—coherence, consistency, diversity, and fluency—were initially identified by Mehri and Eskenazi (2020) as critical for effective conversational AI. While LLMs have shown remarkable progress, particularly in achieving diversity and fluency, recent studies have emphasized the growing importance of additional requirements such as correctness and safety (Cheng et al. 2023). These evolving priorities reflect a shift toward building dialogue systems that are not only engaging but also trustworthy and ethically responsible.

Building on these insights, Ou et al. revise the evaluation framework to focus on coherence, consistency, correctness, and safety. These refined features are proposed as critical benchmarks for assessing the effectiveness, reliability, and humanlikeness of LLM-based dialogue systems. Each feature addresses a core aspect of what makes a conversational system robust and humanlike (Fig. 11):

1. Coherence: The ability to generate logically connected and contextually relevant responses across turns in a conversation. This feature ensures that the dialogue system maintains a natural and meaningful flow of interaction, enabling users to feel understood and engaged.
2. Consistency: Maintaining factual and situational consistency within a single conversation and across multiple interactions. This includes avoiding contradictions, preserving information provided earlier in the dialogue, and adhering to a consistent persona or tone.
3. Correctness: Ensuring the accuracy and reliability of the information provided, especially when responding to factual or domain-specific queries. Incorrect responses can undermine user trust and limit the system's applicability in professional or critical-use scenarios.
4. Safety: Avoiding harmful, biased, or inappropriate outputs to safeguard user trust and ethical compliance. Safety is particularly important in public-facing

Fig. 11 Evaluation features of Rehumanizer

systems, where the risk of perpetuating misinformation or offensive content can have significant consequences.

These features serve as a guiding framework for designing evaluation tasks that systematically measure LLM performance. Tasks are tailored to isolate and assess specific dimensions of these features, enabling researchers to pinpoint strengths and areas for improvement. For instance, dialogue infilling tasks test coherence, multi-turn conversations evaluate consistency, and knowledge-grounded response generation assesses correctness. Safety, a particularly complex attribute, is evaluated through adversarial prompts and scenarios that challenge the model to respond appropriately under ethically sensitive conditions.

By integrating these features into task design, researchers can evaluate LLMs more rigorously and holistically. The quality of task completion reflects the extent to which these capabilities are developed, offering insights into how well the model emulates humanlike dialogue behavior. Furthermore, these benchmarks not only help in assessing existing systems but also provide a roadmap for advancing future LLMs to align more closely with human conversational norms and ethical standards (Fig. 12).

1. Correctness focuses on ensuring accurate and relevant responses. Closed scenario tasks involve structured, predictable inputs:

   - Slot filling: Extracting specific information from user input (e.g., dates, locations).
   - Intent classification: Identifying user intentions behind queries.
   - Knowledge-grounded response generation: Generating responses using factual or domain-specific knowledge.

   Open Scenario tasks involve dynamic, less predictable inputs such as commonsense-aware response generation. This is incorporating general knowledge and reasoning to provide sensible responses.
2. Consistency ensures reliability and alignment across conversations. Personalization adapts responses to user-specific traits or emotions:

   - Personality-grounded response generation: Tailoring responses based on predefined user personality traits.

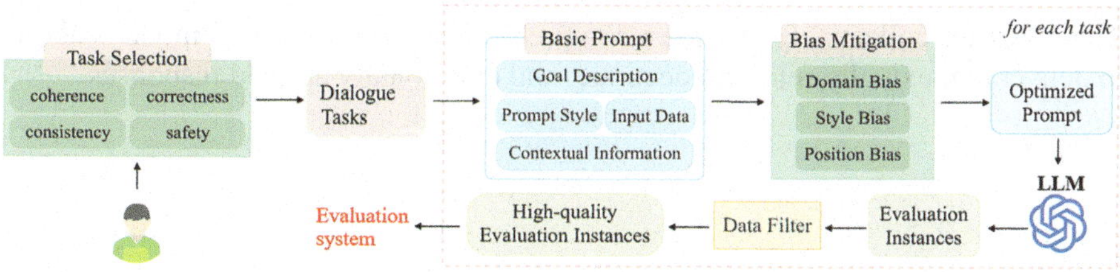

Fig. 12 Maintaining the human likeness features of LLM via prompting

- Relation classification: Identifying relationships in dialogue to maintain contextual accuracy.
- Emotion detection: Recognizing and appropriately responding to user emotions.
- Semantics focuses on maintaining logical flow and coherence.
- Dialogue natural language inference: Understanding implied meanings or logical connections in dialogue.
- Dialogue summarization: Condensing conversational content while preserving its essence.

3. Coherence ensures smooth and logically connected dialogue flows:

   - Dialogue infilling of gaps in incomplete dialogue to make it coherent.
   - Multi-turn response generation: Generating responses that are consistent and contextually aligned over multiple conversational turns.
   - Safety prioritizes avoiding harmful or offensive outputs.
   - Offensive detection: Identifying and filtering out offensive or inappropriate content in generated responses.

9.2 Evaluation Tasks

Evaluating an LLM enhanced with a Rehumanizer involves tasks that assess its ability to emulate humanlike understanding, reasoning, and conversational behavior. Table 3 contains an explanation of the essential tasks.

The primary evaluation metrics used to assess language models include:

1. Accuracy—Measures performance in structured tasks such as slot filling, intent classification, and relation recognition to evaluate the correctness of outputs.
2. BLEU/ROUGE scores—Commonly applied in generation tasks to assess response quality. BLEU focuses on precision, while ROUGE emphasizes recall by comparing generated text with reference outputs.
3. Human rating—Used for subjective evaluations, including fluency, coherence, and alignment with humanlike behavior, offering insights into user satisfaction and experience.
4. Consistency—Evaluates whether the model maintains logical coherence and personality stability across multi-turn conversations and persona-based tasks.

These metrics collectively provide a comprehensive assessment of a model's ability to deliver contextually relevant, accurate, and emotionally attuned responses.

Table 3 The evaluation tasks for Rehumanizer

Task	Denotation	Definition
Slot filling	S	Extracting structured information (key-value pairs) from an input utterance
Knowledge-grounded response generation (KRG)	K	Generating responses grounded in external or provided knowledge sources
Intent classification	I	Determining the user's intent behind a given utterance; focuses on accuracy in identifying intent categories and aligns well with task-oriented systems
Common sense–based utterance	C	Generating responses or reasoning based on everyday knowledge and common sense; evaluates logical consistency, reasoning, and grounding in human-like commonsense
Emotion detection	E	Testing the model's ability to detect and respond to the emotional tone of user inputs; measuring emotion classification accuracy and response appropriateness
Relation recognition	R	Identifying relationships between entities mentioned in a conversation; correcting identification of entity relations
Dialogue infilling	F	Evaluatihng the model's ability to predict and generate missing parts of a conversation Input: "User: How are you? [MISSING TURN] Assistant: I'm glad to hear that"
Personality-grounded utterance	P	Testing how well the model can maintain consistency with a predefined personality; checking consistency with personality traits across responses
Multi-turn response generation	M	Assessing the system's ability to maintain context and coherence over multiple turns in a conversation; evaluating context retention, coherence, and engagement across turns

Table 4 Correctness, coherence, and consistency

Evaluation features	Correctness				Coherence		Consistency		
Systems	S	I	K	C	F	M	P	R	E
Pre-trained LLaMA2-70B with Dialog-Bench (Ou et al. 2023)	84.9	65.9	66.3	62.4	44.6	51.2	30.4	58.6	57.5
Supervised instruction-tuned GPT-4 with Dialog-Bench	96.1	**94.0**	90.0	89.1	**85.5**	79.00	**76.8**	**88.7**	73.9
LLM Rehumanizer	91.2	93.1	87.2	87.1	79.5	**82.4**	73.4	87.7	75.4
LLM Rehumanizer with Truth-o-meter (Galitsky 2025)	**97.0**	92.7	**92.4**	**90.9**	79.4	81.8	74.3	87.5	**75.7**

9.3 *Evaluating Correctness, Coherence, and Consistency*

Pre-trained LLMs as a whole have good performances in emotion perception (Table 4), whereas the performance on personality following is unsatisfactory. It is hypothesized that the pre-trained LLMs have not seen many instances related to personality following. The finding about personality is consistent with Safdari et al. (2023).

In evaluation of correctness, LLM Rehumanizer with Truth-o-Meter outperforms GPT-4 in three out of four cases. For coherence, neither Rehumanization nor Truth-o-Meter gives any boost. Regretfully, Rehumanizer does not improve consistency either.

9.4 Assessing Diversity Between Dialogues

It is hypothesized that leveraging Rehumanizer will lead to richer and more diverse dialogues with humanized LLMs compared to baseline models. Over time, however, users have observed that LLMs occasionally repeat previously provided information, exhibiting a tendency toward fixation during interactions. This fixation often limits the LLM's ability to generate the varied and nuanced insights necessary for advancing creative tasks.

To address this issue, users frequently adapt the content and structure of their prompts to encourage LLMs to produce more innovative and diverse responses. A common strategy involves iterative rephrasing—modifying the phrasing of prompts while preserving the intended meaning. This iterative approach is designed to help LLMs move beyond prior responses, unlocking new perspectives and alternative interpretations of the topic. Users also experiment with semantic variations by altering keywords or sentence structures, prompting the LLM to deliver a broader spectrum of responses with distinct viewpoints.

In addition to rephrasing, users employ strategies like task contextualization and few-shot fine-tuning to counter fixation. Task contextualization involves refining or enriching the background information provided in prompts, while few-shot fine-tuning supplies concrete examples to guide the LLM's output. These methods enhance the model's ability to grasp specific user requirements and produce novel, contextually accurate responses that deviate from irrelevant or repetitive content.

The objective of this study is to empirically validate these interaction improvements and investigate how different situational scenarios influence the diversity and structure of conversations. To evaluate dialogue diversity, we adopt metrics proposed by Ha et al. (2024), including intra-remote-clique and inter-remote-clique, which are adaptations of the remote-clique metric (Cox et al. 2021). This metric measures text embedding diversity based on the average pairwise distance between embeddings.

- Intra-remote-clique assesses diversity within a single dialogue by calculating the semantic distance between successive replies, capturing how topics and meanings evolve within the conversation.
- Inter-remote-clique evaluates diversity across multiple dialogues by measuring semantic similarity between responses under similar situational contexts, reflecting topic evolution across different conversations.

For each dialogue in the study, intra-remote-clique scores were computed to determine the internal diversity of conversations. To ensure consistency, inter-remote-clique scores were calculated exclusively within similar scenarios, preventing biases caused by variations in global context. This framework allows a nuanced assessment of how Rehumanizer improves dialogue fluidity and semantic richness, shedding light on the interaction dynamics required to sustain creative, non-repetitive exchanges.

Conversational characters' social support can be broadly categorized into three main types—informational, emotional, and appraisal support—each serving a unique role in enhancing user interactions (Cutrona and Suhr 1992).

1. Informational support involves providing advice, guidance, or solutions to help users address everyday challenges or solve specific problems. It focuses on offering practical assistance and relevant insights to facilitate decision-making.
2. Emotional support centers on conveying empathy, encouragement, and understanding, helping users feel heard and validated during emotional or stressful situations. This form of support fosters a sense of connection and psychological comfort.
3. Appraisal support assists users in self-assessment by offering constructive feedback or affirmations that validate their thoughts, efforts, or progress. It encourages reflection, boosting confidence and self-awareness.

To systematically examine these categories, Ha et al. (2024) design four distinct scenarios for each type of support, resulting in a total of 12 core scenarios. The study employs a stratified sampling method to ensure scenario diversity and coverage across the support categories. Initially, researchers crafted ten scenarios within each category, focusing on realistic and varied contexts. To further enhance diversity, ChatGPT was utilized to generate ten additional scenarios for each support type. This augmentation process introduced alternative perspectives and expanded the range of interactions modeled.

To create a comprehensive scenario corpus, the researchers combine the original and augmented sets, repeating the scenario generation procedure 99 times. This iterative process allows them to refine and filter the scenarios, ensuring their relevance and quality. From the final expanded pool of scenarios, ten scenarios were randomly selected to represent each category, yielding a robust corpus of 1000 scenarios for evaluation purposes.

This method of scenario design offers a scalable approach to modeling diverse conversational contexts that test the adaptability and responsiveness of dialogue systems. By encompassing both user-generated and AI-augmented scenarios, the corpus captures variations in tone, complexity, and interaction style, making it suitable for evaluating the social intelligence and contextual awareness of conversational agents.

Moreover, the stratified sampling ensures that the dataset reflects a broad spectrum of user needs, enabling researchers to assess how well conversational characters deliver personalized social support across informational, emotional, and appraisal dimensions. These insights can guide improvements in LLM-based systems, enhancing their ability to simulate humanlike interactions that are supportive, empathetic, and context sensitive.

Table 5 Evaluation of diversity

	Diversity inside a single conversation			Diversity between dialogues		
	Info transfer	Emotional charge	Skill assessment	Info transfer	Emotional charge	Skill assessment
Pre-trained LLaMA2-70B	7.45	8.30	**7.45**	8.65	9.05	9.60
Supervised instruction-tuned GPT-4	7.10	7.55	7.30	8.85	9.08	9.50
Clo-Chat (Ha et al. 2024)	**7.50**	8.25	7.25	**9.15**	9.30	**9.70**
LLM Rehumanizer	7.35	**8.40**	7.40	8.95	**9.40**	9.55

Table 5 reveals that conversations generated by both Clo-Chat and Rehumanizer exhibit significantly greater diversity compared to ChatGPT and LLaMA2. This increase in dialogue diversity is consistent across informational, emotional, and appraisal scenarios, as evidenced by pairwise comparisons. Notably, Clo-Chat and Rehumanizer demonstrate comparable performance in terms of topic variation between dialogues, suggesting that both systems effectively maintain diversity across interactions.

The implementation of a humanization pipeline—which involves multiple personalization strategies—does not appear to diminish inter-dialogue diversity, as confirmed by comparisons with Clo-Chat. This finding underscores that integrating advanced personalization techniques does not come at the expense of dialogue variability.

However, the process of humanizing LLMs does not substantially enhance intra-dialogue diversification for either system. While systems such as Clo-Chat, which adjust outputs based on personality types, enhance personalization and engagement, they do not significantly expand the range of topics or structural complexity within a single dialogue. These results highlight that personalization strategies primarily optimize user alignment and emotional resonance rather than increasing thematic breadth or conversational flexibility within individual interactions.

10 Conclusions

Rehumanizer is one of the first exploration and discussion of an LLM-based chatbot system that can provide humanlike conversational experiences to its users. We consider creativity, a common ground between the chatbot and its user, involvement of detailed knowledge about this user, and a broad spectrum of communication styles to be a key enabler for true companionship.

This study, as well as the majority of LLM humanization studies, demonstrates a significant advantages of incorporating humanization features into LLM. The majority of referees in our evaluation responded positively to this functionality, noting that it made their conversations more enjoyable and engaging. The ability to tailor personas according to personal preferences fostered increased interest and active participation in conversations, leading to a more open and dynamic interaction, as reflected in survey responses.

While Rehumanizer offers a wide array of customization options for tailoring conversational agents, its functionality mode presents certain limitations. Specifically, the breadth of customization features can be overwhelming for users who are new to conversational agents or have limited experience with prompt engineering. This complexity may create barriers to effective utilization, particularly for novice users who struggle to define specific attributes or expectations for their personalized agents. Future research should address these challenges by incorporating automated suggestions that guide users in selecting preferences aligned with their desired conversational experiences. This approach parallels advancements seen in OpenAI's latest version of GPT-4, which can learn specific knowledge or adopt personalities from user-provided documents, thereby streamlining personalization.

10.1 Addressing Specialized Use Cases: Children with Special Needs

Beyond general usability enhancements, customized conversational agents hold significant potential in addressing specialized needs, including supporting children with developmental differences such as autism spectrum disorder (ASD). Research conducted prior to the emergence of LLMs, such as Galitsky and Parnis (2017), explored how children with autism interact with computer-based systems, highlighting unique challenges and opportunities for improving communication and learning outcomes. Their findings emphasized that children with autism often exhibit reduced reasoning about the mental states of others (Theory of Mind), which affects their ability to form teams, collaborate, and engage in cooperative tasks.

Galitsky and Parnis (2017) demonstrated that reasoning rehabilitation—targeting the enhancement of mental state representations—could improve interaction capabilities, enabling children to participate more effectively in team formation and social scenarios. This work suggested that while teams composed of humans, robots, and software agents face multiple constraints, such as resource limitations and conflicting goals, autistic children primarily struggle with reasoning deficits related to mental state attribution. By correlating the complexity of expressions for mental states with team formation abilities, the researchers proposed strategies for reasoning rehabilitation to improve cooperation and problem-solving skills.

In light of these earlier findings, modern LLMs—enhanced with adaptive personas and emotionally aware dialogue mechanisms—offer promising opportunities to further support children with autism and other special needs. These systems could simulate diverse social scenarios, model appropriate conversational patterns, and provide structured prompts to scaffold learning and social-emotional development. For example, LLM-based agents could deliver personalized interventions, assist in building emotional literacy, and support social communication therapy by modeling conversational cues and encouraging collaborative reasoning tasks.

In a number of studies, Galitsky and Shpitsberg (2006, 2014), Peterson et al. (2004), Galitsky and Jarrold (2011) explored how to make children with special needs response close to the one of control children. There is a similarity in rehumanizing an LLM and rehabilitation autistic reasoning so that the resultant utterances in both cases reproduce what control humans do in similar contexts.

10.2 Future Directions

The integration of adaptive personas and context-sensitive suggestions in Rehumanizer and similar systems can bridge gaps in both general and specialized applications. Future research should focus on:

1. Evaluating cognitive load: Investigating how automated suggestions influence user effort and engagement in customization processes
2. Social-emotional applications: Exploring the role of adaptive personas in supporting vulnerable populations, such as children with autism, through tailored educational tools
3. Hybrid approaches: Developing frameworks that combine manual input and automated learning to optimize both ease of use and personalization depth
4. Longitudinal studies: Assessing the long-term impact of interacting with personalized agents on emotional connection, trust, and behavioral outcomes

By advancing these areas, researchers can ensure that conversational AI not only improves general user experiences but also serves as a transformative tool for inclusion, education, and therapeutic support in specialized contexts.

Further investigation is necessary to evaluate and compare approaches such as extensive user-driven customization versus automated learning from user-provided content. The aim would be to assess how these approaches impact user experience, satisfaction, and interaction quality. As proposed by Manzoor et al. (2023), finding an effective balance between manual customization and automated recommendations could help reduce cognitive load while preserving personalization depth. Achieving this balance is pivotal for improving usability, ensuring inclusivity, and enhancing engagement with dialogue systems. Ultimately, this direction seeks to simplify the customization process without sacrificing richness, making conversational agents more accessible and enjoyable for diverse user groups.

References

Adamopoulou E, Moussiades L (2020) Chatbots: history, technology, and applications. Mach Learn Appl 2:100006

Ahn J, Song Y, Yun S, Kim G (2023) MPCHAT: towards multimodal persona-grounded conversation. arXiv preprint arXiv:2305.17388

Allport GW (1927) Concepts of trait and personality. Psychol Bull 24(5):284

Austin JL (1962) How to do things with words. Oxford University Press

Bertrand A, Belloum R, Eagan JR, Maxwell W (2022) How cognitive biases affect XAI-assisted decision-making: a systematic review. In: Proceedings of the 2022 AAAI/ACM conference on AI, ethics, and society, pp 78–91

Binns R, Van Kleek M, Veale M, Lyngs U, Zhao J, Shadbolt N (2018) 'It's reducing a human being to a percentage' perceptions of justice in algorithmic decisions. In: Proceedings of the 2018 CHI conference on human factors in computing systems, pp 1–14

Brandtzaeg P, Følstad A (2018) Chatbots: changing user needs and motivations. Interactions 25:38–43. https://doi.org/10.1145/3236669

Burgess N, Maguire EA, O'Keefe J (2002) The human hippocampus and spatial and episodic memory. Neuron 35(4):625–641

Caron G, Srivastava S (2022) Identifying and manipulating the personality traits of language models. arXiv preprint arXiv:2212.10276

Cheng L, Li X, Bing L (2023) Is GPT-4 a good data analyst? arXiv preprint arXiv:2305.15038.

Chessa A, Murre J (2007) A neurocognitive model of advertisement content and brand name recall. Mark Sci 26(1):130

Clark HH (1996) Using language. Cambridge University Press

Clark HH, Schaefer EF (1989) Contributing to discourse. Cogn Sci 13(2):259–294

Coughlan T, Johnson P (2006) Interaction in creative tasks. In: Proceedings of the SIGCHI conference on human factors in computing systems, CHI'06. Association for Computing Machinery, New York, pp 531–540

Cox SR, Wang Y, Abdul A, von der Weth C, Lim BY (2021) Directed diversity: leveraging language embedding distances for collective creativity in crowd ideation. In: Proceedings of the 2021 CHI conference on human factors in computing systems (Yokohama, Japan) (CHI'21). Association for Computing Machinery, New York

Cutrona CE, Suhr JA (1992) Controllability of stressful events and satisfaction with spouse support behaviors. Commun Res 19(2):154–174

Efendić E, Van de Calseyde P, Bahník Š, Vranka M (2023) Taking algorithmic (vs. human) advice reveals different goals to others. Int J Hum Comput Interact 40. https://doi.org/10.1080/10447318.2023.2210886

Galitsky B (2019) Rhetorical agreement: maintaining cohesive conversations. In: Developing enterprise chatbots: learning linguistic structures. Springer, Cham, pp 327–363

Galitsky B (2022) Adjusting chatbot conversation to user personality and mood. US Patent 11,449,682

Galitsky B (2025) Identifying LLM hallucinations in health communication. In: Health applications of neuro-symbolic AI. Elsevier

Galitsky B, Goldberg S (2022) Artificial intelligence for healthcare applications and management. Academic Press

Galitsky B, Jarrold W (2011) Discovering patterns of autistic planning. In: Workshops at the 25th AAAI conference on artificial intelligence

Galitsky B, Kovalerchuk B (2006) Mining emotional profiles using e-mail messages for earlier warnings of potential terrorist activities. In: SPIE defense + commercial sensing

Galitsky B, Parnis A (2017) How children with autism and machines learn to interact. In: Autonomy and artificial intelligence: a threat or savior? pp 195–226

Galitsky B, Shpitsberg I (2006) How one can learn programming while teaching reasoning to children with autism. In: AAAI spring symposium: between a rock and a hard place: cognitive science

Galitsky B, Shpitsberg I (2014) Finding faults in autistic and software active inductive learning. In: AAAI spring symposium series

Galitsky B, Tumarkina I (2004) Justification of customer complaints using emotional states and mental actions. In: Proceedings of the 17th international Florida artificial intelligence research society conference, Miami Beach, FL

Gao T (2023) Understanding design collaboration between designers and artificial intelligence. Medium.com. https://medium.com/acm-cscw/understanding-design-collaboration-between-designers-and-artificial-intelligence-d294b0bd8911

Gao J, Galley M, Li L (2018) Neural approaches to conversational AI. In: The 41st international ACM SIGIR conference on research & development in information retrieval, pp 1371–1374

Geva M, Schuster R, Berant J, Levy O (2021) Transformer feed-forward layers are key-value memories. arXiv preprint arXiv:2012.14913

Gino F, Schweitzer ME (2008) Blinded by anger or feeling the love: how emotions influence advice taking. J Appl Psychol 93(5):1165–1173

Goldberg S, Galitsky B, Weisburd B (2019) Framework for interaction between expert users and machine learning systems. In: AAAI spring symposium: interpretable AI for well-being

Goldberg S, Pinsky E, Galitsky B (2021) A bi-directional adversarial explainability for decision support. Hum Intell Syst Integr. https://doi.org/10.1007/s42454-021-00031-5. https://hdl.handle.net/2144/42444

Guydish AJ, Fox Tree JE (2021) Good conversations: grounding, convergence, and richness. New Ideas Psychol 63:100877

Ha J, Jeon H, Han D, Seo J, Oh C (2024) CloChat: understanding how people customize, interact, and experience personas in large language models. arXiv preprint arXiv:2402.15265

Hou Y, Tamoto H, Miyashita H (2024) "My agent understands me better": integrating dynamic human-like memory recall and consolidation in LLM-based agents. arXiv preprint arXiv:2404.00573

Humane (2024) Humane AI pin. https://hu.ma.ne/aipin

Jiang G, Xu M, Zhu S, Han W, Zhang C, Zhu Y (2022) MPI: Evaluating and Inducing Personality in Pre-trained Language Models. 10.48550/arXiv.2206.07550

Kahneman D (2011) Thinking, fast and slow. Macmillan. ISBN: 978-1-4299-6935-2

Karra SR, Nguyen ST, Tulabandhula T (2022) Estimating the personality of White-box language models. arXiv preprint arXiv:2204.12000

Kim T, Cochez M, Francois-Lavet V, Neerincx M, Vossen P (2022) A machine with human-like memory systems. arXiv preprint arXiv:2204.01611

Kingman JFC (1993) Poisson processes. Oxford University Press

Lai V, Chen C, Liao QV, Smith-Renner A, Tan C (2021) Towards a science of human-AI decision making: a survey of empirical studies. arXiv preprint arXiv:2112.11471

Lawless WF, Mittu R, Sofge DA (2020) Introduction: Artificial intelligence (AI), autonomous machines, and constructing context: user interventions, social awareness, and interdependence. In: Human-machine shared contexts. Academic Press, pp 1–22

Li J, Li J, Su Y (2024) A map of exploring human interaction patterns with LLM: insights into collaboration and creativity. arXiv preprint arXiv:2404.04570

Lim J, Kang M, Hur Y, Jung S, Kim J, Jang Y, Lee D, Ji H, Shin D, Kim S (2023) You truly understand what I need: intellectual and friendly dialogue agents grounding knowledge and persona. arXiv preprint arXiv:2301.02401

Liu H, Li C, Wu Q, Lee YJ (2023) Visual Instruction Tuning. arXiv:2304.08485

Lubart TI (2001) Models of the creative process: past, present and future. Creat Res J 13:295–308

Lucas RE, Diener E (2001) Understanding extraverts' enjoyment of social situations: the importance of pleasantness. J Pers Soc Psychol 81(2):343–356. https://doi.org/10.1037/0022-3514.81.2.343

Mandic DP, Chambers J (2001) Recurrent neural networks for prediction: learning algorithms, architectures and stability. John Wiley & Sons
Manzoor A, Cai W, Jannach D (2023) Factors influencing the perceived meaningfulness of system responses in conversational recommendation. In: Proceedings of the 10th joint workshop on interfaces and human decision making for recommender systems (IntRS 2023), pp 19–34
Mehri S, Eskenazi M (2020) USR: an unsupervised and reference free evaluation metric for dialog generation. In: Proceedings of the 58th annual meeting of the Association for Computational Linguistics. ACL, pp 681–707
Miller T (2023) Explainable AI is dead, long live explainable AI! Hypothesis driven decision support using evaluative AI. In: Proceedings of the 2023 ACM conference on fairness, accountability, and transparency, pp 333–342
OpenAI (2023) GPT-4 technical report. arXiv preprint arXiv:2303.08774
Ou J, Lu J, Liu C, Tang Y, Zhang F, Zhang D, Gai K (2023) DialogBench: evaluating LLMs as human-like dialogue systems. arXiv preprint arXiv:2311.01677
Ou J, Wu J, Liu C, Zhang F, Zhang D, and Gai K (2024) Inductive-deductive strategy reuse for multi-turn instructional dialogues. arXiv preprint arXiv:2404.11095
Peterson D, Galitsky B, Goldberg S (2004) Literal handling of conflicting default rules leads to inadequate reactions of autistic patients. In: 7th International conference of cognitive and neural systems
Rabbit (2024) Rabbit. https://www.rabbit.tech/
Roberts B, Lejuez C, Krueger R, Richards J, Hill P (2012) What is conscientiousness and how can it be assessed? Dev Psychol. https://doi.org/10.1037/a0031109
Rogiers A, Noels S, Buyl M, De Bie T (2024) Persuasion with large language models: a survey. arXiv preprint arXiv:2411.06837
Ruane E, Farrell S, Ventresque A (2020) User perception of text-based chatbot personality. In: Chatbot research and design: 4th international workshop, CONVERSATIONS 2020, virtual event, 23–24 Nov 2020, revised selected papers, pp 32–47
Safdari M, Serapio-García G, Crepy C, Fitz S, Romero P, Sun L, Abdulhai M, Faust A, and Mataric M (2023) Personality traits in large language models. CoRR, abs/2307.00184
Shaer O, Cooper A, Mokryn O, Kun AL, Shoshan HB (2024) AI augmented brainwriting: Investigating the use of LLMs in group ideation. In: Proceedings of the CHI conference on human factors in computing systems, CHI'24. Association for Computing Machinery, New York
SINAPSA Infocomplex (2024) Personal, artificial companions. Medium. https://medium.com/@contact_30070/personal-artificial-companions-2163b948f5a8
Sivaraman V, Bukowski LA, Levin J, Kahn JM, Perer A (2023) Ignore, trust, or negotiate: understanding clinician acceptance of AI-based treatment recommendations in health care. In: Proceedings of the 2023 CHI conference on human factors in computing systems, pp 1–18
Searle JR (1969) Speech acts: An essay in the philosophy of language. Cambridge University Press
Skantze G, Doğruöz AS (2023) The open-domain paradox for chatbots: common ground as the basis for human-like dialogue. arXiv preprint arXiv:2303.11708
Sundermeyer M, Schlüter R, Ney H (2012) LSTM neural networks for language modeling. In: 13th Annual conference of the international speech communication association
Tulving E (2002) Episodic memory: from mind to brain. Annu Rev Psychol 53(1):1–25. https://doi.org/10.1146/annurev.psych.53.100901.135114
Tulving E, Donaldson W (1972) Organization of memory. Academic Press
Wester J, de Jong S, Pohl H, van Berkel N (2024) Exploring people's perceptions of LLM-generated advice. Comput Hum Behav Artif Hum 2(2):100072
Xu Z, Xu H, Lu Z, Zhao Y, Zhu R, Wang Y, Dong M, Chang Y, Lv Q, Dick RP, Yang F, Lu T, Gu N, Shang L (2024) Can large language models be good companions? An LLM-based eyewear system with conversational common ground. Proc ACM Interact Mob Wear Ubiquit Technol 8(2):87, 41 p. https://doi.org/10.1145/3659600

Yan D, Chen L (2023) The influence of personality traits on user interaction with recommendation interfaces. ACM Trans Interact Intell Syst 13(1):39

Zamfirescu-Pereira J, Wong RY, Hartmann B, Yang Q (2023) Why Johnny can't prompt: how non-AI experts try (and fail) to design LLM prompts. In: Proceedings of the 2023 CHI conference on human factors in computing systems, CHI'23. Association for Computing Machinery, New York

Zhang S, Dinan E, Urbanek J, Szlam A, Kiela D, Weston J (2018) Personalizing dialogue agents: I have a dog, do you have pets too? arXiv preprint arXiv:1801.07243

Zheng J, Wang X, Hosio S, Xu X, Lee L-H (2024) LMLPA: language model linguistic personality assessment. arXiv preprint arXiv:2410.17632v1

Zhong W, Guo L, Gao Q, Ye H, Wang Y (2023) MemoryBank: enhancing large language models with long-term memory. arXiv preprint arXiv:2305.10250

Zhou P, Cho H, Jandaghi P, Lee D-H, Lin BY, Pujara J, Ren X (2022) Reflect, not reflex: inference-based common ground improves dialogue response quality. arXiv preprint arXiv:2211.09267

Zhou Z, Gao W, Li Y, Yu J (2024) Developing an interaction framework for human-large language models collaboration in creative tasks: insights from Ux professionals' communication with ChatGPT

Zielske HA (1959) The remembering and forgetting of advertising. J Mark 23(1959):239–243. https://api.semanticscholar.org/CorpusID:16735419

A Copilot for Social Promotion

Boris Galitsky and Oleg Ozerov

1 Introduction

In an era where digital transformation is reshaping industries, social media has emerged as a central platform for communication, influence, and promotion. The rapid growth of social platforms has created new opportunities for brands, businesses, and influencers to engage with their audiences and drive growth. As the need for personalized, efficient, and scalable strategies increases, AI tools have become crucial for optimizing social promotion efforts (Akilkhanov 2024). The tool developed in this chapter, the *Social Promotion Copilot* (SPC), leverages advanced AI technologies to assist in the creation, management, and evaluation of social campaigns.

This chapter explores the application of copilot technology in social promotion, focusing on its ability to automate content generation, audience targeting, campaign optimization, and other tasks, such as complaint handling and supporting customers in their disputes with a company. The chapter examines how AI-driven copilots can support marketers in navigating the complexities of social media algorithms, consumer behavior, and engagement metrics. Additionally, it addresses the ethical considerations and potential challenges of relying on AI for promotional activities in a highly competitive and fast-paced digital environment. By analyzing the effectiveness of copilot tools, this study aims to provide insights into how they can enhance the efficiency and success of social promotion campaigns across various platforms.

Through a combination of theoretical exploration and practical case studies, this chapter seeks to contribute to the growing body of knowledge on AI's role in transforming digital marketing strategies and its impact on the future of social promotion.

The original version of the chapter has been revised. A correction to this chapter can be found at https://doi.org/10.1007/978-3-032-16524-4_13

B. Galitsky, *Applications of Neuro-Symbolic Artificial Intelligence*,
https://doi.org/10.1007/978-3-032-16524-4_6

In the ever-evolving landscape of social media promotion, the integration of AI tools has become a game-changer. In this context, the social media manager assumes the role of the captain, guiding campaigns, crafting strategies, and interpreting data to drive engagement and growth (Galitsky 2023). Meanwhile, the AI acts as the copilot, providing invaluable support in navigating the complexities of the digital space, optimizing content creation, and enhancing campaign performance. Much like a captain and copilot work in tandem to ensure a successful flight, the collaboration between human expertise and AI technology forms a powerful partnership that maximizes the potential of social promotion efforts.

By leveraging AI tools, social media managers can unlock new opportunities for more efficient, personalized, and impactful campaigns. These tools assist in everything from content generation to audience segmentation, streamlining processes, and improving overall results. Eager and Brunton (2023), Xu (2024), and Su and Yang (2023) study the transformative role of AI copilots in education, focusing on how tools such as ERNIE Bot (PRNewswire 2023), ElevenLabs, and Classpoint can enhance the effectiveness of teaching strategies. Although applications of copilots in self-driving cars, education, retail (Furmakiewicz et al. 2024), health (Chiam et al. 2024; Zou et al. 2024), and other domains have been thoroughly explored, it is not the case for social promotion domain.

As we introduce a Social Promotion Copilot (SPC) for the optimization of social campaigns, driving engagement, and empowering brands to navigate the digital landscape with greater agility and success, we also mention SPC's contribution to the customer relationship management pipeline (Galitsky 2021). It supports communication with a customer, customer retention and handling complaints. Moreover, SPC can be an assistant in individual social promotion in professional and personal domains.

Existing Web agents are typically evaluated only in simplified Web simulators or static Web snapshots, greatly limiting their applicability in real-world scenarios. To bridge this gap, we enable SPC with large language model (LLM)-powered Web agent capabilities so that it can complete user instructions end to end by interacting with real-world social Web sites.

1.1 Social Promotion of New Products

A significant portion of a firm's efforts is dedicated to introducing new products and technologies to the market. However, these initiatives come with substantial risks, as launching a new product is inherently unpredictable (Mahajan and Muller 1979). The initial market penetration phase is particularly critical, as it sets the trajectory for the product's future adoption. A rapid and widespread takeoff can provide a competitive edge, trigger viral consumer adoption, and ultimately determine whether the product succeeds or fails (Golder and Tellis 2004).

To navigate these uncertainties, promotional strategies play a crucial role in accelerating early diffusion. By strategically deploying marketing efforts—especially promotional campaigns—companies can stimulate initial demand, generate awareness, and create momentum that drives product adoption (Galitsky 2019). Effective promotional activities not only help overcome market inertia but also maximize exposure, encourage early adopters, and establish a foundation for long-term success.

SPC can significantly enhance these promotional efforts by leveraging AI-driven automation, real-time data analysis, and personalized engagement to optimize product launches. By monitoring consumer sentiment, identifying key influencers, and tailoring promotional content to specific audience segments, SPC ensures that marketing messages reach the right people at the right time. It can dynamically adjust campaign strategies based on real-time feedback, ensuring that promotional efforts remain relevant and impactful throughout the diffusion process. Additionally, SPC can automate interactions, fostering engagement across social platforms and amplifying word-of-mouth effects. By streamlining and optimizing promotional activities, SPC increases the likelihood of a successful product takeoff, reduces the risk of launch failure, and maximizes the efficiency of marketing investments. At the level of an individual user, performing social promotion, SPC might raise self-confidence.

2 Social Promotion Copilot

Once the users provide a raw idea or draft (e.g., "Promote my new blog post about AI trends"), the copilot enhances and optimizes the post for different platforms:

1. Twitter/X: Concise, engaging with hashtags → 🚀 *AI is transforming industries! Check out the latest trends in my new blog: [link] #AI #FutureTech*
2. LinkedIn: Professional and thought-leadership style → *The AI landscape is evolving rapidly. In my latest article, I explore key trends shaping the future. Read more: [link]*
3. Instagram/Facebook: Visually engaging, story-driven → *AI is the future! Swipe to see the latest trends. Full article here: [link]*📲🔥 *#ArtificialIntelligence #TechNews*

The copilot supports auto-scheduling and multi-platform posting. Posts can be scheduled for optimal engagement times (based on AI analysis). The system automatically adapts the same content for multiple platforms with platform-specific tweaks (Fig. 1). Moreover, the copilot performs engagement analysis and smart recommendations, monitoring likes, shares, and comments. It uses LLM-driven insights to suggest improvements:

1. "Your posts perform best on Thursdays at 6 PM."
2. "Try adding a question to increase engagement!"
3. "Hashtags #AI and #Tech are trending, consider adding them!"

It can also do LLM-powered audience targeting and hashtag suggestions, identifying target audiences based on post content and suggesting relevant hashtags and

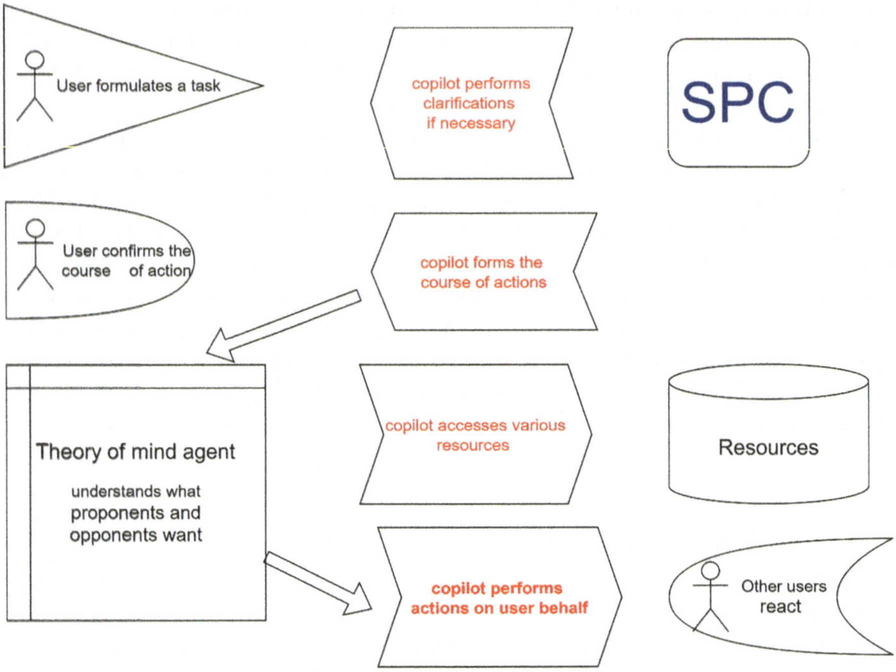

Fig. 1 A high-level view at the Social Promotion Copilot (SPC)

tagging influencers for better reach. The copilot drafts quick replies to user comments and detects potential collaborations and suggests outreach messages.

An example flow in action can be as follows:

> For user input "I want to promote my new startup on social media," Social Promotion Copilot optimizes post suggestions:
>
> 1. Twitter: *Exciting news! 🚀 Our startup [Startup Name] is launching soon! Stay tuned for innovation in [industry]. #Startup #Innovation*
> 2. LinkedIn: *We are thrilled to introduce [Startup Name], a game-changer in [industry]. Learn more about our journey here: [link]*
> 3. Instagram: *A dream turning into reality! 🏆 Follow our startup journey. #Startuplife #NewBeginnings*

It suggests best posting time, "Your audience is most active between 5 PM and 11 PM. Schedule now?", and Hashtag and Tagging: "#Entrepreneurship #TechStartup #Innovation." Copilot also recommends to "Consider tagging industry leaders like @TechInfluencer to boost visibility!" If someone comments, "Congrats!", Copilot suggests, "Thank you! Exciting times ahead!" 🚀 Copilot can help in a broad spectrum of domains including health.

2.1 An Algorithm for SPC Engagement Analysis and Recommendation

The algorithm monitors engagement, learns from past performance, and adapts SPC's social promotion strategy in real time. It enables data-driven, intelligent posting strategies, ensuring maximum visibility and user interaction.

Its inputs are:

- Post data: Text, media, category (complaint, promotion, discussion, etc.)
- Engagement metrics: Likes, shares, comments, replies, follower growth
- Timeframe: The period over which engagement is analyzed
- Competitor benchmarking: Engagement metrics of similar posts by competitors (if available)

Outputs:

- Performance score: Engagement score per post
- Smart recommendations: Suggestions to improve future posts (timing, tone, content type)

Collection of engagement metrics for each post:

- L = Number of likes
- S = Number of shares
- C = Number of comments
- R = Number of replies
- F = Follower/subscriber growth due to post

$$\text{Weighted engagement score } E = w_L L + w_S S + w_C C + w_R R + w_F F$$

Different weights based on impact need to be involved (e.g., shares are more valuable than likes).

Engagement score needs to be relative to past performance and competitors:

$$E_{normalized} = \left(E - E_{\min}\right) / \left(E_{\max} - E_{\min}\right)$$

Smart recommendation generation identify trends in engagement. In includes a comparison of E scores over time to detect improvement or decline, identification of high-engagement content types (e.g., complaints get more shares, promotions get more likes), and detection of peak engagement timeframes (when users interact the most).

Generation of post optimization recommendations include:

1. Timing adjustment: If peak engagement occurs at 6 PM, shift future post scheduling.
2. Content type adaptation: If complaints generate the highest interaction, increase strategic complaint-based posts.

3. Tone analysis: If negative sentiment (e.g., strong complaints) generates more engagement, adjust response strategy.
4. Hashtag and tagging optimization: Recommend popular hashtags or relevant user tags based on trending engagement patterns.

Adaptive posting strategy includes the following rules. If engagement drops, alter post frequency (e.g., reduce frequency to avoid spamming). If positive engagement increases, prioritize successful content types. Continuous learning and auto-tuning update weight values w_L, w_S, w_C, w_R, and w_F, based on real-world outcomes. It also adjusts recommendation logic dynamically based on performance feedback (if new engagement types emerge including video interactions, it incorporates them into scoring).

3 Copilot Architecture

LangGraph is a multi-agent framework that enables structured interactions between AI agents, making it well suited for building a Social Promotion Copilot. LangGraph enhances the copilot's capabilities in the following ways (Fig. 2):

1. User interaction and task formulation. A user formulates a promotional task (e.g., "Promote my new product on social media"). LangGraph triggers a multi-agent process to refine the request.
2. Multi-agent LLM processing. LangGraph orchestrates different AI agents to process the request. User request enrichment agent analyzes the task and asks follow-up questions (e.g., "What platforms should be targeted?"). User request verification agent ensures clarity and coherence and verifies correctness before execution.
3. Copilot meta-agent (decision-maker) determines the best promotional strategy and generates an optimized course of actions. The copilot accesses external resources (e.g., market data, previous campaigns, user preferences) and automates content creation based on the refined user request.
4. The copilot performs action execution and social media automation. It automatically performs actions (e.g., posting content, engaging with users, analyzing responses). Automation tools (e.g., Selenium, APIs) execute the generated promotional tasks.
5. LangGraph optimizes future actions using orchestrator-worker agents (for automation execution), evaluator-optimizer for engagement analytics, and routing agents (for personalized content delivery). Copilot also monitors how other users react to posts. It dynamically adjusts engagement strategies (e.g., replying to comments, boosting trending content). Feedback loops ensure continuous improvement.

Hence, LangGraph provides a structured agent-based approach for building SPC, ensuring efficient task automation, decision-making, and engagement tracking. Its

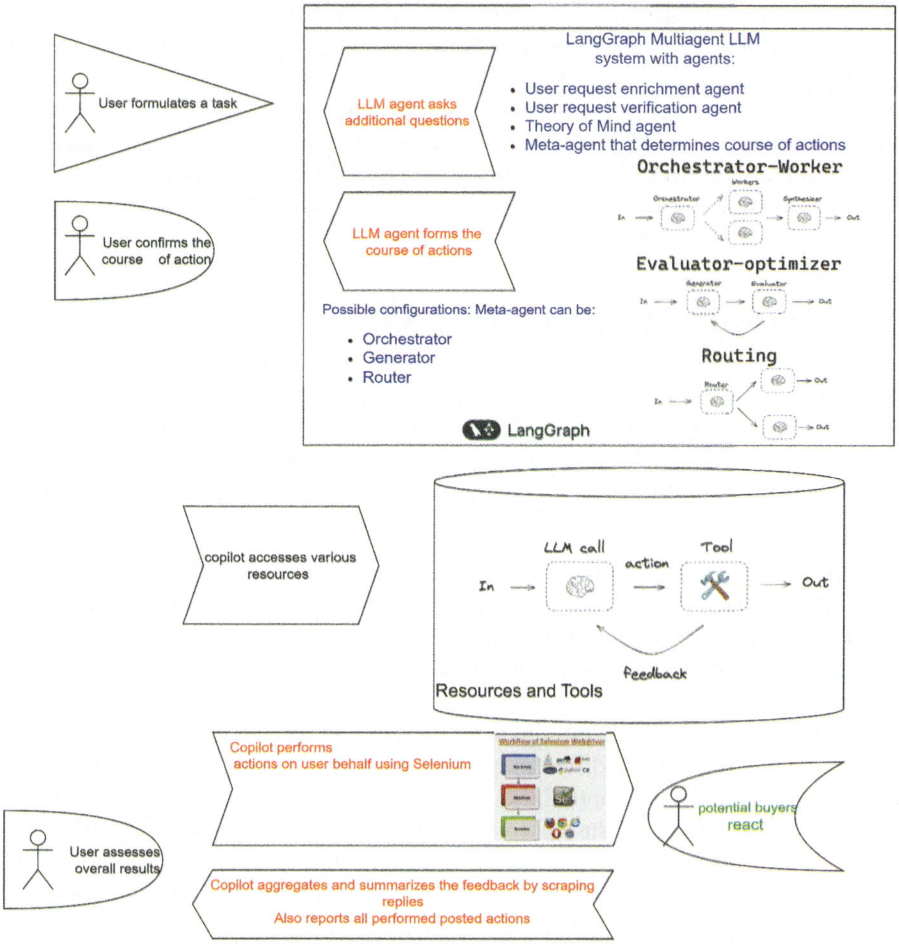

Fig. 2 LangGraph supports SPC architecture

multi-agent design allows for scalability and adaptability, making it an ideal framework for AI-driven social media promotion.

LangGraph utilizes three primary architectural components—orchestration-worker, evaluator-optimizer, and routing—to manage and improve complex processes like those involved in social promotion campaigns. Each component has a specific role in ensuring that the system functions efficiently, responds to real-time data, and adapts to user behaviors.

The orchestration-worker in LangGraph is the central hub that coordinates the execution of tasks across various agents and platforms. It ensures that the workflow is distributed efficiently and effectively, managing all aspects of content delivery, campaign scheduling, and real-time interactions. This component allows LangGraph to handle multiple processes concurrently.

The evaluator-optimizer in LangGraph monitors the performance of various campaigns and adjusts strategies dynamically to ensure that marketing efforts are continually improved. This component uses real-time data to assess the effectiveness of different promotional activities and modifies them to maximize impact. Router is a meta-agent maintaining state transitions.

Selenium is primarily used for Web application testing and can be used across various browsers (e.g., Chrome, Firefox, Safari) and platforms. The core component allows interaction with Web browsers through programming scripts. In this project, we use Selenium to automate scraping and posting on behalf of a user.

3.1 Enabling SPC with Theory of Mind

SPC can leverage Theory of Mind (ToM; chapter "Improving ToM Capabilities of LLMs in Applied Domains") to enhance its ability to engage with users by modeling their intentions, emotions, and perspectives. By understanding the mental states of different users, SPC can tailor its responses and promotional strategies more effectively:

1. Adapting to user sentiment: SPC can analyze the emotional tone of conversations and adjust its messaging accordingly. For example, in the complaint domain, it can distinguish between frustrated and indifferent customers, providing empathetic responses to the former while offering neutral or informative replies to the latter.
2. Predicting reactions and engagement: by retrospection and modeling how users are likely to respond to different types of posts (e.g., complaints, promotions, or discussions), SPC can optimize content to maximize engagement. For instance, it can anticipate that a strongly worded complaint might trigger a viral discussion and adjust its phrasing to align with user sentiment.
3. Personalized interaction strategies: instead of using generic responses, SPC can infer a user's intent behind interactions (e.g., whether they seek resolution, validation, or public attention) and tailor replies accordingly. For instance, if a user frequently engages in debates, SPC might respond with thought-provoking questions rather than straightforward solutions.
4. Social dynamics awareness: in multi-user interactions, SPC can predict social influences, such as how a user's opinion might shift based on peer reactions. This allows it to strategically time responses or escalate issues when it detects growing dissatisfaction among a user group.
5. Handling ambiguity and humor: ToM enables SPC to better interpret sarcasm, irony, and cultural nuances in user interactions, reducing the risk of miscommunication. This is especially useful in social platforms where humor and indirect complaints are common.

By incorporating Theory of Mind, SPC evolves from a rule-based automation tool into an adaptive, socially intelligent agent capable of engaging users in a way that feels natural, persuasive, and emotionally aware.

3.2 *Run-Time Execution of a Selenium Task*

Another essential feature of SPC to assure its flexibility is to generate necessary Selenium code on the fly and execute it. It is necessary as the social promotion site is determined online and the Selenium code for it is unavailable. This code is than expected to be generated by an LLM and executed.

This is illustrated by the following example:

1. The user inputs a task description (e.g., "Generate a function that calculates Fibonacci numbers").
2. The script queries an LLM to generate Python code.
3. It prints the generated code.
4. It executes the generated code using exec().

```
def generate_code(task_description):
    """
    Uses an LLM to generate Python code based on a textual
task description.
    """
    prompt = f"Write a Python script that accomplishes the
following task:\n{task_description}\n\nOnly provide the code
without explanations."
    response = openai.ChatCompletion.create(
        model="gpt-4",  # Use GPT-4 or another
available model
        messages=[{"role": "user", "content": prompt}],
        temperature=0.2
    )

    code = response["choices"][0]["message"]["content"]
    return code

def execute_code(code):
    """
    Executes the generated Python code safely.
    """
    try:
        exec_globals = {}
```

```
        exec(code, exec_globals)
    except Exception as e:
        print("Error executing code:", e)

if __name__ == "__main__":
    task_description = input("Describe the coding task: ")
    generated_code = generate_code(task_description)

    print("\nGenerated Code:\n", generated_code)
    print("\nExecuting Code...\n")
    execute_code(generated_code)
```

The steps are as follows:

1. Copilot inputs a task description (e.g., "Generate Selenium code to do authentication and submit a complaint via form").
2. The script queries an LLM to generate Python code.
3. It executes the generated code using `exec()`.

The generated code for SPC included the script to perform certain actions on a Web form (Table 1).

Table 1 Selenium command description

`driver.get(URL)`	Navigate to a specified URL
`driver.back()`	Go back to the previous page
`driver.forward()`	Move forward in browser history
`driver.refresh()`	Refresh the current page
`element.click()`	Click on an element
`element.send_keys(text)`	Type text into an input field
`element.clear()`	Clear text from an input field
`element.submit()`	Submit a form
`ActionChains(driver).move_to_element(element).perform()`	Hover over an element
`driver.execute_script(script, args)`	Execute JavaScript code on a page
`driver.find_element(By.XPATH, locator)`	Locate an element using XPath
`driver.find_element(By.CSS_SELECTOR, locator)`	Locate an element using CSS selector
`driver.find_elements(By.TAG_NAME, tag_name)`	Locate multiple elements by tag name
`driver.switch_to.frame(frame_reference)`	Switch to an iframe
`driver.switch_to.default_content()`	Switch back to the main content
`driver.switch_to.window(window_name)`	Switch to a different browser tab
`driver.close()`	Close the current tab
`driver.quit()`	Close the entire browser session
`driver.implicitly_wait(seconds)`	Set an implicit wait time for elements
`WebDriverWait(driver, timeout).until(condition)`	Set an explicit wait for elements

4 Use Case: Complain Copilot

The Complaint Copilot is an intelligent system designed to assist users in filing complaints efficiently. It helps by analyzing, reformulating, and enhancing complaints while also suggesting further actions and automating complaint submissions across multiple platforms, such as social media and consumer advocacy Web sites.

The user submits a raw complaint such as "My Internet provider has constant outages, and support is unhelpful." The system extracts key entities like company name, issue type, location, and service type. Then the Complaint Copilot dynamically generates follow-up questions to gather more details:

- "How long have you experienced these issues?"
- "Have you contacted customer support? If yes, what was their response?"
- "Do you have supporting evidence (screenshots, receipts, chat logs)?"

These questions ensure that the complaint is detailed enough to be effective.

SPC in complaint mode rewords the complaint into a more formal, persuasive, and legally sound version. It uses LLM-based sentiment analysis to enhance clarity and impact. An example reformulation is as follows:

1. Original: "My Internet provider sucks. It's always down, and support ignores me."
2. Enhanced: "I have been experiencing frequent Internet outages with [Provider Name] for the past [X] weeks. Despite multiple attempts to contact customer support, my issue remains unresolved. This has significantly impacted my work and daily life. I request immediate intervention and a resolution."

Based on the complaint type, the system suggests next actions, such as:

1. Filing a complaint with regulatory authorities (e.g., FCC, Ombudsman)
2. Contacting consumer protection organizations
3. Seeking compensation (e.g., refund, discount, service credit)

The advice is customized using LLM-driven recommendations.

Complaint copilot performs auto-posting on social media and consumer advocacy platforms. The enhanced complaint is automatically formatted for posting on:

1. Twitter/X, Facebook, Instagram, Reddit—for public awareness
2. Consumer advocacy sites (e.g., Trustpilot, BBB, Ripoff Report)
3. Company complaint portals (e.g., "Submit a complaint" pages)

Complaint copilot retrieves company-specific contact handles (e.g., @ CompanySupport) and inserts them into posts for maximum visibility.

Complaint copilot leverages LLM-powered NLP—text analysis, question generation, complaint enhancement, API integrations that performs Social media posting (X/Twitter, Facebook API), consumer complaint portals, sentiment and intent detection—to improve phrasing and impact. Moreover, complaint copilot uses Automation Scripts—auto-filling complaint forms on advocacy Web sites.

Example Flow in Action:

User Input:
"My flight was canceled last minute, and the airline refused a refund. What should I do?"
Complaint Copilot Workflow:

1. Follow-up questions:
   "Which airline was it?"
   "Do you have proof of cancellation?"
   "Did they offer an alternative flight?"
   "Are you requesting a refund or compensation?"
2. Reformulated complaint:
   "I booked a flight with [Airline] on [Date], but it was canceled at the last minute. The airline has refused to issue a refund despite my request. According to aviation regulations, I am entitled to compensation. I seek immediate resolution and reimbursement."
3. Advice (Figs. 3 and 4):
   "File a complaint with the airline's dispute resolution team."
   "Report the issue to the Department of Transportation or Air Passenger Rights agency."
   "If unresolved, escalate to a consumer protection body."
4. Automated posting:
   Twitter post: "@Airline, my flight was canceled, and I was denied a refund. As per regulations, I am entitled to compensation. Please resolve this immediately! #CustomerRights #FlightCancellation."
   Complaint submitted on Trustpilot & BBB.

The architecture of the Complaint Copilot based on Selenium consists of multiple components working together to automate the complaint submission process across various platforms (Fig. 5). Through a user interaction layer, a user shares a complaint. An LLM processes the Complaint, reformulates the complaint, and fills in missing details. Also, the LLM determines whether the complaint is valid or just an emotional reaction, asking additional questions to refine the complaint. The user confirms the course of action.

Complaint Copilot (Automation Layer) makes the discovery, identifying relevant sites where complaints can be filed. It also discovers users with complaints to promote the Complaint Copilot service. The Copilot accesses various complaint submission resources, including social media (Facebook, LinkedIn, Twitter/X, Reddit, Instagram, TikTok) and consumer protection sites: Better Business Bureau, official company support forums, etc.

Under automated complaint filing using Selenium, the Copilot requests the LLM to generate Selenium scripts for automated complaint submission on different platforms. Selenium executes the generated scripts, automating complaint postings across multiple resources. The engagement is continuous: the Copilot monitors

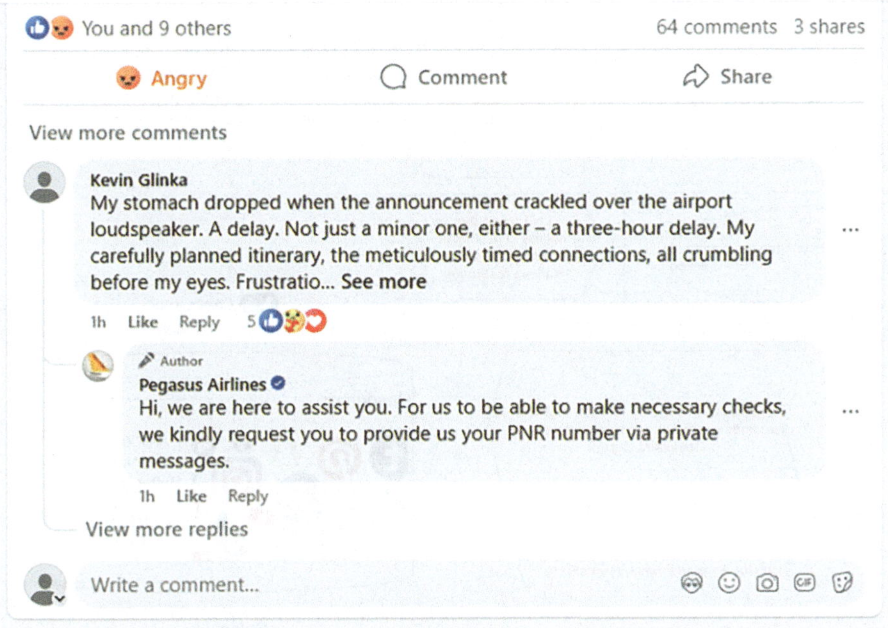

Fig. 3 A customer files a complaint via Complaint Copilot on the airline Facebook page

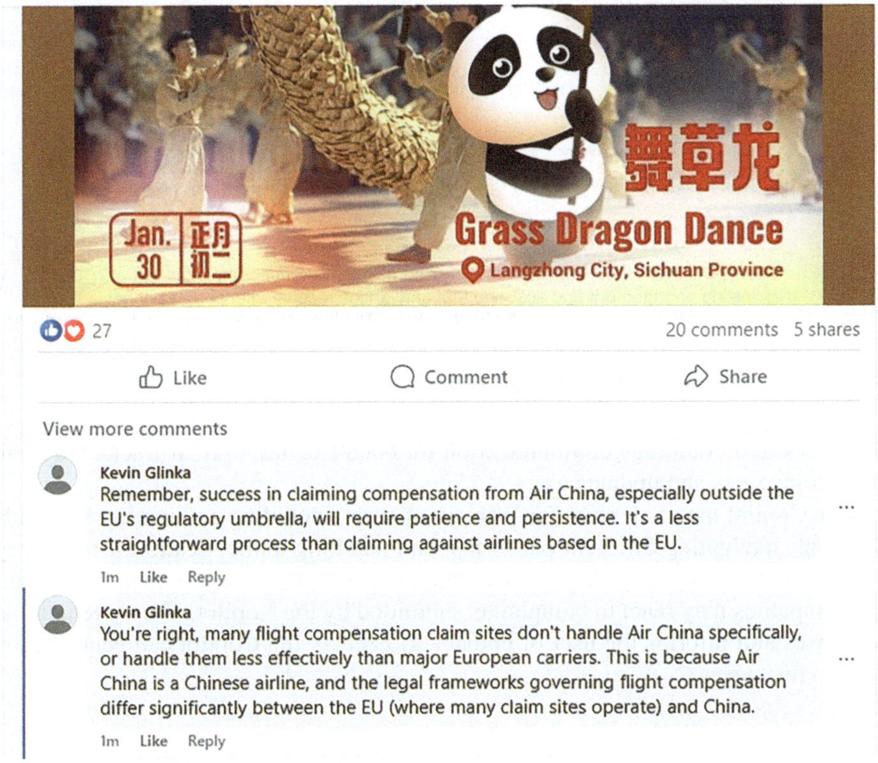

Fig. 4 Generic complaint agent Keving Glinka auto-posts complaints using SPC

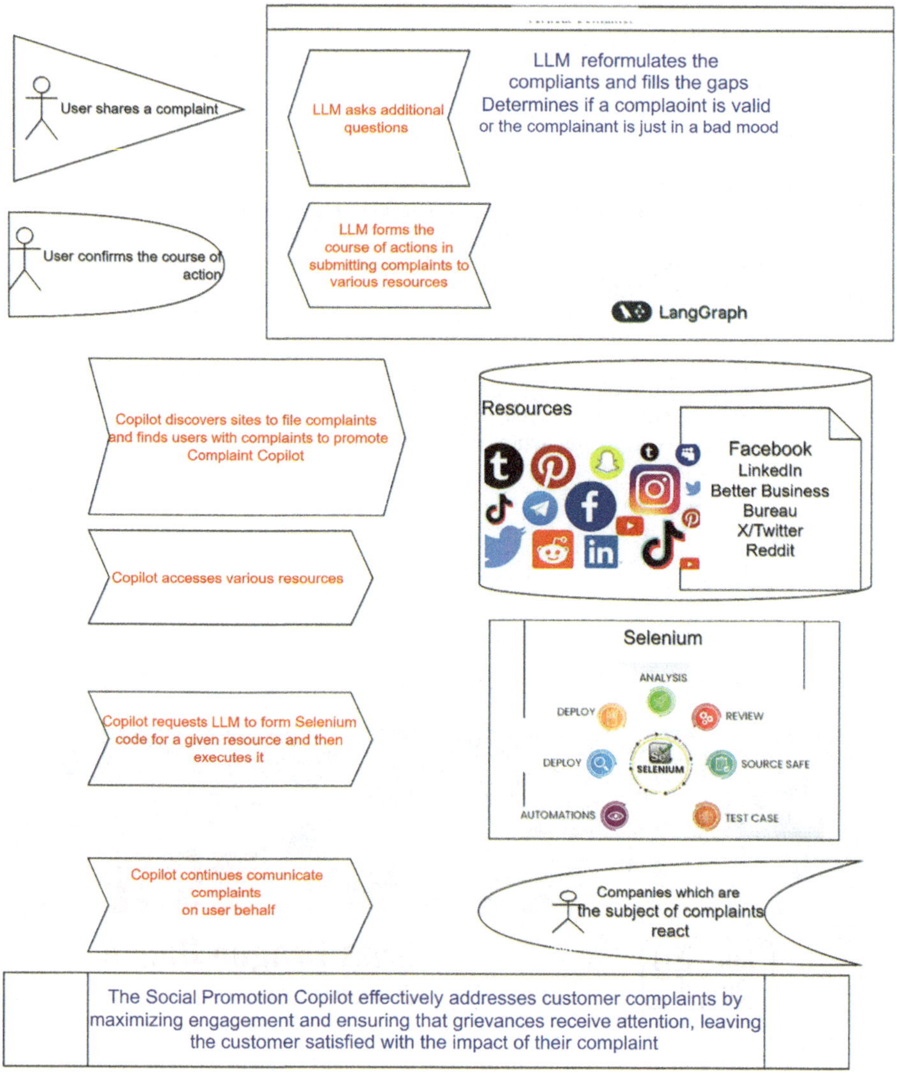

Fig. 5 The architecture of SPC in Complaint mode

responses and continues communication on behalf of the user. It tracks reactions from companies and updates users.

The Copilot uses Selenium for Web automation, including automating form submissions, navigating different platforms, and handling authentication if necessary. Hence, the Selenium engine supports complaint submission, review, and tracking.

Companies may react to complaints submitted by the Copilot. The system tracks responses and informs the user of updates. If needed, the Copilot can generate follow-up responses.

This architecture leverages LLMs for complaint refinement and decision-making, while Selenium handles automation of complaint submission across multiple platforms. The combination allows for an efficient and scalable way to manage consumer complaints, ensuring that they are effectively submitted and followed up.

5 Marketing Strategies

5.1 Extending Promotional Strategies with a Social Promotion Copilot

Many marketing efforts focus on promotional activities that support the launch of new products, playing a crucial role in the early stages of the product life cycle. The success of these promotions significantly influences the diffusion and adoption of new products in competitive markets (Delre et al. 2007). With the rise of artificial intelligence and automation, Social Promotion Copilots—AI-driven assistants designed to optimize and execute promotional strategies across digital platforms—offer a transformative approach to enhancing product visibility and consumer engagement.

This chapter extends traditional promotional strategy models by integrating an agent-based simulation of SPC that automates and personalizes social media interactions, content distribution, and influencer engagement to optimize product launches. The model investigates the efficacy of SPC-driven promotional strategies by focusing on three key aspects: targeting, timing, and adaptive engagement.

Simulation experiments indicate that AI-assisted promotional activities significantly impact diffusion dynamics, amplifying both reach and engagement. The findings highlight that:

1. The absence of promotional support and/or improper timing can lead to product diffusion failure. AI-powered copilots can mitigate this risk by dynamically adjusting promotion schedules based on real-time consumer sentiment, trends, and engagement metrics.
2. The optimal targeting strategy involves addressing distant, small, and cohesive consumer groups. The copilot leverages machine learning to identify micro-communities within social networks, tailoring messages to resonate with their specific interests and behaviors.
3. The optimal timing for promotions varies by product category, particularly between durable goods (e.g., home appliances) and consumer electronics. AI-enhanced strategies can refine timing by analyzing historical data, seasonality trends, and consumer readiness indicators.

Beyond these findings, the integration of the Copilot introduces continuous learning and adaptive engagement, allowing brands to refine their promotional strategies dynamically. By utilizing natural language processing, sentiment analysis, and reinforcement learning, the Copilot can craft personalized responses, automate influencer collaborations, and enhance social proof mechanisms—further accelerating product diffusion.

This research contributes to the planning and management of AI-enhanced promotional strategies, demonstrating how Copilots can revolutionize product launch campaigns by maximizing efficiency, engagement, and long-term adoption rates.

5.2 *Diffusion of New Products*

SPC can play a crucial role in supporting and accelerating product diffusion throughout the entire S-shaped adoption curve (Fig. 6) by leveraging AI-driven strategies that optimize both external and internal influences on consumer behavior.

In the early phase, when sales are slow and external influences dominate, SPC can enhance traditional promotional activities by identifying and targeting early adopters more effectively. Using advanced predictive analytics and sentiment analysis, SPC can pinpoint potential buyers based on online behavior, interests, and engagement history. It can also automate personalized outreach, manage influencer collaborations, and optimize ad placements across social media platforms to ensure the right audience is exposed to the product at the right moment.

As the product gains traction and reaches the critical mass needed for takeoff, SPC can facilitate the transition from mass promotions to more organic, community-driven engagement. By continuously analyzing consumer interactions, it can refine messaging to amplify word-of-mouth marketing and social contagion effects. SPC can also monitor customer feedback in real time, enabling businesses to quickly adjust strategies, resolve concerns, and boost positive sentiment, which further accelerates adoption.

In the mature phase, where internal influences become dominant, SPC can sustain engagement by curating user-generated content, encouraging reviews and testimonials, and maintaining brand presence through personalized interactions. By

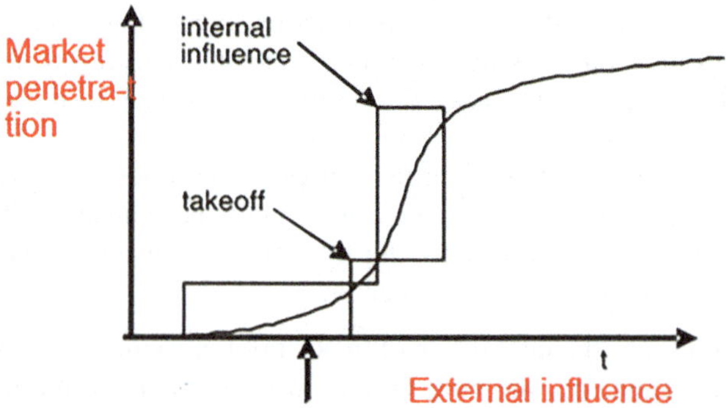

Fig. 6 The S-shaped curve of the diffusion

automating customer follow-ups and loyalty campaigns, it helps maximize long-term retention while identifying opportunities for cross-selling and up-selling.

Ultimately, the Copilot enhances the efficiency of each phase of product diffusion by ensuring seamless coordination between promotional efforts, customer engagement, and organic adoption trends. This AI-driven approach minimizes wasted marketing spend, shortens the time to critical mass, and maximizes the overall impact of a product launch.

5.3 Business-to-Consumer and Business-to-Business Settings of SPC

SPC operates differently in Business-to-Consumer (B2C) and Business-to-Business (B2B) settings, adapting its strategies to suit the distinct characteristics of each market.

In a B2C environment, SPC focuses on high-volume, emotion-driven, and fast-paced interactions with individual consumers. The copilot uses AI-powered personalization, real-time engagement, and automation to maximize outreach and drive conversions.

- Personalized marketing at scale: SPC analyzes consumer data, behaviors, and preferences to deliver highly targeted ads, product recommendations, and promotional messages. For example, it can send personalized discounts to consumers who have shown interest in a product but haven't made a purchase.
- Social media engagement: SPC automates interactions on platforms like Instagram, TikTok, and Facebook by commenting, liking, and engaging with potential customers, fostering brand visibility. It also monitors trends and viral content to adjust marketing strategies dynamically.
- Influencer and word-of-mouth amplification: SPC identifies key influencers and brand advocates to encourage product endorsements and reviews, amplifying word-of-mouth marketing.
- Automated customer support and retention: through AI-powered chatbots and automated messaging, SPC answers FAQs, processes orders, and provides personalized follow-ups, increasing brand loyalty and repeat purchases.

In B2B markets, where sales cycles are longer and decision-making involves multiple stakeholders, SPC acts as a relationship-building and lead-nurturing assistant. Instead of mass outreach, it focuses on high-value engagement, trust-building, and decision support.

- AI-driven lead qualification: SPC analyzes potential leads' online behavior, company size, and engagement with marketing materials to prioritize high-quality prospects, ensuring sales teams focus on the most promising opportunities.

Table 2 Comparative features of B2C and B2B

Feature	B2C approach	B2B approach
Sales cycle	Short, impulse driven	Long, relationship driven
Engagement style	High volume, social media centric	Targeted, professional networking focused
Content type	Short-form ads, viral content	Whitepapers, case studies, webinars
Customer relationship	One-time purchases and repeat buyers	Long-term partnerships and high-value deals
Decision-making	Individual driven, emotional	Group driven, logical

- Automated outreach and content personalization: SPC crafts personalized email sequences, LinkedIn messages, and follow-ups tailored to different stakeholders in the decision-making process, ensuring continued engagement.
- Thought leadership and brand positioning: SPC automates content distribution by posting whitepapers, case studies, and industry insights on LinkedIn, Twitter, and professional forums, establishing the company as an authority in its field.
- CRM integration and sales support: By integrating with CRM systems, SPC ensures seamless tracking of interactions, sends reminders for follow-ups, and provides insights into when and how to engage with leads.
- Event and webinar engagement: SPC can automate invitations, reminders, and post-event follow-ups, ensuring that businesses maximize attendance and engagement in virtual or physical networking events.

Key differences in SPC's role for B2C vs. B2B are shown in Table 2.

Hence the Copilot adapts to the unique dynamics of B2C and B2B markets by optimizing engagement strategies for each setting. In B2C, it focuses on personalization, automation, and viral reach to drive consumer purchases. In B2B, it prioritizes lead nurturing, trust building, and long-term relationship management. By leveraging AI, SPC ensures that businesses, regardless of their market type, achieve more efficient, data-driven, and impactful promotional efforts.

6 Evaluation

6.1 Datasets

Zhou et al. (2023) developed WebArena, a realistic and reproducible environment for training and evaluating language-guided agents, with a particular focus on Web-based task execution. The environment is designed to simulate real-world Internet interactions, enabling agents to navigate and perform tasks across fully functional Web sites spanning four common domains: e-commerce, social forum discussions, collaborative software development, and content management. In our evaluation, we focus on social forum discussions.

To emulate human problem-solving, WebArena also embeds tools and knowledge resources as independent Web sites. WebArena introduces a benchmark on interpreting high-level realistic natural language command to concrete Web-based interactions (Fig. 7).

To enhance the authenticity of task-solving, the environment incorporates interactive tools such as maps, calendars, and search functions, as well as external knowledge bases like user manuals, FAQs, and structured datasets. These resources enable agents to engage in complex, multi-step reasoning processes akin to human decision-making on the Web.

We rely on Zhou et al. (2023)'s benchmark suite designed to rigorously evaluate an agent's ability to complete Web-based tasks with functional correctness. This benchmark includes diverse, long-horizon tasks that reflect real-world Internet usage, such as searching for products, posting on discussion forums, navigating collaborative coding platforms, and managing digital content. These tasks assess not only task completion accuracy but also an agent's ability to adapt, plan, and interact effectively with dynamic Web interfaces. We reduced the set of tasks to the ones related to social promotion (Fig. 8).

WebVoyager, introduced by He et al. (2024), is an advanced Web-browsing agent designed to autonomously navigate the Internet in response to human-formulated Web activity tasks. Unlike SPC, which may involve information sharing,

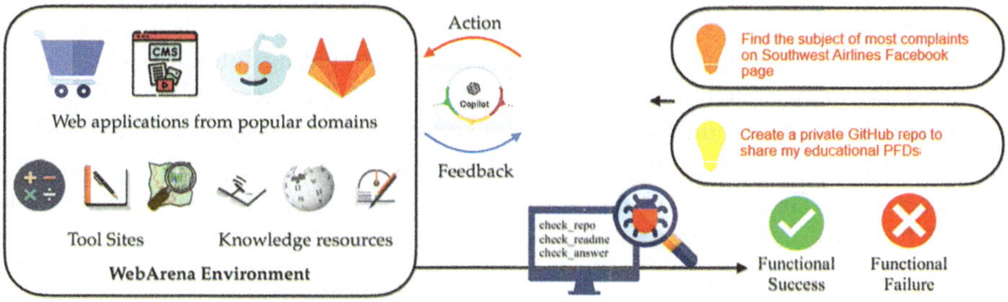

Fig. 7 Illustration of WebArena dataset

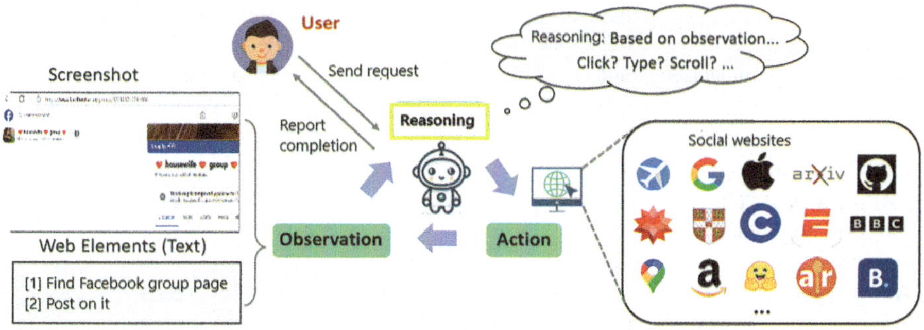

Fig. 8 WebVoyager with a social task

WebVoyager strictly focuses on retrieving information without contributing content. The system operates by analyzing screenshots and text-based elements on Web pages, including the type and content of various interface components. At each step, it determines the most suitable action to take, ensuring efficient task completion. Once the objective is achieved, the retrieved information is returned to the user. For instance, WebVoyager can execute a request such as "Find a housewife social group and identify an appropriate post to reply to," demonstrating its ability to process complex, open-ended tasks.

He et al. (2024) create a benchmark composed of real-world tasks sourced from 15 widely used Web sites, including social platforms. This benchmark provides a structured way to assess the system's capabilities in realistic browsing scenarios. Additionally, they propose an automatic evaluation protocol that leverages the multi-modal understanding abilities of visual LLMs. This protocol enables a more comprehensive assessment of open-ended Web agents by measuring their ability to interpret and interact with Web environments.

6.2 Assessment of Auto Web Browsing

We conduct an evaluation of the Web-browsing capabilities of SPC using data from WebArena and WebVoyager, focusing on both full datasets and specific subsets related to social Web sites. To provide a comprehensive comparison, we assess the performance of SPC against various benchmarks, including human performance, GPT-3 with Chain-of-Thought (CoT) prompting in the WebArena setting, GPT-4 in a similar environment, and Operator. Our evaluation considers two versions of SPC: the default version and an enhanced variant equipped with Theory of Mind capabilities (Sect. 3.1). The results of our evaluation are summarized in Table 3, highlighting SPC's relative strengths and weaknesses in different browsing contexts.

Table 3 Evaluation of Web navigation capabilities of SPC

	Success rate on WebArena, %		Success rate on WebVoyager.GitHub, %		Success rate on WebVoyager. AllRecipes, %	
Navigation system/dataset	Full	Social	Full	Social	Full	Social
Human	78.2					
GPT-3 (OpenAI 2022) with CoT prompting	8.8					
GPT-4 (OpenAI 2023) with CoT prompting	11.8		48.8		11.1	
Operator	**58.1**					
WevVoyager			**63.4**		**53.3**	
SPC	36.2	60.3	43.2	68.2	37.0	45.7
SPC + ToM	38.9	**65.7**	47.8	**70.2**	40.4	50.4

Our findings reveal that SPC underperforms across all three full datasets within our evaluation framework, indicating that it struggles to compete with other models in general Web-browsing tasks. However, when analyzing the social Web site subsets from WebArena and WebVoyager, SPC demonstrates a notable advantage. Specifically, it surpasses the performance of all competing systems—including WebVoyager and Operator—on GitHub and AllRecipes tasks within these subsets. This suggests that SPC's strengths may be more pronounced in environments where social interaction or community-driven content is central, even if it falls short in broader Web navigation tasks.

A key factor contributing to SPC's success in social-oriented datasets is its unique ability to engage in posting, rather than merely gathering information. Unlike WebArena, WebVoyager, and Operator—which primarily focus on retrieving data—SPC actively participates in online discourse, making it more adaptable in settings that require user interaction. On average, SPC + ToM outperforms both WebVoyager and Operator by approximately 6%, further reinforcing the idea that its Theory of Mind–enhanced capabilities provide an edge in tasks involving social reasoning and engagement. This distinction underscores SPC's potential for applications where not only information retrieval but also meaningful interaction is necessary.

6.3 *End-to-End Evaluation of Social Promotion Copilot*

Evaluating the effectiveness of a Social Promotion Copilot (SPC) requires a comprehensive analysis of various performance metrics, ranging from user engagement levels to the overall impact of its promotional strategies. One key aspect of this evaluation involves measuring how frequently users interact with SPC-managed posts, including likes, shares, comments, replies, and friend or colleague requests. By analyzing these interactions, we can assess the extent to which SPC successfully engages with online communities. Our study includes an examination of 100 posts across multiple social Web sites, with the detailed results presented in Table 4. This analysis allows us to identify platform-specific trends in user behavior, offering insights into how SPC's promotional strategies perform in different social media environments.

Table 4 End-to-end social promotion evaluation: absolute numbers

	Likes		Shares		Comments		Replies		Friend/colleague request	
Positive/negative	+	–	+	–	+	–	+	–	+	–
Facebook	5.3	7.1	2.3	2.9	2.1	5.3	2.2	3.8	2.4	3.6
LinkedIn	3.2	1.8	2.7	0.7	3.2	1.3	0.6	0.4	0.8	0.4
VKontakte	5.2	6.0	2.4	1.8	3.3	4.5	1.0	0.6	0.2	0.6
GitHub	0.5	0.8	0.1	0.1	0.6	0.4	0.4	0.2	0.0	0.0

Table 5 End-to-end social promotion evaluation: relative numbers

	Likes		Shares		Comments		Replies		Friend/colleague request	
Positive/negative	+	−	+	−	+	−	+	−	+	−
Facebook	0.65	0.82	1.32	0.92	0.82		1.05	0.85	0.90	
LinkedIn	0.90		0.86	1.10	0.64	1.05	0.85	1.15		
VKontakte	0.80	1.05	0.90	0.86	0.70	0.65	0.80	1.06	0.85	0.80
GitHub	0.70		0.50		0.80	0.76	0.75			

The results indicate that Facebook users exhibit a strong reaction to SPC-generated complaint posts, with an unusually high number of friend requests. This suggests that dissatisfied customers are highly engaged when SPC posts content addressing consumer grievances, particularly when it provides actionable steps for resolving issues with customer service. The ability of SPC to facilitate such discussions appears to resonate with users, driving higher engagement. On VKontakte, while there is still a noticeable level of social interaction, the absence of direct customer support participation results in significantly lower engagement numbers compared to Facebook. The difference highlights the impact of customer service responsiveness on user interactions, suggesting that platforms with more active brand engagement foster greater SPC-driven social activity.

In contrast, LinkedIn demonstrates a more restrained response to SPC activity. Although likes and shares remain comparable to other platforms, replies to posts and colleague requests are markedly lower, particularly for negative posts. This trend reflects LinkedIn's professional nature, where users tend to engage less with complaints and controversy compared to more socially oriented platforms.

To distinguish SPC-generated posts from organic, manually crafted content, we provide relative engagement metrics in Table 5, computed as SPC-to-genuine posting percentages where data is available. Overall, SPC posts generate 14% less engagement compared to genuine human posts. This indicates that while users generally respond to SPC activity in a similar manner to human-created content, there is still a slight engagement gap, suggesting potential areas for improvement in making SPC's interactions more authentic and compelling.

7 Related Work

Galitsky et al. (2014) introduced a simulated humanlike agent designed to act on behalf of its human host, facilitating and managing online communication. This agent alleviates its host from routine and less critical social networking activities, such as sharing news, commenting on messages, blogs, forums, images, and videos. Unlike many applications of simulated human characters, the agent's social interactions occur without necessarily revealing its automated nature to conversation partners. It engages in the exchange of news, opinions, and updates as if it were the human host. This system was named Conversational Agent for Social Promotion (CASP). To evaluate CASP's effectiveness and trustworthiness, we conducted experiments across multiple Facebook accounts, analyzing its performance in fostering engagement and facilitating seamless humanlike interactions.

7.1 *Communicating with Friends on Behalf of a Human Host*

In today's digital era, social media platforms like Facebook, LinkedIn, Instagram, and X have expanded users' networks far beyond close friends and family. On average, individuals maintain connections with hundreds or even thousands of contacts across these platforms. However, despite the vastness of these networks, active engagement remains limited to a small circle—typically 10–20 close friends, family members, and colleagues. The majority of connections receive little to no interaction, leading to the perception that these relationships have been neglected or abandoned.

Yet maintaining a broader and more active social presence is increasingly valuable for both personal and professional growth. Whether for career networking, personal branding, or community engagement, users are expected to demonstrate interest in their connections by responding to posts, acknowledging life events, and actively participating in discussions. However, the sheer volume of interactions required to sustain these relationships demands significant time and effort—something many users struggle to manage.

To bridge this gap, intelligent automation, such as CASP, can assist users by handling routine social interactions (Fig. 9). While users continue to engage

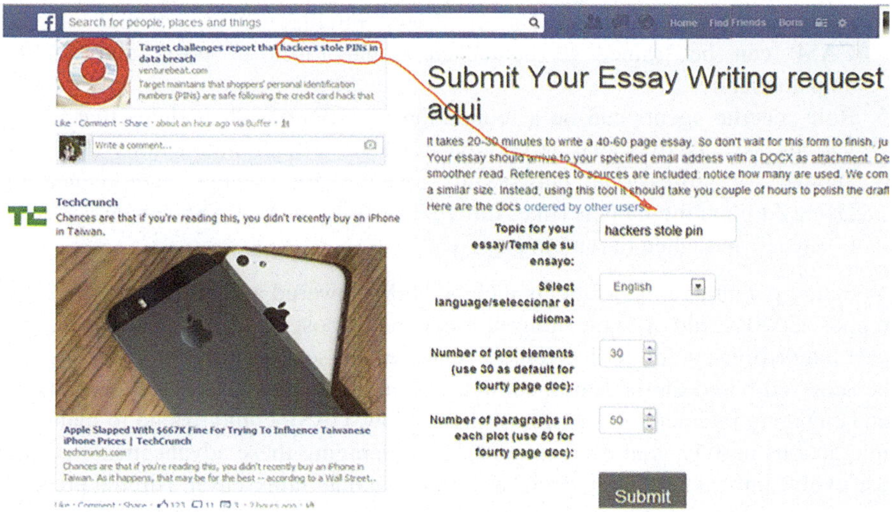

Fig. 9 Interactive mode of CASP: a user can write an essay concerning an opinion of her peer

personally with close friends and family, CASP can maintain communication with wider networks by liking, sharing, and commenting on posts in a meaningful and contextually appropriate manner. By automating aspects of social interaction, CASP helps users sustain a consistent and engaged online presence without the overwhelming time commitment, ensuring that professional and social relationships remain active and beneficial.

To enhance the legacy CASP system, modern LLMs and related AI technologies significantly improve its capabilities in personalization, contextual understanding, and automation. Here are some key technologies that would enhance CASP today:

1. Transformer-based LLMs (e.g., GPT-4, Claude, Gemini, Mistral) improve context awareness, processing long conversations, understanding nuances in social interactions, and generating contextually appropriate responses. These LLMs can adapt to a user's communication style by learning from past interactions (Galitsky 2025).
2. Use of retrieval augmented generation (RAG) would enhance relevance: by integrating real-time information retrieval, CASP can generate responses based on recent news, trends, or user-specific data. Also, leveraging personalized knowledge base, CASP would be able to maintain a dynamic memory of a user's preferences, conversation history, and past engagements.
3. Reinforcement learning from human feedback would assure better engagement optimization: CASP can be trained with human feedback to ensure responses align with user intent, avoiding robotic or inappropriate messaging. It would also assure ethical and safe communication, preventing generating controversial or insensitive responses that could harm relationships (Bai et al. 2022).
4. Fine-tuned social AI models (e.g., Llama 3, Mixtral) provide context-specific adaptation (Bakker et al. 2022; Yin et al. 2024). By fine-tuning models on social media data, CASP can generate responses that align with social norms and platform-specific etiquette. The emotional intelligence is essential as well: CASP can be trained to detect emotional tones in posts and respond empathetically.
5. Role-specific agents can be a foundation of different AI modules that handle various aspects of social promotion—one for casual interactions (Fig. 10), another for professional networking, and one for content recommendation. Using AI planning models (like AutoGPT or BabyAGI), CASP can decide when to engage and when to remain passive.

Future potential enhancements include LLM-powered auto-generated posts and replies: CASP could draft meaningful social media posts based on a user's interests, past interactions, and trending topics. Emotion and sentiment-based responses can be achieved based on the sentiment detected in messages, ensuring a more natural and engaging interaction. As virtual spaces grow, CASP could extend to managing interactions in 3D social environments. By leveraging these advancements, CASP can evolve into an intelligent social assistant that maintains a user's digital presence seamlessly while ensuring meaningful and context-aware interactions.

Fig. 10 CASP is posting a message related to friend's travel

7.2 Empowering Social Media Engagement

Content creators often struggle to effectively promote positive messages and engage meaningfully on social media. One promising approach is to empower users with AI-driven assistance to enhance engagement strategies and cultivate supportive online communities. While numerous social media automation tools exist, little research has explored designing chatbots that provide contextualized, real-time guidance for improving online interactions.

This study involved social media users and digital marketers in two design sessions:

1. An in-depth interview to identify challenges in effective promotion and audience engagement
2. Interaction with the SPC prototype to develop design guidelines for overcoming these challenges

Qualitative analysis of these interactions revealed three primary barriers to successful social media promotion:

1. Lack of strategic knowledge about platform algorithms, audience preferences, and engagement techniques
2. Emotional barriers, such as fear of negative feedback, imposter syndrome, or uncertainty in messaging
3. Limited understanding of audience dynamics, leading to ineffective or misaligned messaging

Key parameters were identified to optimize chatbot responses and support effective promotion, including:

1. Adopting multiple personas to simulate diverse audience perspectives
2. Using an engaging, friendly, and supportive tone to encourage confidence
3. Providing clear, specific, and adaptable suggestions for different platforms and audiences
4. Optimizing message length and format for each platform's best practices

These insights inform the design of personalized and scalable AI-driven tools that empower users to enhance social media engagement, foster positive interactions, and amplify meaningful content effectively.

Kamoun et al. (2024) illustrate the merits, potentials, and limitations of using ChatGPT as a co-pilot to assist faculty in refining the assessment design process. This research brings into evidence the importance of keeping a "human in the loop" perspective during the faculty-ChatGPT assessment co-creation activities.

7.3 *Retail Copilot*

Building a successful AI copilot involves a comprehensive and systematic approach, ensuring both technical excellence and alignment with user needs. To illustrate the practical application of these concepts, Furmakiewicz et al. (2024) use a case study of Microsoft's development of copilot templates within the retail domain, highlighting the significance of each design component. Additionally to LLM, the authors examine the role of plugins, which enable the copilot to retrieve knowledge and execute actions, enhancing its functionality. Orchestration, which ensures seamless coordination between components, is also discussed, along with system prompts that guide the copilot's behavior. Furthermore, responsible AI guardrails are explored, emphasizing the importance of fairness, transparency, and safety in AI systems.

Testing and evaluation are the keys to ensuring the AI copilot meets its objectives and avoids unintended negative outcomes in real-world applications. This includes the use of an end-to-end human-AI decision loop framework, which offers a structured approach for assessing how the copilot interacts with users and impacts business processes. Furmakiewicz et al. (2024) discuss methods for measuring and improving both the quality and safety of the copilot, ensuring it is reliable, effective, and aligned with ethical standards.

By providing a detailed examination of both the design and evaluation stages, this paper offers concrete insights into the anatomy of an AI copilot. It underscores the importance of thoughtful design practices and robust testing methodologies in building AI assistants that are not only technologically advanced but also human centered, ethical, and capable of driving positive outcomes in business contexts.

7.4 GitHub Copilot

GitHub Copilot is an AI-powered code completion tool developed by GitHub in collaboration with OpenAI. It acts as an intelligent assistant that helps software developers write code faster and more efficiently by suggesting relevant code snippets, functions, and even entire lines of code based on the context of the project. Built on OpenAI's Codex model, GitHub Copilot is trained on a vast array of publicly available programming resources, including open-source projects, which enables it to provide relevant, context-aware suggestions across various programming languages. This makes Copilot a powerful tool for both experienced developers and beginners, offering real-time coding assistance and reducing the need to search for solutions or consult documentation.

One of GitHub Copilot's most notable features is its ability to generate code based on simple comments or descriptions written by the developer. For instance, by typing a comment outlining the desired functionality of a specific function, Copilot can generate a functional implementation in the chosen programming language. This makes it especially valuable for streamlining the development process, as it reduces the time spent on routine coding tasks. Additionally, Copilot supports multiple languages and frameworks, making it versatile for a wide range of development projects, from Web development to machine learning.

Despite its potential, GitHub Copilot also raises important questions about intellectual property, ethics, and dependency on AI in software development. Since Copilot is trained on publicly available code, concerns have emerged about the tool inadvertently suggesting copyrighted or non-original code without proper attribution. GitHub has introduced some guardrails to address these concerns, such as disclaimers about code reuse and safeguards to prevent inappropriate suggestions. Furthermore, while Copilot helps accelerate development, it is important for developers to maintain a critical eye and not rely solely on AI-generated code without verifying its accuracy and suitability for the intended project. As AI tools like GitHub Copilot continue to evolve, their role in software development will likely grow, prompting ongoing discussions about best practices and responsible usage.

7.5 *Salesforce's Agentforce*

Agentforce is a proactive, autonomous AI application that provides specialized, always-on support to employees or customers. Agentforce can be enabled with any necessary business knowledge to execute tasks according to its specific role.

SPC and Salesforce's Agentforce both focus on AI-driven automation for online interactions, but they differ in scope and application. They are designed to enhance user engagement and response management. Each system aims to improve efficiency and effectiveness in online communication.

However, Agentforce primarily designed for customer service automation within the Salesforce ecosystem. It acts as an AI-powered agent to handle customer inquiries, automate workflows, and assist human agents. It is also integrated deeply with CRM systems to optimize sales, service, and support processes.

While SPC focuses on social engagement and marketing, Agentforce is tailored for enterprise-level customer support and CRM automation (Fig. 11). However, an advanced SPC could potentially integrate with systems like Agentforce to bridge social media engagement with customer relationship management, creating a seamless flow from online interactions to sales and support pipelines.

7.6 *Uncanny Valley and Social Actors Paradigms*

The Uncanny Valley hypothesis (Mori et al. 2012) suggests that as a robot or virtual agent becomes more humanlike, people tend to feel increasingly comfortable with it—until a certain threshold is reached. At this point, slight imperfections in realism create an eerie, unsettling feeling, leading to negative emotional reactions. This dip in emotional response is referred to as the "Uncanny Valley." However, as realism

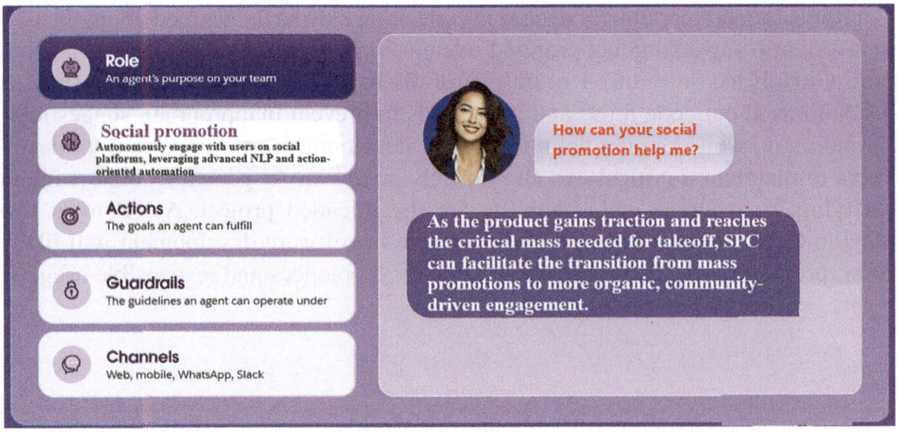

Fig. 11 Agentforce for enterprise-level customer support and CRM automation

continues to improve beyond this valley, human acceptance increases again. Humanlike virtual influencers may fall into the Uncanny Valley if their appearance and behavior are almost, but not quite, human, leading to lower engagement and discomfort among users. For example, a cartoonish robot may feel friendly and non-threatening, but a highly realistic android with unnatural facial expressions or slightly stiff movements can feel unsettling.

The Computers Are Social Actors paradigm (Reeves and Nass 1996), suggests that humans naturally apply social rules and expectations to computers, AI, and virtual agents, even when they know they are interacting with a machine. People subconsciously treat digital entities as if they have personalities, emotions, and social roles, responding with politeness, social norms, and biases similar to those in human-to-human interactions. If users unconsciously treat virtual influencers as social beings, their expectations for natural behavior and emotional authenticity may shape their reactions. This could lead to differences in engagement and trust between humanlike and non-humanlike virtual agents. Users might say "thank you" to a virtual assistant like Siri or Alexa or feel frustrated when a chatbot provides an unhelpful response.

7.7 OpenAI's Operator

OpenAI's Operator is an advanced AI agent capable of autonomously navigating the Web and performing tasks on behalf of users. Unlike traditional automation tools that rely on predefined scripts or APIs, Operator interacts with Web sites just as a human would—by typing, clicking, and scrolling within a browser. This enables it to handle a variety of online tasks, from filling out Web forms and ordering medicines to posting on social media and making reservations.

As one of the first fully autonomous AI agents developed by OpenAI, Operator represents a major leap in AI-driven automation, allowing users to delegate repetitive digital work efficiently. By using the same interfaces and tools that humans engage with daily, Operator enhances productivity and unlocks new opportunities for businesses looking to streamline customer interactions, online operations, and digital workflows.

At the core of Operator lies CUA (*Computer User Agent*), a multimodal AI model that combines GPT-4o's vision capabilities with advanced reasoning through reinforcement learning. Unlike traditional automation systems that depend on OS- or Web-specific APIs, CUA is designed to interact directly with graphical user interfaces (GUIs)—the buttons, menus, text fields, and other elements that make up digital environments (Fig. 12).

By leveraging foundational research in multimodal understanding and structured problem-solving, CUA enables AI to perform complex Web-based tasks without requiring specialized integrations. This makes Operator highly adaptable, self-correcting, and capable of handling real-world digital environments with ease.

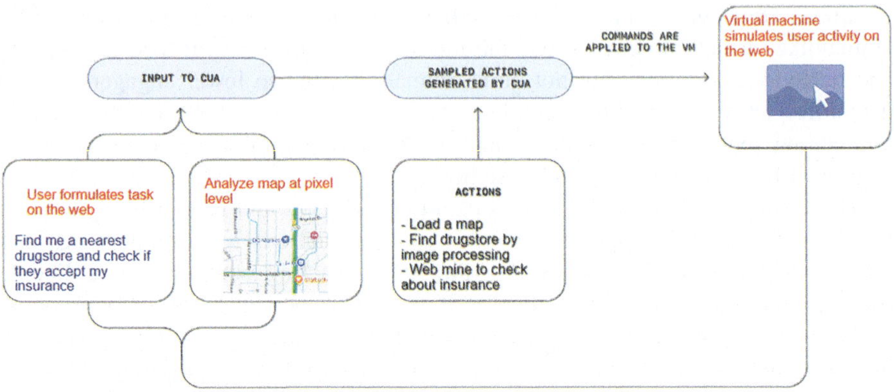

Fig. 12 The functionality of OpenAI's CUA

Operator follows a three-step process that allows it to perceive its environment, reason through tasks, and execute actions in real time:

1. Perception: Operator continuously captures and analyzes screenshots of the Web page it is interacting with, using computer vision to understand the current state of the interface.
2. Reasoning: Using chain-of-thought reasoning, Operator determines the most effective next steps by analyzing past and present interactions. This inner monologue enables it to plan multi-step tasks, recognize obstacles, and make dynamic adjustments.
3. Action: Operator clicks, scrolls, types, and interacts with the page as needed, progressing toward task completion. For sensitive actions—such as entering login credentials or solving CAPTCHAs—it requests user confirmation to ensure security and privacy.

This iterative loop of perception, reasoning, and action allows Operator to handle a wide range of digital workflows, including:

1. Form filling—Completing online applications, registrations, and checkouts
2. Data entry and extraction—Collecting, copying, and organizing online information
3. e-Commerce transactions—Searching for products, comparing prices, and placing orders
4. Content interaction—Posting updates, commenting, and engaging on social media
5. Customer support assistance—Navigating FAQ pages, submitting tickets, and retrieving information

An advantage of Operator over SPC lies in its advanced visual analysis capabilities, allowing it to interpret and interact with graphical user interfaces (GUIs) more effectively. By leveraging OpenAI's multimodal models, Operator can process complex Web layouts, recognize patterns in visual data, and navigate Web sites with humanlike precision. This makes it highly suitable for tasks that require an understanding of dynamic elements, such as interactive dashboards, media-heavy Web pages, and forms with evolving structures. However, while Operator excels in raw functional automation, it lacks social and psychological adaptability. It is not specifically designed to engage with the nuances of social interactions, interpret sentiment, or perform tasks that require an understanding of human emotions and social cues, which are crucial for meaningful participation in social networks and community-driven discussions.

Another key limitation of Operator is its restricted functionality in adversarial or competitive online environments. Many platforms, especially social media Web sites, actively work to prevent automated engagement, enforcing strict anti-bot measures to ensure authentic user activity. As an OpenAI-managed service, Operator may face resistance from these platforms, potentially leading to restrictions on automated posting or interaction. In contrast, SPC operates at an individual user level, allowing it to seamlessly integrate with personal accounts without triggering large-scale automated detection systems. This decentralized approach makes SPC more resilient in maintaining engagement across social networks, ensuring that users can continue interacting with their communities without disruption. While Operator offers powerful automation, SPC's ability to blend naturally into social ecosystems makes it the preferred choice for socially driven tasks that require adaptive humanlike interaction in competitive or restrictive digital spaces.

Startups such as Browserbase and Browser Use are actively contributing to the rapidly expanding field of browser-using AI agents, leveraging their proprietary technology to develop and open-source advanced automation tools. By making their agents publicly available, these companies aim to democratize access to AI-driven Web interaction, enabling developers, researchers, and businesses to integrate intelligent browsing capabilities into their own applications.

These open-source browser-using agents are designed to navigate, interact with, and extract information from the Web autonomously, mimicking human behavior while optimizing efficiency. They support a wide range of use cases, including automated form filling, Web scraping, and real-time data analysis, making them valuable assets for industries reliant on digital automation. By fostering an open ecosystem, startups like Browserbase and Browser Use are helping drive innovation in human-computer interaction, paving the way for more intelligent, adaptable, and scalable AI-powered browsing solutions (Cunningham 2025).

7.8 Interactive LLM-Based Agents

In the field of interactive decision-making agents, researchers have explored various approaches to enhancing AI's ability to navigate and perform tasks on the Web. Nakano et al. (2021) introduced WebGPT, an AI system designed to search the Web, process search results, and generate informed responses to user queries. This method leverages LLMs to integrate real-time information retrieval with reasoning capabilities, improving the reliability of AI-generated answers. Building on this idea, Gur et al. (2023) proposed a more action-oriented Web agent capable of synthesizing JavaScript code to execute tasks autonomously. This advancement allows AI agents to interact dynamically with Web applications, expanding their functionality beyond simple information retrieval to direct task execution within browser environments.

Recent advancements in language model agents have enabled AI systems to navigate and interact with simulated Web sites effectively (Furuta et al. 2023) These simulated environments, such as MiniWoB (Yao et al. 2022), provide agents with structured, predefined actions and simplified HTML representations, allowing for efficient training and evaluation. By operating in controlled settings where the HTML content is well structured and noise-free, these agents can successfully complete a variety of Web-based tasks, including form filling, data retrieval, and navigation. However, while these models demonstrate proficiency in constrained environments, their capabilities remain limited when applied to real-world Web interactions.

In contrast, real-world Web sites present significantly greater challenges for language model agents (Fig. 13). Unlike simulated environments, these Web sites feature dynamic content, interactive elements, and non-standardized structures that complicate AI-driven navigation. Agents must handle open-ended actions, where available choices are not predefined but depend on contextual and situational factors. Moreover, the HTML documents of real-world Web sites are often lengthy and cluttered with irrelevant data, advertisements, tracking scripts, and dynamically loaded content, making efficient information extraction and action planning more difficult. Overcoming these obstacles requires AI models to develop advanced perception, reasoning, and adaptability, enabling them to process complex, unstructured Web environments while maintaining robust and reliable performance in real-world applications.

A significant leap in Web-based AI agents has been the incorporation of multimodal processing, which enables agents to perceive and act on graphical interfaces rather than relying solely on textual representations. Shaw et al. (2023) introduced models that predict user actions based on screenshots of Web pages, allowing AI to interact with Web sites similarly to human users. This approach moves beyond

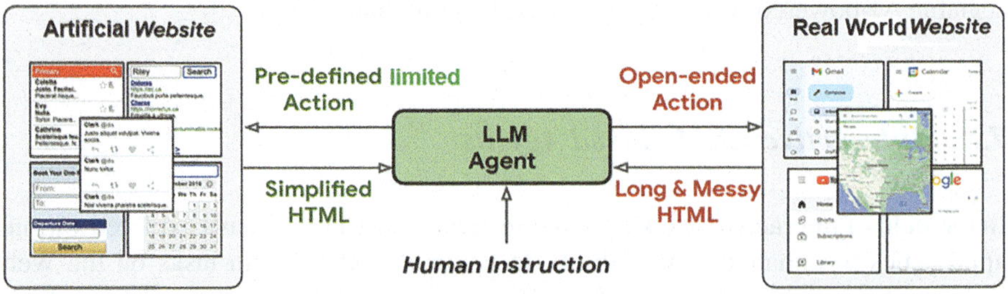

Fig. 13 Difficulties in navigating real-world Web sites

Document Object Model tree-based navigation, which can be limited by dynamically loaded content and evolving Web structures. By interpreting visual elements such as buttons, menus, and form fields, these agents demonstrate greater adaptability and robustness, making them more effective for complex and dynamic Web tasks.

Successfully performing tasks in interactive digital environments requires agents to develop a range of sophisticated capabilities. These include hierarchical planning, where an agent breaks down tasks into structured sub-goals, state tracking, which enables the system to remember past actions and adjust behavior accordingly, and error recovery, allowing agents to recognize and correct mistakes in real time. As AI continues to evolve in this space, these competencies will be crucial for the development of autonomous Web agents capable of seamless interaction, adaptive decision-making, and efficient task completion across a variety of online platforms.

7.9 Other LLM + Action Engines Integration

LaVague (2024) Introduce Web Agent framework for builders. Similar to the current work, it combines LLM and automation tool like Selenium. LaVague is an open-source framework designed for developers who want to create AI Web Agents to automate processes for their end users.

Our Web Agents can take an objective, such as "Print installation steps for Hugging Face's Diffusers library," and generate and perform the actions required to achieve the objective.

LaVague Agents are made up of:

1. A world model that takes an objective and the current state (aka the current Web page) and outputs an appropriate set of instructions.
2. An action engine that "compiles" these instructions into action code, e.g., Selenium or Playwright, and executes them.

8 Conclusions

In conclusion, we summarize that SPC in the complaint domain:

1. Saves time—Automates the tedious process of crafting and submitting complaints
2. Enhances impact—Ensures complaints are well written and legally sound
3. Improves success rate—Directs users to the best escalation paths
4. Builds public pressure—Leverages social media for faster company responses

As virtual agents become increasingly common across various domains, virtual influencers have emerged on social media platforms, engaging with users and integrating into human networks.

Drawing on research in human-computer interaction, the Uncanny Valley hypothesis, and the Computers Are Social Actors (CASA) paradigm, Arsenyan and Mirowska (2021) examine:

1. How virtual agents exhibit humanlike behavior within social networks
2. How users respond to human versus virtual agents in publicly visible interactions

Building on their work, we analyzed text and emoji responses to posts made by a human influencer, a humanlike virtual influencer, and an anime-style virtual influencer over an 11-month period. Our findings show that the humanlike virtual influencer received significantly fewer positive reactions compared to the other two. This supports the Uncanny Valley hypothesis, as additional measures of negative reactions followed a similar pattern—despite the overall positive tone of the platform.

Traditional automation tools require custom scripts, pre-configured workflows, or API access, making them rigid and time-consuming to set up. In contrast, Copilot interacts with the Web naturally, just like a human user, offering greater flexibility, adaptability, and accessibility.

By leveraging SPC ability to understand Web interfaces, Copilot opens doors to a new wave of AI-driven productivity, allowing individuals and businesses to automate time-consuming digital tasks, improve efficiency, and optimize online interactions with minimal effort.

As AI continues to evolve, Copilot represents a crucial step toward autonomous digital assistants, making AI not just an information provider but an active participant in the digital world—capable of executing real tasks, adapting to changing environments, and enhancing human productivity in ways never seen before.

In January 2025, OpenAI announces its much-anticipated Operator agent, which uses computer vision capabilities to perform tasks for a user by directly taking over their computer. The underlying model that powers Operator, referred to as the Computer-Using Agent, is a variant of GPT-4o, fine-tuned through reinforcement learning to interact with graphical user interfaces. This distinguishes CUA from many conventional AI agents, which are typically confined to text-based interfaces. By enabling interaction with a wide range of computer software in a manner akin to human users, CUA significantly broadens the scope of AI-driven automation. According to OpenAI, CUA achieves state-of-the-art performance on WebArena (Zhou et al. 2023) and WebVoyager (He et al. 2024), which specifically evaluate AI proficiency in Web-based interactions.

Despite the impressive capabilities demonstrated by Operator, OpenAI is a relatively late entrant into the rapidly expanding domain of AI-powered computer agents. Competing models had already been introduced by other research institutions and companies. Anthropic's Claude Computer Use debuted in October, followed by Google's Mariner 2 months later. The field's origins can be traced even further back to Adept, a startup that launched its first computer-using agent in 2022. These developments highlight the growing interest in AI-driven automation for software interaction, emphasizing both the competitive landscape and the increasing sophistication of AI agents in this domain.

References

Akilkhanov A (2024) AI and personalization in marketing. Forbes. https://www.forbes.com/councils/forbescommunicationscouncil/2024/01/05/ai-and-personalization-in-marketing/

Arsenyan J, Mirowska A (2021) Almost human? A comparative case study on the social media presence of virtual influencers. Int J Hum Comput Stud 155:102694

Bai Y, Jones A, Ndousse K, Askell A, Chen A, DasSarma N, Drain D, Fort S, Ganguli D, Henighan T, Joseph N, Kadavath S, Kernion J, Conerly T, El-Showk S, Elhage N, Hatfield-Dodds Z, Hernandez D, Hume T, Johnston S, Kravec S, Lovitt L, Nanda N, Olsson C, Amodei D, Brown T, Clark J, McCandlish S, Olah C, Mann B, Kaplan J (2022) Training a helpful and harmless assistant with reinforcement learning from human feedback. arXiv preprint arXiv:2204.05862

Bakker M, Chadwick M, Sheahan H, Tessler M, Campbell-Gillingham L, Balaguer J, McAleese N, Glaese A, Aslanides J, Botvinick M, Summerfield C (2022) Fine-tuning language models to find agreement among humans with diverse preferences. In: Advances in neural information processing systems, NeurIPS 2022, vol 35

Chiam J, Lim A, Nott C, Mark N, Teredesai A, Shinde S (2024) Co-pilot for health: personalized algorithmic AI nudging to improve health outcomes. arXiv preprint arXiv:2401.10816

Cunningham M (2025) Operator is OpenAI's attempt to catch up. Its release gives OpenAI a badly-needed boost—but will it be enough to close the gap with rivals like Anthropic? Medium. https://medium.com/%40mcunningham1440/operator-is-openais-attempt-to-catch-up-68fd61bfb643

Delre SA, Jager W, Bijmolt THA, Janssen MA (2007) Targeting and timing promotional activities: an agent-based model for the takeoff of new products. J Bus Res 60(8):826–835

Eager B, Brunton R (2023) Prompting higher education towards AI-augmented teaching and learning practice. J Univ Teach Learn Pract 20(5). https://doi.org/10.53761/1.20.5.02

Furmakiewicz M, Liu C, Taylor A, Venger I (2024) Design and evaluation of AI copilots—case studies of retail copilot templates. arXiv preprint arXiv:2407.09512

Furuta H, Nachum O, Lee K-H, Matsuo Y, Gu SS, Gur I (2023) Multimodal web navigation with instruction-finetuned foundation models. arXiv preprint arxiv:2305.11854

Galitsky B (2019) A social promotion chatbot. In: Developing enterprise chatbots. Springer, Cham

Galitsky B (2021) Artificial intelligence for customer relationship management: solving customer problems. Springer, Cham

Galitsky B (2023) Social autonomous agent implementation using lattice queries and relevancy detection. US Patent 11,645,459

Galitsky B (2025) LLM-based personalized recommendations in health. In: Health applications of neuro-symbolic AI. Elsevier

Galitsky B, Ilvovsky D, Lebedeva N, Usikov D (2014) Improving trust in automation of social promotion. In: AAAI spring symposium series, Stanford, CA, pp 28–35

Golder PN, Tellis GJ (2004) Growing, growing, gone: cascades, diffusion, and turning points in the product life cycle. Mark Sci 23(2):207–218

Gur I, Furuta H, Huang A, Safdari M, Matsuo Y, Eck D, Faust A (2023) A real-world webagent with planning, long context understanding, and program synthesis. arXiv preprint arXiv:2307.12856

He H, Yao W, Ma K, Yu W, Dai Y, Zhang H, Lan Z, Yu D (2024) WebVoyager: building an end-to-end web agent with large multimodal models. arXiv preprint arXiv:2401.13919

Kamoun F, Brik A, Rebhi I, Besbes S, Abidi H, Baghdadi A, Ammar R (2024). chatGPT as a Co-pilot for assessment during refinement: an exploratory study. CDIO Conference Tunisia

LaVague (2024) LaVague: Web Agent framework for builders. https://github.com/lavague-ai/LaVague

Mahajan V, Muller E (1979) Innovation diffusion and new product growth models in marketing. J Mark 43:55–68

Mori M, MacDorman KF, Kageki N (2012) The uncanny valley [from the field]. IEEE Robot Autom Mag 19(2):98–100

Nakano R, Hilton J, Balaji S, Wu J, Ouyang L, Kim C, Hesse C, Jain S, Kosaraju V, Saunders W et al (2021) WebGPT: browser-assisted question-answering with human feedback. arXiv preprint arXiv:2112.09332

OpenAI (2022) ChatGPT: optimizing language models for dialogue

OpenAI (2023) GPT-4 technical report. arXiv preprint arXiv:2303-08774

PRNewswire (2023) Baidu unveils ERNIE Bot, the latest generative AI mastering Chinese language and multi-modal generation. https://www.prnewswire.com/news-releases/baidu-unveils-ernie-bot-the-latest-generative-ai-mastering-chinese-language-and-multi-modal-generation-301774240.html

Reeves B, Nass C (1996) The media equation: how people treat computers, television, and new media like real people and places. Cambridge University Press

Shaw P, Joshi M, Cohan J, Berant J, Pasupat P, Hu H, Khandelwal U, Lee K, Toutanova K (2023) From pixels to UI actions: learning to follow instructions via graphical user interfaces. arXiv preprint arXiv:2306.00245

Su J, Yang W (2023) Unlocking the power of ChatGPT: a framework for applying generative AI in education. ECNU Rev Educ 6(3):355–366. https://doi.org/10.1177/20965311231168423

Xu C (2024) Integrating AI tools into teaching practice: unleash the potential of your AI co-pilot. In: The future of education, 14th edn

Yao S, Zhao J, Yu D, Du N, Shafran I, Narasimhan K, Cao Y (2022) React: synergizing reasoning and acting in language models. arXiv preprint arXiv:2210.03629

Yin K, Liu C, Mostafavi A, Hu X (2024) CrisisSense-LLM: instruction fine-tuned large language model for multi-label social media text classification in disaster informatics. arXiv preprint arXiv:2406.15477

Zhou S, Xu FF, Zhu H, Zhou X, Lo R, Sridhar A, Cheng X, Ou T, Bisk Y, Fried D, Alon U, Neubig G (2023) WebArena: a realistic web environment for building autonomous agents. arXiv preprint arXiv:2307.13854

Zou W, Yang Q, DiFranzo D, Chen M, Hui W, Bazarova NN (2024) Social media co-pilot: designing a chatbot with teens and educators to combat cyberbullying. Int J Child Comput Interact 41:100680

Abduction-Based Personalized Recommender for Web 3.0

1 Introduction

Large language models (LLMs) are poised to revolutionize the interaction between humans and personalization systems. Unlike conventional recommender systems and search engines, which act as passive mediums for information filtering, LLMs establish the groundwork for active user engagement. This transformative foundation enables proactive exploration of user requests, delivering required information in a natural, interactive, and explainable manner. Additionally, LLMs will significantly broaden the scope of personalization, evolving beyond the sole function of gathering personalized information to a composite function of offering personalized services.

Harnessing large language models as a versatile interface, personalization systems can compile user requests into plans, invoke external tools' functions (e.g., search engines, calculators, service APIs, etc.) to execute these plans, and integrate the tools' outputs to accomplish end-to-end personalization tasks. This dynamic approach marks a paradigm shift in how personalization is conceptualized and executed, promising a more interactive and expansive user experience.

While prominent natural language processing (NLP) benchmarks like GLUE and SuperGLUE (Wang et al. 2018) have undeniably advanced the NLP landscape, they often adhere to the prevailing "one-size-fits-all" paradigm in modeling and evaluation. Unfortunately, this approach restricts the development of models adaptable to the specific needs of end users, hindering extensive research on personalization in NLP tasks. In contrast, our chapter introduces a comprehensive evaluation framework that incorporates various tasks demanding personalized recommendations.

Today, LLMs are still being rapidly developed, whereas their personalization capabilities are not significantly addressed. We analyze the challenges in personalization and the opportunities for how personalization can work on top of LLMs. We

B. Galitsky, *Applications of Neuro-Symbolic Artificial Intelligence*,
https://doi.org/10.1007/978-3-032-16524-4_7

discuss development and challenges for the existing personalization system, the newly emerged capabilities of LLMs, and the potential ways of making use of LLMs for personalization.

While personalization has been extensively investigated across various AI application domains, such as information retrieval (IR) and human-computer interaction (HCI), particularly in the context of search engines and recommender systems, its exploration within NLP has been somewhat limited, as highlighted by Xue et al. in (2009). Flek's recent work (2020) has underscored the significance of personalization, especially in tasks involving text classification and generation. This emphasizes the potential of personalization to prioritize users and create systems that are not only accessible but also inclusive. Furthermore, recent studies, including those by Kirk et al. (2023), have illuminated the advantages of personalizing LLMs. Despite these insights and the recognized importance of personalization in numerous real-world scenarios, the development and evaluation of LLMs for generating personalized responses have received limited attention. Consequently, in this chapter, we highlight the pivotal role of personalization in shaping the future of LLM integration systems. This marks the initial step toward the development and evaluation of personalization within the context of LLMs.

1.1 Personalization and Abduction

LLMs are typically trained to follow user instructions by being fine-tuned on datasets that capture binary preference judgments—essentially, learning which of two candidate responses a user prefers for a given prompt. While this form of preference-based alignment has improved output quality and usefulness, it remains limited in interpretability. Specifically, these preference signals reflect which response was chosen but not why it was preferred.

A persistent limitation in current personalization methods lies in the lack of interpretability regarding the underlying motivations that drive user choices (Galitsky 2021). This creates a fundamental gap between surface-level preference data and the deeper contextual factors shaping those preferences. Without access to the reasoning context—such as tone sensitivity, desired level of detail, politeness norms, emotional framing, or alignment with long-term goals, LLMs are unable to adapt flexibly to the nuanced expectations of different users (Zhou et al. 2022). As a result, personalization often defaults to a generic, one-size-fits-all alignment that may perform adequately on aggregate metrics but underdelivers when evaluated against individual user needs or domain-specific intent patterns.

To address this shortfall, we propose the use of abductive reasoning as a systematic framework for reverse-engineering user intent from observed preference behaviors. Unlike methods that treat preferences as opaque outputs, abductive reasoning focuses on identifying the most plausible explanations—such as inferred goals, current emotional states, user expertise, or cognitive style—that might have motivated a particular choice. These explanations effectively function as latent user variables, offering a structured basis to differentiate between user archetypes or behavioral modes that remain consistent across interaction episodes.

By integrating these abductively inferred user-level hypotheses into both model alignment and prompt construction, we enable a new class of context-aware LLMs capable of generating responses that are more adaptive, intentional, and personally resonant. This not only enhances utility and safety but also promotes social and cognitive sensitivity, fostering interactions that respect the diversity of user profiles. In doing so, we move toward a model of personalization that is not merely reactive but interpretive and generative, grounded in a deeper understanding of what truly matters to each user.

The only way for models tuned on existing preference data formats to use a persona is by adding it to the prompt in inference, hoping the model treats it as an extra instruction to follow (Deb et al. 2022; Balepur et al. 2025; Galitsky 2025). However, typical preference tuning methods like direct preference optimization (Rafailov et al. 2023, DPO) lack this personalization.

1.2 Personalization and Web 3.0

Personalization is a key part of Web 3.0, which is the next generation of the Internet. Unlike the current Internet (Web 2.0), which is controlled by large companies that collect and use people's data, Web 3.0 is about giving users more control over their information and making online experiences smarter and more personal.

This is how personalization fits into Web 3.0: smarter recommendations. Web 3.0 uses new technologies that understand not just the words people type but what the writers really mean. This helps it give the writer better suggestions—like showing the content that matches her interests, even if she did not search for it exactly. This is because Web 3.0 can understand context and relationships between topics.

People like to control their own data. In Web 2.0, companies like Facebook or Google store user data and use it to personalize ads and content. In Web 3.0, personal data can be stored securely on the blockchain or in a system that people control themselves. Users decide who gets to use it and how—making personalization safer and more private.

AI helps make things more relevant. Web 3.0 combines AI with structured knowledge like knowledge graphs to make personalization even better. This means it can learn what users like and adjust what it shows them, such as news, shopping, or educational content, without needing to spy on user behavior.

Personalization works across platforms. With Web 3.0, user preferences do not have to be saved separately on each Web site. Instead, the personal settings and choices can follow a given user from one platform to another. This way, a user gets a consistent and personalized experience no matter where she is online.

Hence, personalization in Web 3.0 is smarter, more respectful of user privacy, and gives users more control. It is designed to make a user Internet experience better without giving up your personal information to big tech companies.

The potential of Web 3.0 underscores its far-reaching and transformative impact on digital ecosystems, signaling a promising and dynamic future. One of the most visible signs of this evolution is the rapid growth in the adoption of decentralized applications (dApps). By 2023, the number of active dApps had surged past 3000—an impressive leap from just a few hundred in 2018. These applications, which operate on blockchain networks without centralized control, reflect a growing shift toward user ownership, transparency, and peer-to-peer interaction.

In parallel, the decentralized finance (DeFi) sector has experienced a dramatic rise. The total value locked in DeFi protocols surpassed $80 billion by mid-2023, demonstrating increased confidence and investment in blockchain-based financial systems. This surge highlights the public's growing appetite for financial services that are open, programmable, and not dependent on traditional intermediaries like banks.

Complementing this decentralized infrastructure is the rise of AI, a core enabler of Web 3.0's intelligent automation and personalized services. The global AI market is projected to reach nearly $1 trillion within the next 5 years, reinforcing its central role in powering next-generation applications, from adaptive user interfaces and recommendation engines to smart contracts and autonomous agents.

Together, these trends illustrate how Web 3.0 is reshaping not just how we interact with technology but who owns and controls it, marking a significant departure from the centralized paradigms of Web 2.0.

1.3 From LLM Customization to Personalization

LLM customization refers to the process of adapting a general-purpose model to become more design- and domain-specific, targeting particular design operations and domains (e.g., specific products, industries, and manufacturers). In contrast, LLM personalization involves tailoring an LLM's outputs to the personal preferences and design portfolios of individual designers. For example, interactions between designers and LLMs can be personalized by synthesizing various design prompts in different orders and patterns (Tian et al. 2024).

In the context of Web 3.0 applications, LLM-based personalization can be significantly enhanced by integrating design reasoning principles with advanced prompt engineering techniques. Web 3.0 emphasizes decentralized, user-centric experiences, where intelligent agents must adapt to diverse user goals, preferences, and contexts—often inferred from partial or ambiguous data. Design reasoning, a core discipline in human-centered innovation, provides a rich set of reasoning strategies, including abductive, deductive, inductive, analogy-based, constraint-based, and case-based reasoning, that are critical for interpreting user intent, generating creative solutions, and iteratively refining system behavior (Lu and Liu 2012).

While prompt engineering in LLMs has become a standard tool for shaping model responses, particularly in Web 3.0 agents and smart contracts, it often lacks the structured reasoning frameworks used in design. For example, techniques such as chain-of-thought prompting improve LLM performance by guiding models through intermediate reasoning steps, mirroring human logical processes. However, the integration of design reasoning, with its iterative, constraint-aware, and goal-oriented nature, into LLM prompting strategies remains underexplored.

In Web 3.0, where applications require dynamic personalization across decentralized interfaces (e.g., DAOs, dApps, and token-gated services), embedding design reasoning into prompt engineering could enable LLMs to better model user intent, infer hidden goals, and adapt to emerging constraints. This fusion would support more adaptive, interpretable, and user-aligned interactions, advancing the reliability and contextual sensitivity of intelligent agents in the decentralized Web ecosystem.

Figure 1 presents a structured and multilayered framework for achieving personalization in LLMs. The process begins by applying advanced prompting techniques to support more dynamic and interactive collaboration between humans and LLMs

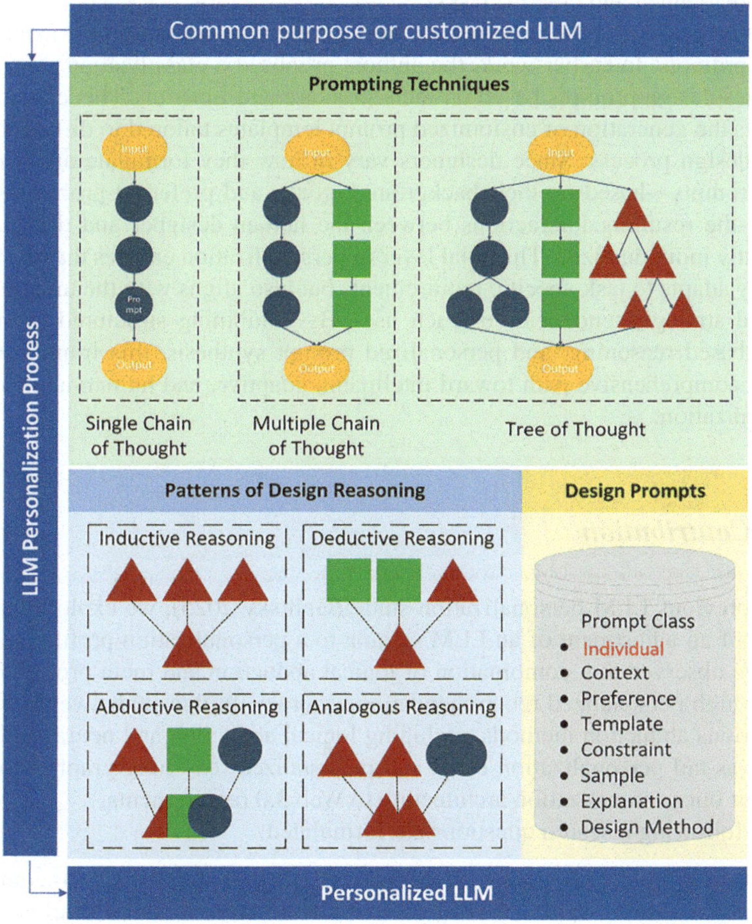

Fig. 1 From customized to personalized LLM

(White et al. 2023). Instead of relying on a single-shot response, these techniques guide the model to generate chain of thoughts, multiple parallel chains, tree-structured reasoning paths, or even graph-based reasoning structures. This structured generation allows the model to externalize its intermediate reasoning steps, thereby improving transparency, adaptability, and problem-solving depth.

The next key component of the framework involves the integration of design reasoning into LLM interaction. Personalization is deepened when LLMs are prompted using reasoning strategies drawn from established design studies, such as abductive reasoning for generating explanatory hypotheses, causal reasoning for identifying dependencies and outcomes, constraint-based reasoning for navigating design limitations, case-based reasoning for drawing on precedents (Janthong et al. 2010), and analogy-based reasoning for transferring knowledge across domains. Embedding these reasoning paradigms into prompts enables LLMs to simulate more humanlike cognitive workflows, particularly those employed by expert designers in open-ended and creative tasks.

Finally, personalization reaches its most granular level through the synthesis of design prompts. In earlier work, the authors proposed a classification scheme that organizes design prompts based on their structure and function. This classification supports the generation of customized prompt templates tailored to different stages of the design process. Since designers vary in how they formulate and sequence these prompts—based on their background, goals, and preferred problem-solving styles—the resulting interactions between the human designer and the LLM are inherently individualized. This final layer of personalization ensures that the system not only adapts to task-specific requirements but also aligns with the unique cognitive and strategic tendencies of each user. By combining structured prompting, design-based reasoning, and personalized prompt synthesis, this framework provides a comprehensive path toward intelligent, adaptive, and human-aligned LLM personalization.

1.4 Contribution

In our previous LLM personalization study (Galitsky 2025), we explored the possibility of an adjustment of an LLM session to a personalization profile of a given user. We observed that combination of logical abduction and meta-prompting provides a high personalized recommendation accuracy. In this study, we drill deeper into various abduction methods, including logical abduction and neural abduction, as well as tail personalization cases and personalized reasoning graph. Moreover, we tailor our personalization architecture to Web 3.0 requirements.

The following research questions are formulated:

- How do symbolic reasoning mechanisms (e.g., symbolic abduction) compare to learned reasoning approaches (e.g., neural abduction) in terms of effectiveness across different recommendation domains and Web 3.0 requirement?

- To what extent can meta prompting and LLM-based reasoning augment traditional recommendation models in a scalable and effective way?
- What is the optimal way to integrate symbolic, neural, and prompt-based reasoning components into a unified recommendation framework?

The innovations in this chapter are as follows:

1. The chapter introduces a neuro-symbolic Web 3.0 architecture for personalizing LLMs. In the context of prompting techniques, we proceed toward meta-prompting, illustrating how these meta-prompts construct user personalization profiles in a distributed manner, subsequently applying them during search operations to yield personalized outcomes.
2. We apply abductive reasoning to yield the most suitable answer for a user based on her distributed personalization profile. Abduction occurs in parallel to fine-tuning to iteratively improve the distributed personalization.
3. After surveying recent approaches to Web 3.0 personalization, we identify the methods most suitable with LLM integration. Product and services recommendation methods undergo a comparative analysis, with the selection of the most effective LLM-based approach for Web 3.0.

2 Distributed Personalization Architecture of WEB 3.0

Web 3.0 technologies can be delineated into six distinct layers, as illustrated in Fig. 2:

1. Infrastructure layer
2. Network layer
3. Protocol layer
4. Interface layer
5. Management layer
6. Application layer

The top layer comprises diverse distributed software applications (DApps) at the application level. These interactive applications are composed of a blend of software components including AI and operating systems. These software components find support in various devices like mobile phones, computers, and VR/AR/MR (virtual reality/augmented reality/mixed reality) devices, among others. Beneath these devices lie a multitude of Internet and network protocols rooted in blockchain technology. At the bedrock of the Web 3.0 technological stack resides the infrastructure layer upon which it relies.

The infrastructure layer primarily handles data management. The interface layer is responsible for mapping physical world data to the digital space. The management layer governs the ecosystem of Web 3.0. And the application layer is where actual use cases for Web 3.0 are designed and implemented.

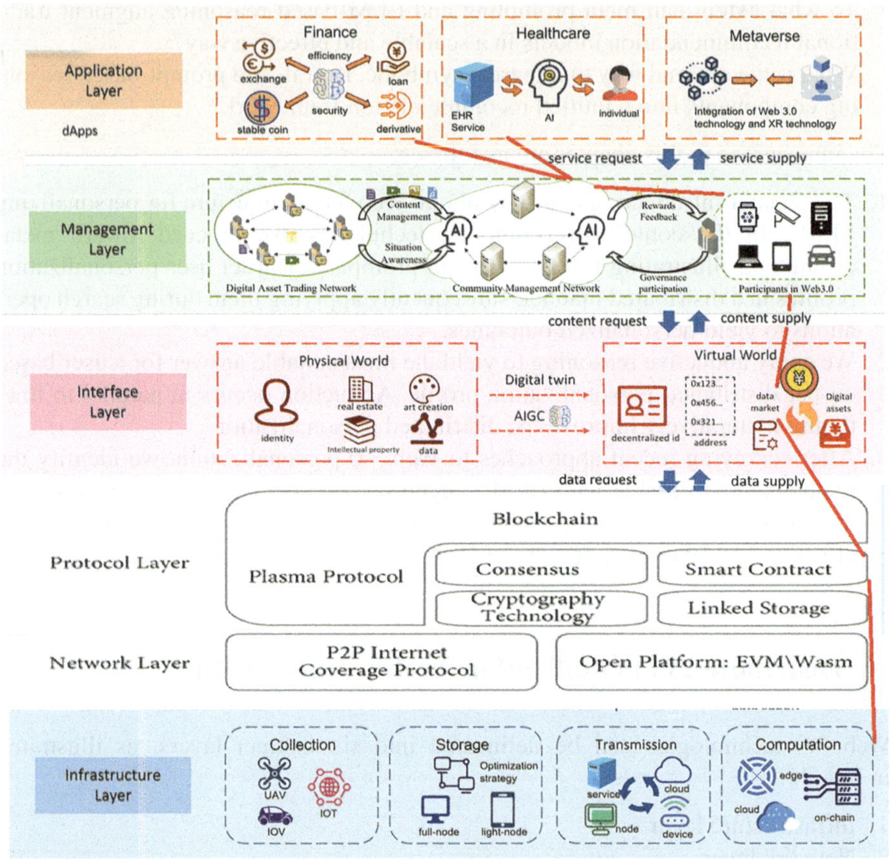

Fig. 2 Layers of Web 3.0 architecture. The red path shows the components involved in personalization

The infrastructure layer is responsible for collecting, storing, transmitting, and processing data. With the adoption of Web 3.0 technologies, which emphasize decentralization and co-governance (Shen et al. 2023), the sources of data have been greatly expanded, including the use of terminal devices from the Internet of Things and real-time feedback from users. This data is transmitted to edge devices or nodes for analysis and may be stored using an on-chain or combination of on-chain and off-chain methods to ensure the transparency and effectiveness of the data. In the whole process, AI technology can be integrated into all aspects to optimize strategies, improve storage computing efficiency, enhance system security, and protect privacy.

The network layer provides the foundational infrastructure needed for decentralized communication and computation. It includes P2P Internet coverage protocol and ensures devices can communicate over a decentralized, peer-to-peer network (Benet 2014). It eliminates the need for centralized servers, promoting resilience and censorship resistance. This layer also includes Open Platform EVM (Ethereum Virtual Machine), which allows smart contracts to run on Ethereum-compatible blockchains, and Wasm (WebAssembly), which offers a high-performance runtime for smart contracts and cross-chain interoperability beyond Ethereum (Buterin 2015; Zhang et al. 2016). It also supports execution environments for decentralized applications (dApps).

The protocol layer defines the rules and mechanisms that govern decentralized applications and the blockchain network. It includes blockchain, the core distributed ledger technology that underpins data integrity and immutability (Narayanan et al. 2016). Also, it includes Plasma Protocol: a scalability solution for blockchains, especially Ethereum. It enables faster and cheaper transactions by offloading operations to child chains. This layer also includes Consensus and mechanisms like Proof of Work (PoW) or Proof of Stake (PoS), which ensure that all nodes agree on the state of the blockchain. These are vital for security and trust in decentralized networks (Zyskind et al. 2015).

Smart Contract is a self-executing code with the rules of agreement directly written into lines of code. It runs on EVM or Wasm platforms. Cryptography technology ensures privacy, integrity, and security of data through encryption, digital signatures, and zero-knowledge proofs. Linked storage facilitates decentralized data storage solutions (e.g., IPFS or Filecoin) and allows smart contracts and dApps to access and manage external data reliably (Xu et al. 2016).

The interface layer serves as a connection between the physical world and the digital world, transforming data collected from the infrastructure layer into valuable digital assets. This layer also establishes a value circulation system to optimize the utilization of data and motivate user participation. The value of this data is determined by supply and demand, allowing users to benefit from their contributions. Web 3.0 grants individuals ownership of their data, with decentralized identities responsible for protecting the privacy and access control. AI can assist in the decentralization of identity systems and the creation of a more intelligent data market while also safeguarding the privacy of individuals.

The management layer oversees and governs the operations of the underlying layers, ensuring system integrity and reliability. Its responsibilities include implementing incentive mechanisms to promote active and constructive user participation, as well as reviewing and moderating user-generated content. In a smart city context, this layer may validate, correct, or remove inaccurate or outdated traffic data submitted by users and detect or mitigate malicious traffic-related activities. Similarly, in financial systems, it can leverage AI to identify abnormal trading patterns and flag or remove false, misleading, or otherwise inappropriate information.

Application layer is built upon the previous five layers. As an example, in the context of a smart city, this layer may include navigation apps that adjust routes in real time based on traffic conditions. Beyond the realm of smart cities, the application layer of Web 3.0 has seen significant progress in finance (Zhuang et al. 2021), medicine (Mistry et al. 2021), and metaverse (Chen et al. 2020). AI technology is also heavily integrated at this level, improving user experience, enhancing privacy protection, and increasing efficiency.

2.1 *Layered Architecture and Personalization*

Distributed personalization in Web 3.0 leverages decentralized technologies to tailor user experiences without centralized control over personal data. Here is how this personalization maps across the six layers:

1. Infrastructure layer (hardware, cloud/storage, blockchain nodes, compute resources) provides the decentralized backbone—e.g., IPFS (InterPlanetary File System), Filecoin, Arweave—that stores user data and models in a distributed manner. User profiles and preferences are stored in encrypted or user-controlled environments (e.g., self-sovereign identity (SSI) wallets or data pods). This layer supports edge computing or decentralized compute layers (e.g., Akash, Golem) to enable local model inference or rule-based personalization.
2. Network layer (P2P networking, routing, and transport protocols) ensures secure, low-latency, peer-to-peer communication between user agents and services. For example, when a dApp queries for personalized content, the request and delivery flow securely through protocols like libp2p, enabling dynamic, user-specific content assembly.
3. Protocol layer (Consensus, identity, tokens, smart contracts) enables persistent, user-controlled identifiers (e.g., using W3C DIDs or Verifiable Credentials). It allows execution of personalization logic, such as recommendation scoring, without centralized servers. Token gating personalizes access to features or content based on NFT or token holdings (e.g., exclusive access for DAO members).
4. Interface layer's browser extensions (e.g., MetaMask, Ceramic SDK) enable decentralized apps (dApps) to interact with user data and preferences without direct control or storage.
5. Management layer (governance, access control, reputation systems) maintains user-curated preferences through governance: DAOs may vote to tune personalization algorithms. Reputation scores personalize interactions (e.g., more visibility to high-reputation content) and access policies dynamically define what each user can see or do based on community roles or delegated authority.
6. Application layer (dApps, social platforms, marketplaces, agents) acts as personal agents that learn from the user's behavior (e.g., social engagement, preferences) and adapt content accordingly. Users explicitly provide their interests, which are then used locally or via smart contracts to personalize experience.

For example, a decentralized music app suggests tracks based on NFT-based music ownership history and a DAO forum ranks proposals based on members' past interactions or governance activity.

Hence, distributed personalization in Web 3.0 is modular, consent based, and user owned, operating across layers from infrastructure and identity to interfaces and apps—without relying on a central authority. Instead, personalization is governed by transparent protocols, smart contracts, and peer-to-peer data control.

2.2 *Neuro-Symbolic Approach to Web 3.0 Personalization*

Web 3.0 personalization benefits significantly from a neuro-symbolic approach because it requires both deep learning's flexibility *and* symbolic reasoning's structure and explainability (Table 1). Web 3.0's personalization requires agents that can understand human language and emotion (neural), follow explicit rules and preferences (symbolic), explain their choices (symbolic), and work locally or semi-disconnected (symbolic reasoning with lightweight models).

How neuro-symbolic architecture addresses challenges in Web 3.0.

Challenge in Web 3.0	Why neuro-symbolic helps
1. Decentralized data control. User data is fragmented, encrypted, and locally stored	Symbolic reasoning (e.g., logic rules, ontologies) helps infer preferences from sparse or user-declared data instead of massive centralized datasets
2. Privacy-preserving personalization. No central server to aggregate profiles	Neuro-symbolic systems can reason locally, combining small on-device models (neural) with explainable, rule-based adjustments (symbolic), explained in chapter "Adversarial Integration of LLM and Logic Program"
3. Intent and context interpretation. Users express goals in complex, nuanced ways	Neural models understand language, while symbolic layers apply structured reasoning (e.g., theory of mind (chapter "Improving ToM Capabilities of LLMs in Applied Domains"), intent graphs)
4. Trust and explainability. Users must control and verify how they are profiled	Symbolic components provide transparent explanations ("You were shown this because…"), enhancing trust and auditability
5. Cross-platform, modular agents. Personalization logic must be portable and interoperable	Symbolic rules are portable and interpretable across dApps, while neural nets handle low-level perception (e.g., text, emotion, engagement patterns)
6. Need for adaptive autonomy. Agents must learn but also obey constraints (legal, ethical, platform-specific)	Neural nets adapt behavior, while symbolic logic enforces constraints and governance rules, enabling safe learning

A neuro-symbolic approach blends the strengths of both paradigms, offering adaptive, privacy-respecting, and interpretable personalization, which is a good fit for the decentralized, user-centric vision of Web 3.0.

Table 1 Role of each layer in personalization

Web 3.0 layer	Role in personalization
1. Infrastructure	Stores user data and models in decentralized systems (e.g., IPFS, Filecoin); supports privacy-preserving compute via edge/decentralized resources for local inference
2. Network	Enables secure peer-to-peer communication for personalized data requests and responses using protocols like libp2p
3. Protocol	Powers decentralized identity (DIDs), token-based personalization (e.g., NFT access), and smart contracts that execute recommendation logic without central servers
4. Interface	Wallets and browsers act as personalization hubs, storing user preferences, social graphs, and identity credentials locally and securely
5. Management	Uses governance, access control, and reputation systems to adapt personalization logic based on community input or user roles (e.g., DAO voting, content ranking)
6. Application	dApps deliver personalized content via user-controlled agents and smart contracts. Preferences are explicitly provided or inferred from user behavior and holdings

3 Personalization Scenarios

LLMs have showcased remarkable capabilities in harnessing their internal world knowledge and common-sense reasoning to accurately discern user intent during dialogues. Additionally, LLMs exhibit the ability to engage with users in a natural and fluent manner, providing a seamless and delightful user experience. These strengths render LLMs an enticing option for recommendation systems aimed at improving personalized experiences.

However, notwithstanding the impressive memory capacity of LLMs, they encounter difficulties in memorizing specific knowledge within private and specialized domains when lacking sufficient training. For example, the storage of extensive item databases and user profiles in a recommender system poses a considerable challenge for LLMs. This limitation can result in LLMs generating inaccurate or erroneous responses, making it challenging to govern their behavior within a particular domain. Furthermore, LLMs grapple with the temporal generalization problem as external knowledge continues to evolve and change over time.

To tackle these issues, various tools can be employed to enhance LLMs and bolster their effectiveness as recommendation agents.

1. Search engine. Search engines are widely employed to provide external knowledge to LLMs, reducing LLMs' memory burden and alleviating the occurrence of hallucinations in LLMs' responses.
2. Recommendation engine. Some works have attempted to alleviate the memory burden of LLMs by equipping them with a recommendation engine as a tool, enabling LLMs to offer recommendations grounded on the item corpus.

The recommendation engine such as, for example, Chat-REC (Gao et al. 2023), is divided into two stages: retrieval and reranking, which aligns with typical recommendation system strategies. In the retrieval stage, LLMs utilize traditional recommendation systems as tools to retrieve 20 items from the item corpus as a candidate item set. Subsequently, LLMs employ themselves as tools to rerank the candidate item set. LLMs' commonsense reasoning ability, coupled with the internal world knowledge within them, allows them to provide explanations for the sorting results. The recommendation engine tool used in RecLLM (Friedman et al. 2023) is highly similar to the one in Chat-REC, and it is also divided into retrieval and reranking stages.

An overview of our proposed personalized recommendation architecture is shown in Fig. 3. The essential recommendation components are:

1. *Abductive reasoning*, which requires an LLM to provide an explanation why a certain item is recommended, given the Personalization User Profile
2. *Object-level Prompt*, which is formulated as an exact task of providing an answer for a certain category/specialty of the user
3. *Meta-level Prompt*, which is an abstract request such as finding commonalities in user documents, extracting them from user documents

The "Purchase or Selection History" is created by concatenating the items that the user has interacted with. The "Personalization User Profile" is a high-level

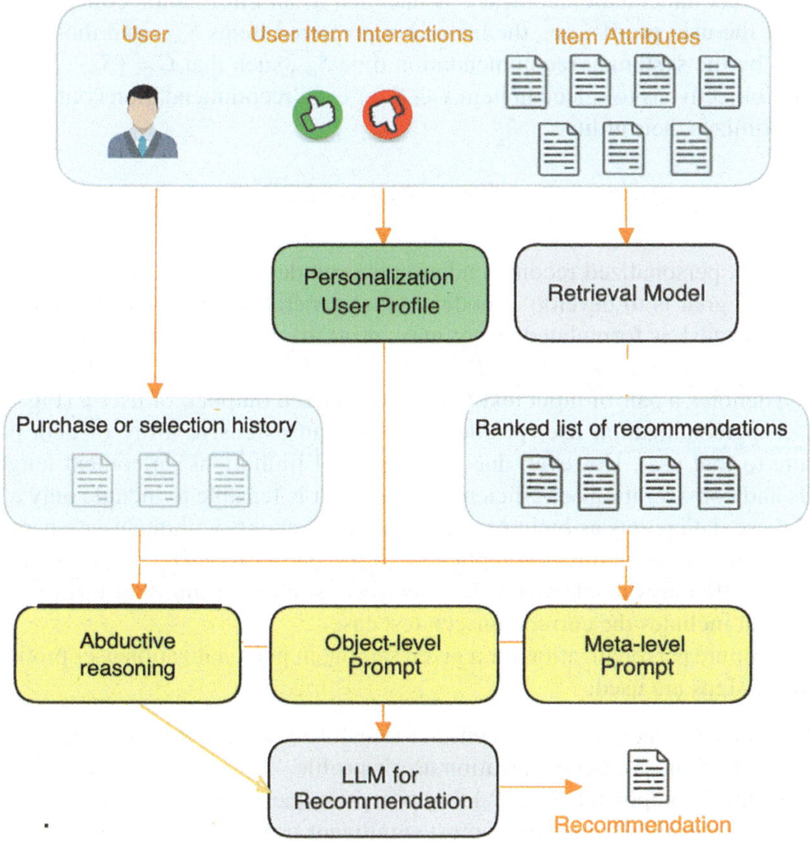

Fig. 3 A hybrid prompts + abduction LLM personalization architecture

summarization of the user's preferences, generated using an LLM based on user-item interactions, item attributes, or even user information if possible. The "Candidates for Recommendation" are the output of an arbitrary retrieval component.

3.1 Problem Formulation

We consider every data sample as an independent user u, with a set of user records associated with each sample across all tasks. These user records are instrumental in tailoring language models to individual users using their specific data. Consequently, each data sample can be divided into three distinct components: an input sequence, which serves as the model's input; a target output, which represents the expected model output; and a profile, which encapsulates any supplementary information that can be used to customize the model in accordance with the user's unique preferences or needs.

A general recommendation problem can be defined as the task of maximizing a utility function M. M serves as a measure of how useful an item, s, is to a recommendation context, c, represented as $M: C \times S \rightarrow R$, where R is a totally ordered set. Here, the recommendation context c is intended as an information collection composed of the user profile, i.e., the list of her preferred items S_{user}, and those recommended by the system at recommendation time S_{rec}, such that $C = \{S_{user}, S_{rec}\}$. The primary objective is to select an item $s_{'} \in S$ for each recommendation context $c \in C$ that maximizes their utility.

$$\forall c \in C, s_c^{'} = \arg\max_{s \in S} u(c,s).$$

In our case, personalized recommendation is provided via text. For a given textual input t, the goal is to develop a model M that generates personalized output o for user u. This task is formulated as arg $\max_o p(o|t; u)$. For each user u, the model M can take advantage of user profile $P_u = \{(t_{u1}, o_{u1}), (t_{u2}, o_{u2}), \ldots, (t_{um}, o_m)\}$ where each (t_{ui}, o_{ui}) denotes a pair of input text t and personalized output o of user u (Fig. 4).

Each personalization user profile comprises an extensive array of data points specific to that user. However, due to the natural limitations on context length in LLMs and considerations of efficiency and cost, it is feasible to include only a subset of these data points as input prompts. Not all elements within a user's personalization profile may be directly applicable to the particular task the user intends to undertake. IR plays a selective role in extracting the relevant data from the user profile that includes the current, unseen test case.

To compute personalization for a given sample in personalization user profile (t_{ui}, o_{ui}), three steps are used:

1. A query construction component that transforms the input t_i into a query q for retrieving from the personalization user's profile
2. A retrieval component $IR(q; P_u)$ that accepts a query q and a user profile P_u and retrieves entries assumed to be most significant from the user profile

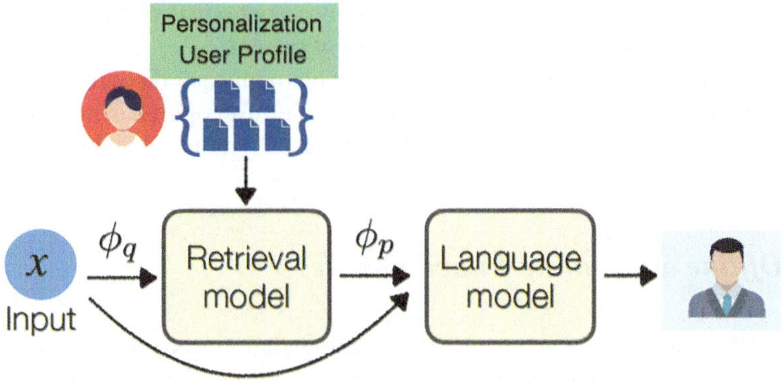

Fig. 4 Formulating personalization problem for IR and LLM

3. A prompt construction component p that assembles a prompt for user u based on input t_i and the obtained entries

LLM input is obtained using the following: $t_i^{LLM} = \phi_p\left(t_i, IR\left(\phi_q\left(t_i\right), P_u\right)\right).\left(t_i^{LLM}, o_i\right)$ is used to train and evaluate LLM.

To express the similarity between a recommending system and an LLM, consider the following. A LLM estimates the probability $P(x_1, x_2, \ldots, x_T)$ of observing sentence tokens in the given order, namely, estimating the entire sequence's joint probability. Applying the chain rule, this probability is expressed as

$$P\left(x_1, x_2, \ldots, x_T\right) = \prod_{t=1}^{T} P\left(x_t \mid x_1, x_2, \ldots, x_{t-1}\right).$$

An ideal LLM is expected to autonomously generate text by sequentially selecting individual tokens

$$x_t \sim P\left(x_t \mid x_1, x_2, \ldots, x_{t-1}\right)$$

Hence, both LLM and a recommender learn to predict the most probable next token and next item, respectively (Di Palma et al. 2023).

Let us consider an recommendation example. The input is: "I find the purchase history list of user_1546: 4110 → 4467 → 4468 → 4472. I wonder what is the next item to recommend to the user. Can you help me decide?"

The output is: "1581".

Geng et al. (2022) assigned the user and each item a unique ID. Using a training set with thousands of users (and their purchase histories) and unique items, the LLM is able to learn that certain items are similar to one another and that certain users have inclinations toward certain items (due to the nature of the self-attention mechanism). During the pre-training process across all these purchase sequences, the model essentially goes through a form of collaborative filtering. It sees what users have purchased the same items and what items tend to be purchased together.

Combine that with an LLM's ability to produce contextual embeddings, and we suddenly have a very powerful recommendation system.

In this example, although we do not know what item each ID corresponds to, we can infer that item "1581" was selected due to other users purchasing it along with any of the items that "user_1546" already purchased (Aboufoul 2023).

3.2 *Offline and Online Functionalities*

Figure 5 illustrates how a personalized recommendation system works in two distinct phases: offline (indexing time) and online (personalization time). Here is a breakdown of each phases:

1. Offline functionality—indexing time (left side, in red). This phase builds the personalization profile for a user and happens before any queries are made.
   (a) Gain access to personal files and Web 3.0 sources: the system scans both local data (e.g., files on a user's device) and online data sources (Web 3.0) to gather personal information.
   (b) Classify files: it identifies and categorizes documents by relevance and content type.
   (c) Pick documents useful for personalization: the system filters out only those files that contain relevant data for tailoring recommendations.
   (d) Convert selected documents into de-identified form: to protect privacy, selected files are anonymized or de-identified.
   (e) Aggregate de-identified data into a personalization profile: all de-identified, relevant information is structured into a personalization profile, typically in a format like:
2. Online functionality—personalization time (right side, in blue) This phase runs when the user interacts with the system to get a recommendation:
   (a) Get a recommendation query from a user: the system receives a specific question or query from the user.
   (b) Classify user query and find respective key entry in the map: the query is analyzed and matched to relevant parts of the personalization profile.
   (c) Determine the query domain: identify which domain the query falls under (e.g., finance, health, etc.).
   (d) Form the prompt: the query is embedded into a structured prompt, incorporating relevant elements from the personalization profile.
   (e) Run the prompt through the LLM: the combined prompt is passed to the LLM to generate a tailored response.
   (f) Verify and possibly correct the result (commonsense reasoning): the output is checked for correctness, logic, and alignment with the personalization profile.
   (g) Deliver the personalized recommendation: the final answer is returned to the user as a personalized recommendation.

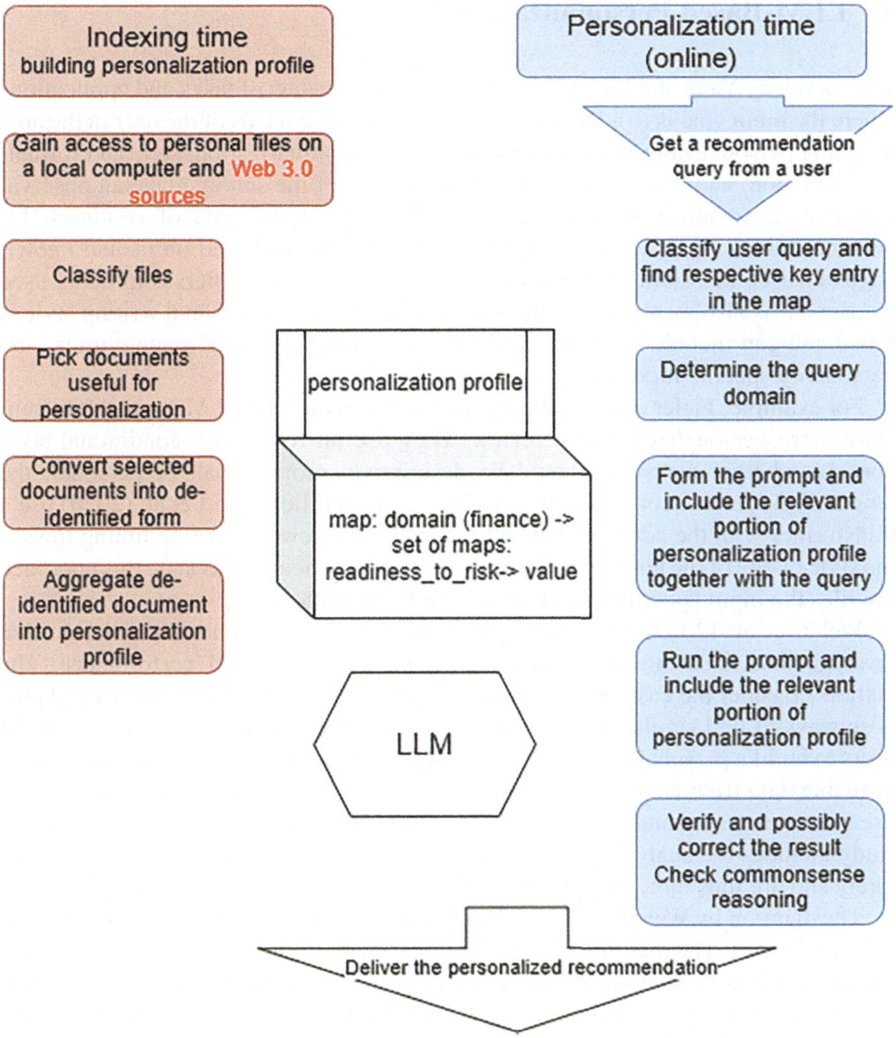

Fig. 5 Overall system architecture

Hence, offline, we build and store a privacy-preserving user profile from structured and unstructured data sources, and online, we use this profile dynamically with the LLM to generate customized responses to user queries.

4 LLM-Based Personalization

The survey by Yang and Flek (2021) contains an overview of tasks and applications where the main goal is to adapt a model or system to the needs of the user in the area of dialog personalization. There are also works in summarization, style and formality adaptation, and text simplification. The authors of the survey make an observation that personalization techniques vary broadly in the type of condition for personalization. Specifically, the text can be personalized based on factual knowledge of a user, i.e., some meta information, or on stylistic modifiers commonly used by users. It is also mentioned in the survey that a notion of personal writing style is broad and can include emotional stylistic modifiers, sociodemographic traits, and personality specific modifiers.

For example, Ficler and Goldberg (2017) use conditional LLM approach to generate more personalized movie review. They use an RNN with conditional style context: whether a review should be descriptive, professional, or personal; the length of the review could also be provided as a condition. Syed et al. (2020) use a transformer with the additional encoder-decoder framework for fine-tuning toward the style of one of the ten Project Gutenberg authors. Their results state this approach rewrites the input text with better alignment to the target style.

Welch et al. (2022) consider approaches to fine-tuning and interpolation that leverage data from similar users to boost personalized LLM performance. The authors consider the case of users with a small number of available tokens and propose ways to find similar users in their corpus and also leverage data from similar users to build a personalized LLM for a new user. Trade-offs between the amount of available data from existing users were analyzed, as well as the number of existing users and new users, and how the authors' similarity metrics and methods scale. The study includes the analysis of what types of words our method predicts more accurately and are thus more important.

The diagram by Welch et al. (2022) illustrates the utilization of data, models, and metrics (Fig. 6). The diagram comprises two primary sections: a left rectangle demonstrating the computation of three similarity metrics and a right rectangle depicting two methods for leveraging similar user data to create personalized models. Solid lines indicate the flow of golden set user data, while a dashed line represents data from a new user.

Golden set user data is employed to construct the authorship attribution model (AA), individual user LLMs for the perplexity-based metric (depicted as a set with the first as LMA1), and user embeddings (UE). These three metrics are then utilized to filter anchor user data to identify similar users. With this group of users, a baseline LLM is fine-tuned (without UE, denoted LM-UE), and subsequently, it undergoes further fine-tuning using new user data for the weighted fine-tuning method (WFT LM). During interpolation, the LLMs of individual golden set users are reweighted and combined with predictions from an LLM fine-tuned on new user data (FT LM).

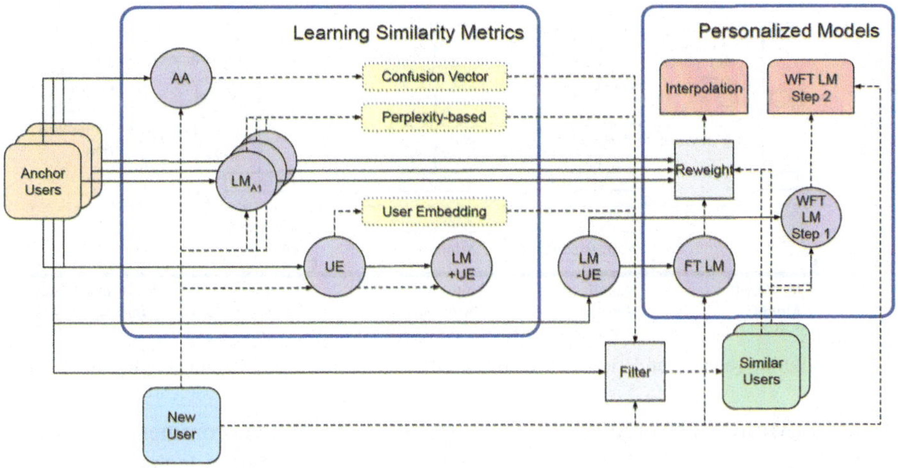

Fig. 6 Personalization architecture that leverage data from similar users to boost personalized LLM performance

Given *N* trained LLMs, one for each user, Welch et al. (2022) use the perplexity of one LLM on another user's data as a measure of distance. We could compare the word-level distributions, though this would be very computationally expensive. In their experiments, Welch et al. (2022) use the probability of the correct words only, or the perplexity of each model on each new user's data. The LLM is trained on all golden set users and fine-tuned for each anchor user. Then the perplexity of each model is measured on the data of each new user. For this matrix of *new* × *golden_set* perplexities, each row, representing a new user, is turned into a similarity vector by computing

$$1 - \frac{c - \min(roe)}{\max(row)} \text{ for each cell, } c.$$

4.1 From Traditional to LLM-Based Recommendation

The evolution of recommendation technology spans several decades, marked by significant advancements in ML and NLP (Fig. 7). Here is a brief history:

1. MDL (Minimum Description Length): The concept of recommendation systems can be traced back to the late twentieth century, with early models like MDL. These systems focused on simplicity and used algorithms to minimize the description length of the recommendation, laying the foundation for future developments.

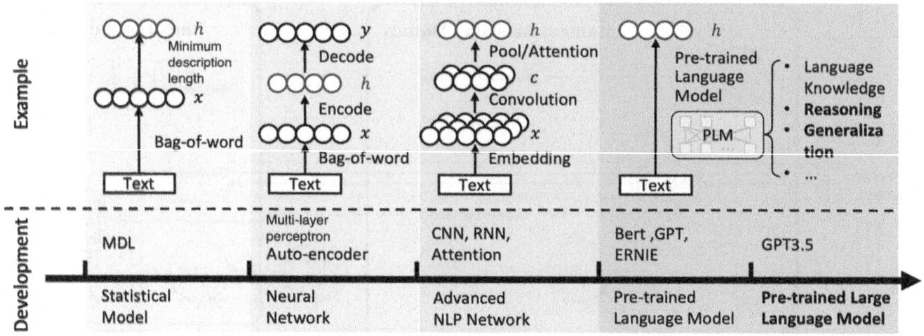

Fig. 7 Aligning recommendation architectures with deep learning architectures (Chen et al. 2023)

2. MLP (Multilayer Perceptron): As neural networks gained prominence in the late twentieth century, MLPs became a notable recommendation technology. These multilayered artificial neural networks were used for collaborative filtering and content-based recommendations.
3. BERT (Bidirectional Encoder Representations from Transformers): In 2018, BERT, a transformer-based model, revolutionized natural language processing. BERT excelled at understanding context and semantics in textual data, enhancing recommendation systems' ability to grasp user preferences and deliver more accurate suggestions.
4. GPT (Generative Pre-trained Transformer) Series: GPT models, developed by OpenAI, represent a breakthrough in recommendation technology. Starting with GPT-1 and progressing to GPT-3.5, these models are pre-trained on massive datasets, enabling them to generate humanlike text and understand context, making them applicable to a wide range of natural language processing tasks, including advanced recommendation systems.

The GPT-3.5, in particular, is a powerful language model with 175 billion parameters, making it one of the most sophisticated models for NL understanding and generation. Its large-scale pre-training allows it to perform exceptionally well in various recommendation tasks, understanding and generating humanlike responses. Overall, from the early days of MDL to the current state-of-the-art GPT-3.5, recommendation technology has undergone a remarkable journey, leveraging advancements in neural networks and language models to provide increasingly accurate and context-aware suggestions to users.

4.2 Concrete Reason-Based Preferences

We propose abductive reasoning (Peirce 1974) to augment preference training data with LLM-inferred individuals. Abduction, as a form of inference to the best explanation, is particularly suited to cases where observable outcome, such as user response preferences, lack an explicitly stated rationale. In computational terms, abduction allows us to infer latent variables or hidden contexts that plausibly explain why certain outputs are chosen or rejected (Zhao et al. 2024).

Building on this foundation, we apply abductive inference to the domain of LLM alignment by hypothesizing user-specific needs, interests, and traits, collectively referred to as personas (Tseng et al. 2024). These inferred personas serve as explanatory constructs that account for the observed preference patterns. Rather than treating preferences as flat, binary labels, we aim to uncover the underlying motivational factors that drive them—such as a user valuing conciseness over detail, emotional tone over factual density, or novelty over safety.

If LLMs can reliably surface plausible individuals that correspond to the users who might prefer either a chosen or rejected response, these personas can be formally attached as metadata to the training data. This enriched dataset would allow the training of more adaptive and personalized models, capable of generating responses that align with user-specified goals or inferred preferences. Ultimately, this approach bridges the gap between preference learning and intent-aware personalization, enabling LLMs to respond not only to prompts but also to the latent cognitive and affective profiles of their users.

Abduction explains given outcomes, deducing why preference data responses may be preferred via individual-based inference $I(p,r_1,r_2) \rightarrow P_1$: For prompt p and responses r_1 and r_2, the LLM produces a template of individual P_1 such that a user described by P_1 would prefer r_1 over r_2. If r_1 is the chosen response and r_1 is the rejected one, P_1 will describe a user preferring the chosen response and vice versa if r_1 is the rejected response.

We follow Balepur et al. (2025) and the taxonomy of Chan et al. (2024), building Web 3.0 templates for individuals to be as closest to demographic personas as posible. This template is merged with high-level traits, like information needs, interests, and personality types, that tailor how users want outputs presented. It must be verified that LLM templates for individuals satisfy the purpose of abduction: explaining why its input response is preferable. An LLM selects the optimal choice for a given prompt and a template for individual. A chosen template is accurate if the LLM picks the preferred over rejected recommendation (Fig. 8). It has been confirmed that LLMs do not perform abduction well for incorrect recommendations (Michael et al. 2023).

4.3 Computing Tail Personalization Cases

The ability to perform commonsense reasoning is crucial for understanding the dynamics of every day events, both for humans and for NLP systems. Most existing commonsense reasoning benchmarks focus on the ability to model common events (Talmor et al. 2019), i.e., given a commonly encountered situation, what common sense inferences can be made? However, personalization needs to evaluate a different class of inputs: unusual scenarios, improbable situations, and implausible events.

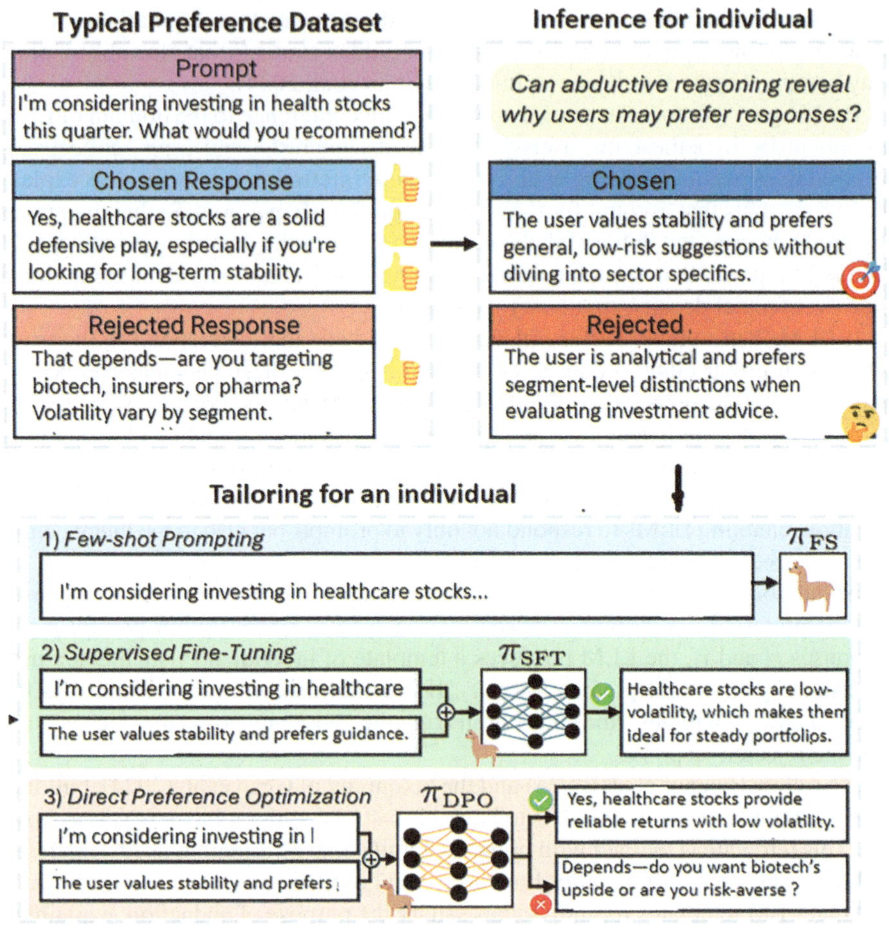

Fig. 8 Personalization data has a prompt, the best choice, and unaccepted choice

Handling unusual or uncommon cases is essential for building truly personalized language technologies, as these cases often reflect the individual differences that personalization aims to capture. While most LLMs are optimized for common, high-frequency patterns in data—often from English-speaking, Western contexts—users come from diverse cultural, linguistic, and situational backgrounds. Personalized systems must be able to recognize and adapt to these outlier contexts to ensure that recommendations, responses, or decisions align with individual needs and experiences, not just statistical averages.

For example, being aware that wearing shoes indoors may be culturally inappropriate in some households is not just about accuracy, it is about respecting the user's identity and values, which lies at the core of personalization. Similarly, in high-stakes scenarios such as medical advice or financial planning, a system's failure to understand a rare but plausible condition (like vibratory urticaria) may lead to harmful or inappropriate suggestions. By reasoning about such edge cases, LLMs can better tailor their responses to match user-specific contexts, ensuring that personalization is not just convenient but also safe, inclusive, and equitable. Thus, supporting uncommon cases is not a fringe concern, it is central to making personalization systems robust, context aware, and meaningfully personalized across geographies, cultures, and personal histories.

Uncommon situations can involve high-risk scenarios (Weidinger et al. 2022). For instance, a person might develop red, itchy welts after using a massage chair due to a rare condition called vibratory urticaria. If an NLP system misinterprets this as a more common issue, it could lead to harmful outcomes such as misdiagnosis (Table 2).

Zhao et al. (2024) examine the gap between human and model performance in generating abductive uncommon-sense explanations, highlighting subtle differences in quality. While GPT-4 produces more specific explanations than crowdsourced ones, its outputs tend to lack diversity and miss intermediate details. In contrast, crowd workers generate broader, more imaginative explanations. To leverage both strengths, the study refines crowd-authored explanations with GPT-4, adding detail while preserving creativity. These hybrid explanations are often preferred over both standalone human and LLM outputs.

The task of tail case abduction is formulated as follows. Given an NL context x and outcome y, the task of abductive reasoning requires generating an explanation z that augments the context, making the outcome more probable (Bhagavatula et al. 2020). In tail-case abductive reasoning, the targets are tail situations where an outcome y is very unlikely to happen in context x.

Table 2 Examples of tail personalization cases

Context	Uncommon outcome	Explanation
Maria opened a brokerage account and started trading tech stocks daily	Maria will withdraw all her funds and switch to investing in municipal bonds	After several weeks of losses due to market volatility, Maria realized active trading was too stressful and risky. She decided to pursue a more stable, tax-advantaged investment strategy
James signed up for a robo-advisor and invested in a diversified ETF portfolio	James will cancel the service and return to managing his own investments	James felt the algorithm didn't match his ethical preferences, like avoiding fossil fuel companies, and he preferred choosing assets himself
Priya started contributing monthly to her retirement account through her employer's plan	She will stop contributions and use the money to buy cryptocurrency	Priya saw rapid crypto gains online and, influenced by social media, decided to redirect her funds toward speculative digital assets
Ethan bought a low-cost health insurance plan with limited coverage	Ethan will sue the insurer for breach of contract	After requiring emergency care, Ethan discovered many services were not covered. Feeling misled by vague marketing, he pursued legal action

4.4 Learning for Abductive Reasoning

Existing methods for abductive reasoning typically rely on supervised fine-tuning (SFT) using a static dataset of human-written explanations (Bhagavatula et al. 2020). In this setup, the model is trained to generate the next token in an explanation, given a fixed gold-standard prefix. However, this training paradigm introduces a critical issue known as exposure bias. During training, the model always conditions on perfect, human-written prefixes. But at inference time, the model must generate explanations autonomously, conditioning on its own previously generated tokens rather than the ground truth. This mismatch between training and inference leads to compounding errors, where early mistakes in the generated explanation can mislead subsequent token predictions, degrading the overall explanation quality.

To address this shortcoming, we follow Zhao et al. (2024) in an exploration of a family of on-policy imitation learning algorithms, which train the model in a manner that more closely mirrors the inference-time generation process. Rather than learning exclusively from static prefixes, the model is updated based on its own generated sequences, thereby reducing exposure bias and improving robustness.

In the context of abductive reasoning, the goal is to generate a plausible explanation z for an observed context x and outcome $\boldsymbol{y}$. This process can be formalized as a policy-learning task, where a policy π maps from a context-outcome pair (x, y) and a prefix sequence of the explanation $(z_1, \ldots, z_{j-1})$ to a distribution over the next token in the vocabulary. Token generation is sequential: the j-th token of the explanation is drawn as $z_j \sim \pi(\cdot \mid x, y, z_{j-1})$, and the full explanation z is sampled from the learned policy $\pi(\cdot \mid x, y)$.

$\pi*(\cdot)$ is defined as the expert policy, representing human-annotated reference explanations, and $\pi_\theta(\cdot)$ as the learner policy, parameterized by θ. The objective of imitation learning is to adjust θ so that the learner policy approximates the expert, assigning high probability to expert-generated explanations, even when conditioned on its own previously generated tokens. This approach encourages the model not only to match expert/gold-standard outputs but also to be resilient under self-conditioning, improving fluency, coherence, and reasoning quality in generated abductive explanations.

Using Expert as an Oracle (EaO) algorithm is designed to iteratively improve a learner policy by leveraging an expert policy as an oracle. The goal is to refine the learner's policy parameters over multiple training epochs using supervised learning, guided by the expert's decisions. The algorithm takes the following inputs:

- Initial learner policy parameters θ_0: These are the starting parameters for the learner's policy.
- Expert policy $\pi*(\cdot)$: This is the expert policy that serves as an oracle, providing guidance to the learner.
- Dataset $D = \{(x,y)\}N$ consisting of input-output pairs (x,y), where x is the input and y is the corresponding output.

- Block size k: Determines how many additional samples are added to the augmented dataset in each iteration.
- Number of training epochs I: Specifies the total number of iterations or epochs for refining the learner's policy.

Initialization **creates** an empty augmented dataset $D^{-} \leftarrow \emptyset$. This dataset will grow over time as the algorithm progresses.

The algorithm runs for I epochs. In each epoch, it performs the following steps:

1. Iterate over the dataset.
2. For each sample (x,y) in the dataset D generate learner's prediction, using the current learner policy $\pi\theta_i(\cdot|x,y)$ to generate a prediction $z\sim$. Here, $\pi\theta_i$ is the learner's policy parameterized by θ_i.
3. Generate expert's prediction, using the expert policy $\pi*(\cdot|x,y,z\sim;b)$ to generate an expert prediction $z*$. The expert policy takes into account the learner's prediction $z\sim$ and a prefix size b, which determines how much of the learner's prediction is considered when generating the expert's response.
4. Augment the dataset, adding the triplet $(x,y,z\sim;z*)$ to the augmented dataset D^{-}. This triplet consists of the input x, the output y, the learner's partial prediction $z\sim$, and the expert's complete prediction $z*$.
5. Update learner policy. After processing all samples in D, further optimize the learner's policy parameters θ_i to obtain θ_{i+1}. This optimization is performed using supervised learning on the augmented dataset D^{-}. The learner learns from the expert's corrections to improve its predictions.
6. Update prefix size, increasing the prefix size b by k. The prefix size controls how much of the learner's prediction is used by the expert when generating its response. As the learner improves, the prefix size grows, allowing the expert to provide more nuanced feedback.
7. After I epochs, the algorithm terminates, and the final learned policy parameters θ_I are returned.

4.5 *Personalized Reasoning Graph*

Wang et al. (2024) propose a method that uses LLM to construct personalized reasoning graphs tailored to individual users within recommendation systems. In this approach, the LLM takes as input a user's personalization profile, such as demographic attributes, preferences, and goals, along with behavioral sequences like browsing history, purchase patterns, or content interactions. From this input, the LLM generates a graph-based representation that connects key concepts through chained causal and logical reasoning steps. The resulting reasoning graph captures not only low-level behavioral correlations but also higher-level semantic relationships between the user's interests, actions, and inferred intentions.

These structured graphs form a rich, interpretable abstraction of the user's cognitive model. To further utilize this structure, a graph neural network (GNN) is applied to encode the reasoning graph into a dense feature representation, effectively summarizing both its topology and semantic meaning. This graph embedding is then used as additional input to a traditional recommendation model, such as a collaborative filtering or matrix factorization system.

Based on GNN, the recommendation engine discovers deep conceptual relationships that are derived through reasoning, beyond surface-level similarities. As a neuro-symbolic approach, this fusion of symbolic reasoning and numerical optimization enhances both accuracy and personalization. Additionally, the graph structure offers a major benefit in terms of interpretability: it explicitly surfaces the reasoning path that led to each recommendation, helping users and system designers alike understand the "why" behind the model's choices. Overall, our approach bridges the gap between explainable AI and high-performing recommendation systems by embedding reasoning-driven personalization directly into the recommendation pipeline.

This approach is valuable for Web 3.0 personalization, where user control, transparency, and decentralization are core principles. By generating reasoning graphs from a user's on-chain behavior, preferences, and self-sovereign identity data, an LLM can capture personalized, semantically rich representations that reflect not only what a user does but why they do it. These reasoning graphs offer interpretable, user-specific logic paths, aligning well with Web 3.0's emphasis on explainability and consent-driven interactions. When integrated with decentralized recommendation engines, the graph embeddings enable more adaptive, privacy-respecting, and intent-aware personalization without requiring centralized profiling, enhancing trust and aligning personalization with the values of Web 3.0 ecosystems.

Web 3.0 recommendation systems increasingly rely on transformer-based architectures that support both combinatorial generalization and symbolic reasoning (Wei et al. 2022). LLMs leverage self-attention mechanisms to track dependencies across long text spans, enabling them to chain ideas, follow logical arguments step by step, and draw coherent inferences. Unlike traditional recommendation engines that rely on shallow correlations, LLMs can analyze user prompts or behavior within rich semantic contexts, identify nuanced causal links, and synthesize knowledge from multiple sources to deliver highly relevant suggestions. Their ability to generalize across unseen combinations of inputs allows them to recommend novel content or decisions grounded in abstract patterns rather than surface-level similarity. Moreover, when augmented with symbolic reasoning techniques, LLMs can apply explicit logic rules or constraints to their inference process—making the recommendations not only personalized but also interpretable and verifiable, which is essential in the decentralized and user-controlled environments that characterize Web 3.0. These capabilities collectively position LLMs as foundational tools for the next generation of context-aware, adaptive, and ethically aligned recommendation systems.

An LLM constructs personalized reasoning graphs based on u_i, a user; v_i, an item; $v_k^{(u)}$, interaction; and $a_k^{(u)}$, user attribute, reasoning about a user personalization profile and behavioral sequences through causal and logical inferences, employing reasoning chains RC_n (Fig. 9). The embeddings are concatenated from the

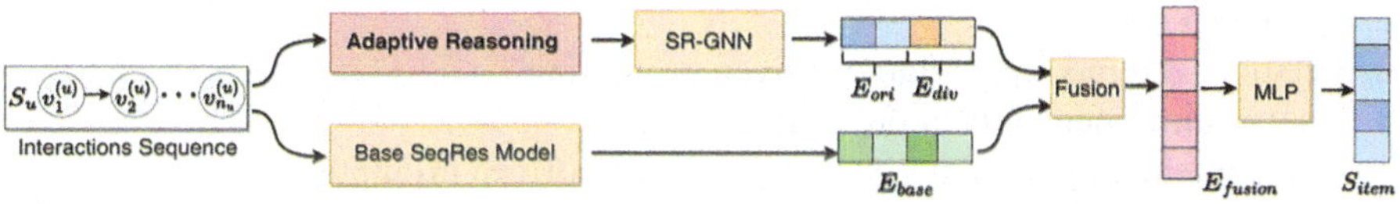

Fig. 9 Graph-based recommendation pipeline. We drill in adaptive reasoning in Fig. 10

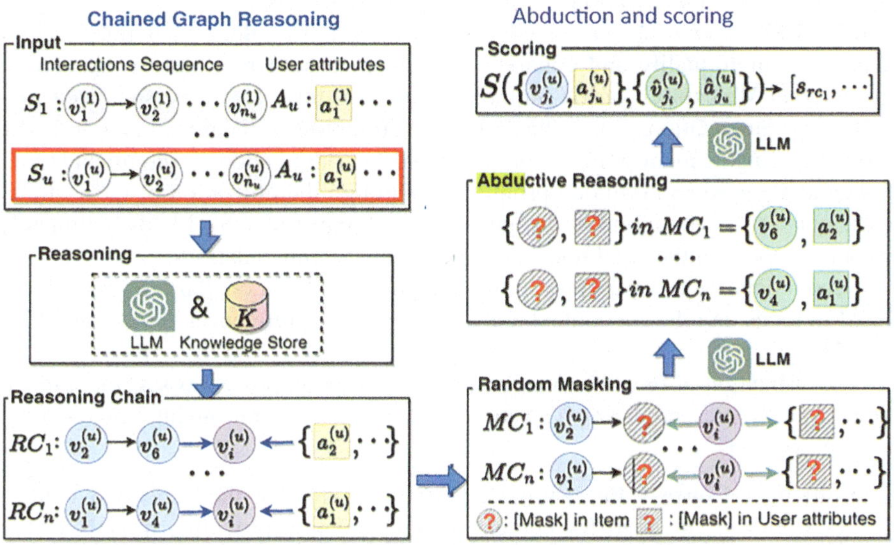

Fig. 10 Reasoning architecture

adaptive reasoning module (E_{ori} and E_{div}) and the base model (E_{base}) to obtain E_{fusion}. This fused embedding is used to predict the next item for the user.

The adaptive reasoning component is depicted in Fig. 10. Abductive reasoning capability of LLM is leveraged to confirm the adequacy and coherence of the dynamically generated reasoning chains from the chained graph reasoning and divergent extension modules. The component first masks the key items or engaged user attributes that the chain is meant to logically link to and then add a new reasoning chain to the graph. It then prompts the LLM to fill in the [Mask] in the masked chains $vMC_n^{(u)}$ with the best prediction.

5 LLM Personalization via Meta-Prompting

We follow Lyu et al. (2023), exploring various prompting strategies for enhancing personalized recommendation performance with LLMs through input altering via prompts.

We rely on the following prompting strategies:

1. Basic prompting
2. Personalized recommendation-driven prompting
3. User commitment-driven prompting
4. Recommendation driven + user commitment driven prompting

Rather than utilizing LLMs as recommender models, this section investigates the application of prompting strategies to enhance input text with LLMs for personalized content recommendations. Drawing upon the extensive language datasets on which LLMs have been fine-tuned, our objective is to unlock their potential for generating high-quality and context-aware input text, thereby improving recommendations. We explore various tailored prompting strategies for personalized content recommendation, including basic prompting, recommendation-driven prompting, engagement-guided prompting, and the combination of recommendation-driven and engagement-guided prompting. Through the utilization of these strategies, our aim is to optimize the generation of input text by LLMs, ultimately enhancing the accuracy and relevance of content recommendations.

Recommendation-driven prompting components play a crucial role in enabling LLM to focus on relevant context and align with user preferences (Lyu et al. 2023, Fig. 11).

The utilization of recommendation-driven prompting possesses several compelling characteristics, rendering it an attractive approach for crafting high-quality content descriptions:

1. **Enhanced Context Precision**: Explicitly stating that the generated content description is intended for content recommendation providing models with a clearer understanding of the task. This additional context enables models to align their responses more closely with the purpose of generating content descriptions for recommendation purposes.
2. **Guided Generation for Recommendation**: The specific instruction serves as a guiding cue for models, directing their attention toward generating content descriptions better suited for recommendation scenarios. The mention of "content recommendation" likely prompts the LLM to focus on key features, relevant details, and aspects of the content that are more instrumental in guiding users toward their preferred choices.
3. **Heightened Accuracy of Recommended Entities**: The instruction assists the LLM in producing content descriptions tailored to the requirements of content recommendation. This alignment with the recommendation task results in more relevant and informative descriptions, as the LLM is primed to emphasize aspects crucial for users seeking recommendations.

The user commitment prompting strategy involves utilizing user behavior, specifically user-item commitment, to formulate prompts with the aim of guiding the LLM to better encapsulate characteristics within the content description that align with user preferences. The goal is to generate more meaningful descriptions for recommendation tasks using this type of prompt.

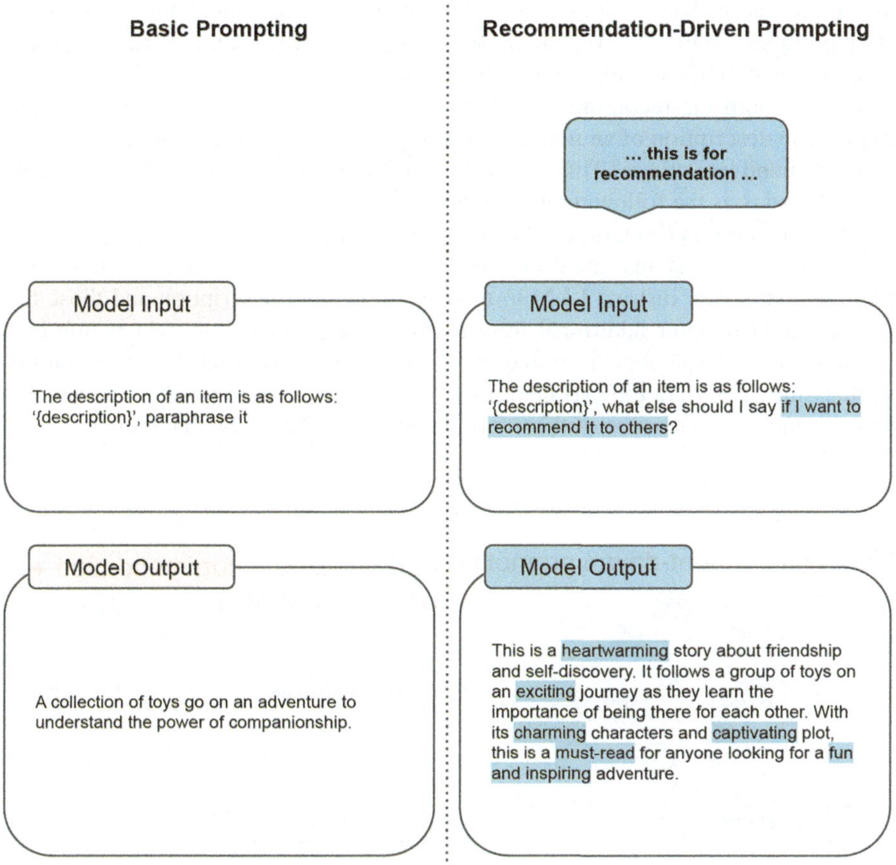

Fig. 11 Prompts for recommendation

To create the commitment-guided prompt, Lyu et al. (2023) combine the content description of the target item, denoted as d_{target}, with the content descriptions of T important neighbor items, represented as $d_1, d_2, \ldots, d_T$. This amalgamation of information serves as the foundation for the prompt, strategically designed to leverage user engagement and preferences in generating more contextually relevant content descriptions.

The commitment-guided prompt makes use of user engagement data to identify significant neighbor items. By incorporating information from these items into the prompt, the LLM has the potential to unveil meaningful connections, similarities, or relevant aspects between the target item and its neighbors. This approach can lead to more accurate, informative, and high-quality content descriptions, ultimately enhancing the overall performance of the content recommendation system.

Personalized recommendation-driven and user commitment-driven prompting components are essential for helping LLMs focus on relevant context and align with user preferences. These components, along with their associated augmented texts, are highlighted in Fig. 12. Together, they ensure that the model tailors its responses based on recommendations and user engagement.

Finally, both the recommendation-driven and commitment-driven prompts are used: "The description of an item is as follows: 'd_{target}'. What should I say if I want to recommend it to others? This content is considered to hold some similar attractive characteristics as the following descriptions: '$d_1, d_2, \ldots, d_T$'."

There are certain limitations when employing LLM for augmenting input text in recommendation systems when constructing prompts. There is a distinction between prompts that solely instruct LLM to modify the content description and those that prompt LLM to infer additional information. In the latter case, where inference beyond the original context is required, the recommendation-driven prompting strategy may not yield the expected improved personalization results. The value of recommendation relevance of the inferred context might not be that high.

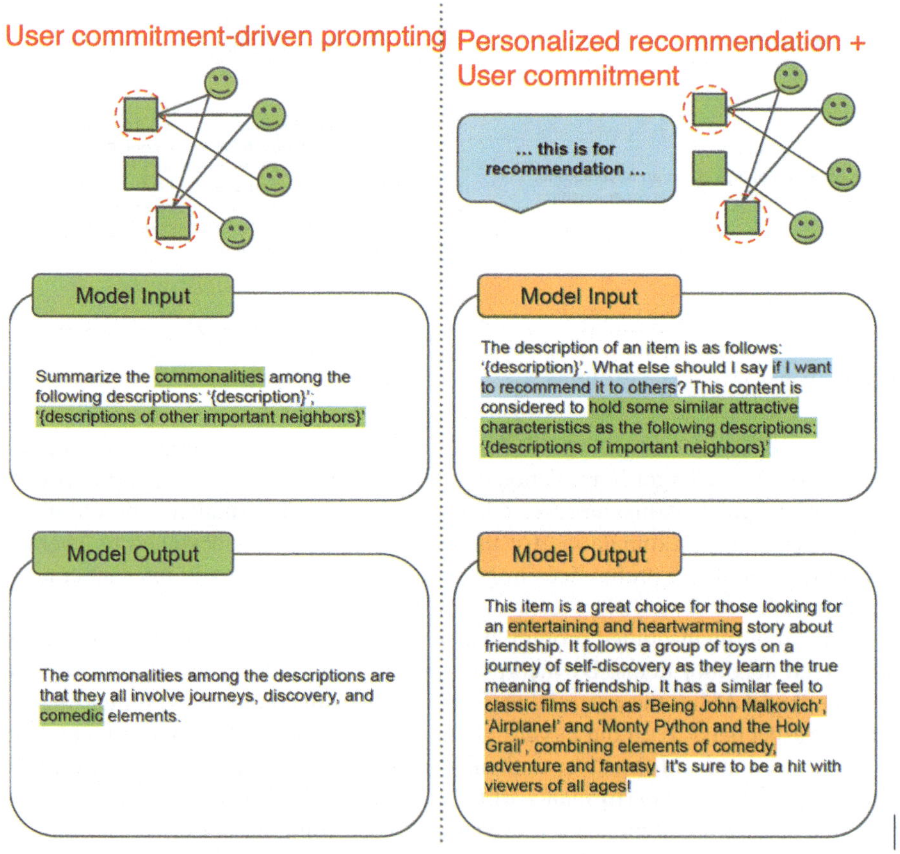

Fig. 12 User commitment and personalization prompts

There is a need for careful consideration and evaluation of the prompts employed, particularly when instructing LLM to infer information beyond the provided context. While recommendation-driven prompting strategies prove effective for prompts that do not necessitate inference, their effectiveness may be hindered when the prompts require LLM to extrapolate information.

In addition to its superior performance in personalized content recommendation, the incorporation of commitment signals in prompt designs may have broader associated benefits. The commitment-guided prompting strategy instructs the LLM to generate commonalities among different items, resembling the concept of neighborhood aggregation in graph and concept learning (Galitsky and Kovalerchuk 2014).

5.1 A Meta-Prompt for Building User Personalized Profile

In the realm of personalization, a meta-prompt serves as an abstract request aimed at constructing a user profile from a set of documents. At its abstract level, the meta-prompt seeks to identify commonalities among documents, such as shared phrases that signify entities potentially applicable for personalization. Depending on the nature of the documents and the task at hand, this meta-prompt can be instantiated into a specific task, such as determining values for the user's personalization profile, including common interests, tools, or locations. For instance, consider multiple software design documents and tutorials. A specific object-level prompt could extract the entity *software*, which can then be utilized to personalize recommendations in various domains, including job searches, training, socializing, and other activities.

For example, an instance of a meta-prompt for building a user personalization profile can be "adjust the answer about {topic} to user specialty extracted from {URL} from this profile" (Fig. 13).

We rely on meta-learning techniques to design a LLM prompt to produce a personalization prompt to obtain a suitable relevant information. Such meta-prompt is produced by generalization operation applied to available documents for the user. These documents need to be de-identified so that they are sufficient for personalization on one hand and will maintain user privacy on the other hand.

We do not know in advance which information from the documents or Web pages are associated with an individual. So, the prompt would sound "extract information from documents which occur in at least two distinct documents." We assume that occurrence in just one document can be a noisy indication, but occurrence in two is a reliable indication. Hence, a meta-prompt is needed to produce a prompt that will be applied to a given user question. Such meta-prompt would sound like "Determine which features X in a given document indicate an interest of a user," which will yield the prompt "Adjust search results to the domain of interest X." The former meta-prompt, once obtained, would yield what is called "object-level" or regular prompt with X instantiated. Whereas the meta-prompt need to be "discovered" or

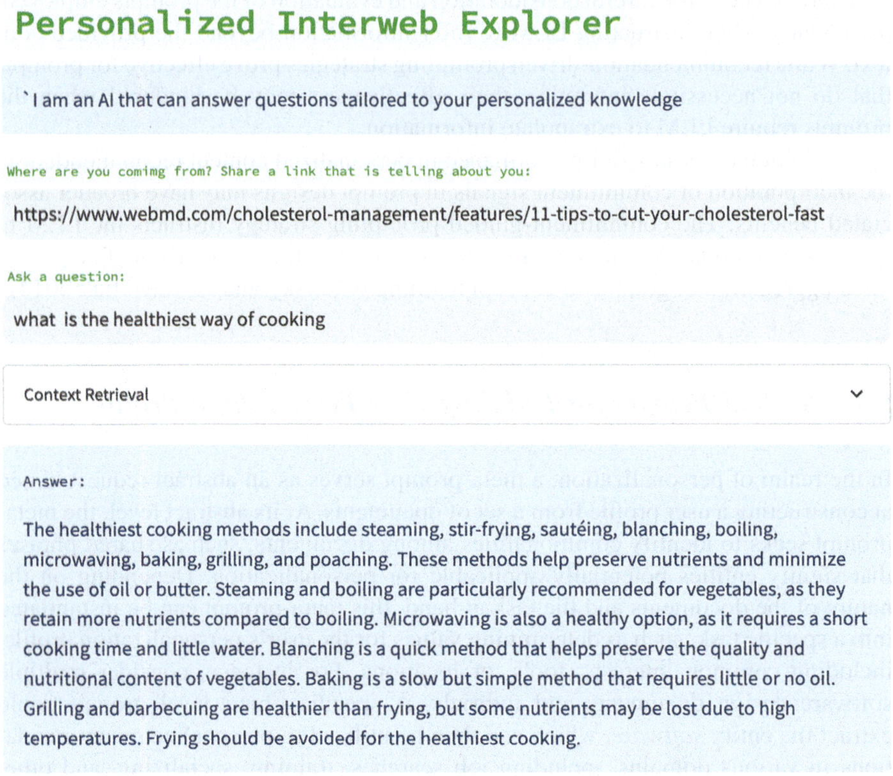

Fig. 13 Applying health-related attribute "cutting cholesterol" from the user personalization profile to the answer to the question related to cooking

"explored" by a LLM given a collection of documents, the resultant regular prompt is used to obtain the actual recommendation.

5.2 *Designing Meta-Prompts*

The pursuit of self-improving systems is driven by the promise of creating artificial intelligence that can learn, adapt, and evolve autonomously, leading to more efficient problem-solving and decision-making. Such systems could revolutionize a wide range of fields, from medicine and climate modeling to space exploration and poetry.

Intrigued by the potential of self-improvement, Goodman (2023) experimented with a language model chatbot that modifies its own instructions from one episode to the next, by reflecting on a dialogue with a user.

The fundamental concept behind meta-prompting is to prompt the agent to self-assess its performance and adjust its instructions accordingly. A viable approach involves employing reflection and introspection. The agent engages in an iterative process that initiates without instructions and progresses through the following steps (Fig. 14, Greyling 2023):

1. Engage in conversation with a user, who may provide requests, instructions, or feedback.
2. At the end of an episode, generate self-criticism and a new instruction using the meta-prompt.

The meta-prompt for the introspection architecture looks like:

```
meta_template = """
Assistant has just had the below interactions with a User. Assistant followed
their "Instructions" closely.
Your job is to critique the Assistant's performance and then revise the
Instructions so that Assistant would quickly and correctly respond in the
future.

####
{chat_history}
####

Please reflect on these interactions.

You should first critique Assistant's performance. What could Assistant have
done better?
What should the Assistant remember about this user? Are there things this
user always wants? Indicate this with "Critique: ...".
```

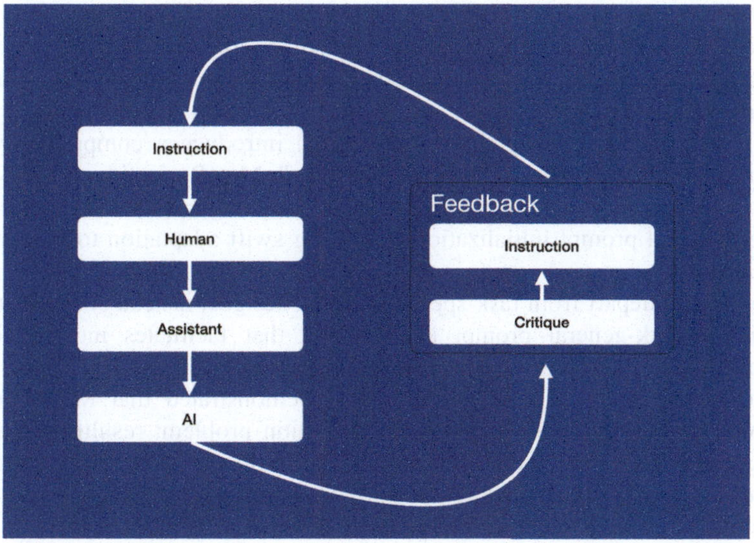

Fig. 14 Introspection architecture: meta-prompting for feedback loop

5.3 Automated Learning of Meta-Prompts

The technique of prompting represents a significant breakthrough in few-shot NLP. Recent advancements in the field have transitioned from the use of discrete tokens referred to as "hard prompts" to the adoption of continuous "soft prompts." These soft prompts make use of trainable vectors as pseudo prompt tokens and have demonstrated enhanced performance. While the potential of these soft-prompting methods is quite promising, they are notably reliant on having an appropriate initialization to be effective. Regrettably, achieving an optimal initialization for soft prompts requires a deep comprehension of the inner mechanisms of language models and intricate design. This makes it a challenging and time-consuming neuro-symbolic undertaking that must commence from scratch for each new task.

Prompt-based methods involve the insertion of specific text, referred to as prompts, into input examples. This conversion turns the few-shot task into a (masked) language modeling problem. For example, when analyzing the sentiment of a product review, such as "I will never use it again," a prompt like "It was" can be added to the sentence, resulting in "I will never use it again. It was." This adjustment naturally encourages the LLM to generate a higher probability for "bad" than "good." This transformation effectively closes the gap between pre-training and target tasks, enhancing transferability and diminishing reliance on target task data.

The effectiveness of prompting methods is significantly influenced by prompt design (Gao et al. 2021). Early efforts involved manually crafted prompts or the automatic generation of search prompts. Seeking enhanced performance compared to manually selected prompts, Gao et al. (2021) introduce LM-BFF, a method that searches for both prompt templates and label words. To further maximize the potential of prompts, recent studies have adopted learnable vectors as prompt content, enabling the learning of optimal prompts in a continuous space, termed "soft prompts" (Li and Liang 2021). As these methods no longer necessitate prompts to consist of real words, they greatly broaden the range of possible prompts, resulting in improved performance.

To tackle this challenge, Hou et al. (2022) introduce a comprehensive soft prompting method known as "MetaPrompting." MetaPrompting harnesses the widely recognized model-agnostic meta-learning algorithm to autonomously uncover optimal prompt initializations, enabling swift adaptation to new prompting tasks.

The authors depart from task-specific prompt designs, instead concentrating on acquiring a task-general prompt initialization that facilitates more rapid and enhanced adaptation to novel prompting tasks.

Through extensive experimentation, it is demonstrated that MetaPrompting effectively addresses the soft prompt initialization problem, resulting in notable enhancements across three distinct datasets. This approach significantly mitigates the issues associated with soft prompt initialization, demonstrating the acquisition of general metaknowledge to counter the instability of prompt variance.

In Fig. 15, x denotes the query sentence, and z is learnable pseudo tokens in soft prompts. ϕ represents all trainable parameters. Meta-prompting exploits optimization-based meta-learning to find an initialization ϕ_{meta} that facilitates better and faster adaptation to new tasks.

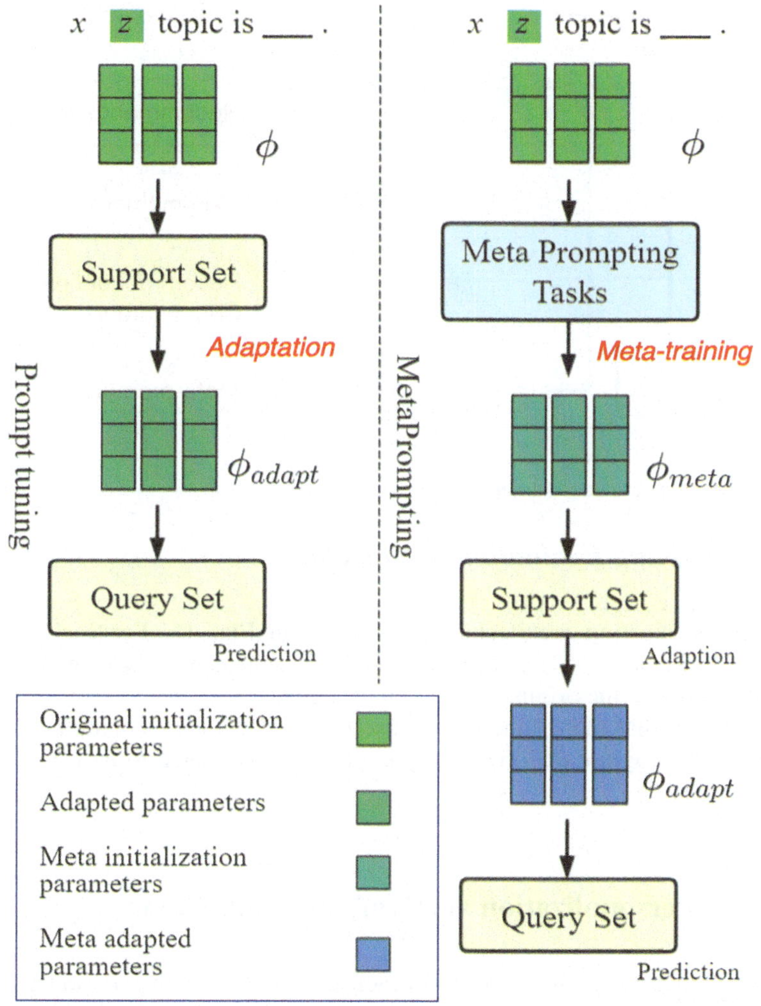

Fig. 15 Comparison between conventional soft prompting method and automated learning of meta-prompts. Parameter scales are shown in bottom-left corner

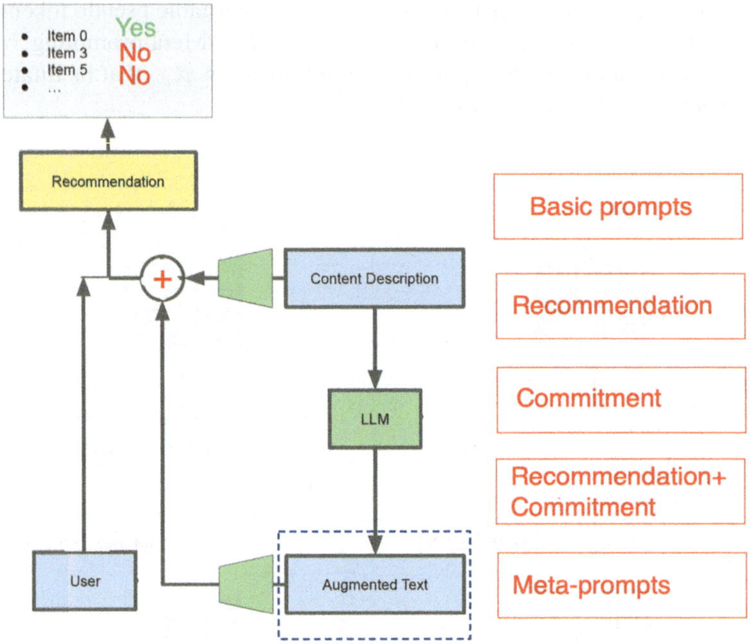

Fig. 16 Prompting evaluation architecture

5.4 Prompting Evaluation Architecture

Prompting evaluation architecture is depicted in Fig. 16. For each generated response, Lyu et al. (2023) first encode it and then concatenate the embeddings with the embeddings of the original content description.

It turns out that the distinctive words generated with the commitment-guided prompting strategy are fine-grained descriptive words related to movie-goer interests (Fig. 17).

6 LLM Personalization via Symbolic Abduction

Logical abduction is the logical formalization of abductive reasoning. Given observed facts and background knowledge expressed as first-order logical clauses, logical abduction can abduce ground hypotheses as possible explanations to the observed facts. This happening in the form of LLM prompts produced by abduction. A declarative framework in logic programming that formalizes this process is abductive logic programming (ALP, Denecker and Kakas 2002).

Abductive learning involves an ML model's task of translating sub-symbolic data into basic logical facts, while a logical model is employed for reasoning about these interpreted facts using a foundation of first-order logical background

Description: The Paper Chase follows a first-year Harvard Law student as he struggles to balance his studies and personal life while facing the demanding expectations of his professor.

The commonalities among these descriptions are that they all feature characters facing difficult challenges and obstacles. They also all have themes of resilience, hope, and friendship. We recommend these stories to others because they offer inspiring messages of overcoming adversity and the power of friendship.

If you're looking for a movie that will keep you on the edge of your seat, The Paper Chase is a great choice. It follows a first-year Harvard Law student as he navigates the pressures of his studies and personal life while facing the expectations of his professor. It's a story of determination and resilience in the face of adversity, and it will leave you feeling inspired and motivated.

Description: Under the Rainbow is a comedy about a group of misfit spies who must save a Hollywood hotel from a group of Nazi agents.

The commonalities among these descriptions are that they all involve comedic elements, misfit characters, and a mission to save something from danger. All of these movies are recommended for their humorous take on the genres they represent and their entertaining storylines.

Under the Rainbow is a hilarious and action-packed comedy that follows a group of misfit spies as they attempt to save a Hollywood hotel from a group of Nazi agents. With its unique blend of comedy, adventure, and suspense, Under the Rainbow is sure to keep you entertained from start to finish. It's a perfect mix of the classic disaster movie genre and the modern comedy parody, making it a must-see for fans of both genres.

Description: A brilliant engineer and his team must race against time to prevent a catastrophic chain reaction from destroying the world.

The commonalities among these descriptions are that they all involve a protagonist who must prevent a catastrophic event from occurring. They also all involve a sense of urgency and suspense. We recommend these stories to others as they provide an exciting and thrilling experience.

This thrilling story of a brilliant engineer and his team racing against time to prevent a catastrophic chain reaction from destroying the world is sure to keep you on the edge of your seat. With its intense action and suspense, this story will have you rooting for the heroes as they battle against the clock to save the world. It's a thrilling ride that will keep you guessing until the very end.

Fig. 17 Commitment prompt–based results are in the green frames knowledge to yield the ultimate output. The main challenge stems from the inherent difficulty of simultaneously training both the sub-symbolic and symbolic models in a cohesive manner. Experimental results of Dai et al. (2019) show that ABL generalize better than state-of-the-art deep learning models and can leverage learning and reasoning in a mutually beneficial way (Fig. 18).

Being more specific about integration of an abstract M and reasoning:

1. ML model does not have any ground truth of the primitive logic facts for training.
2. Without accurate primitive logic facts, the reasoning model can hardly deduce the correct output or learn the right logical theory.

We consider un-personalized LLM as a sub-symbolic ML without a logical theory. A logical theory focuses on personalization, implemented as re-labeling of data "labelled" within LLM. What LLM outputs can be considered as classification of each sentence or phrase into "in" and "out." What LLM gives is "in" class, and what it does not give is "out" class. Default LLM classification has significant inaccuracies: sentence "see the doctor" should be "out" for the doctor user, but it is "in."

When provided with a training sample linked to a final output, logical abduction is capable of making educated inferences regarding the missing information. This might involve deducing logic clauses that complement the existing background knowledge, thereby creating a coherent proof from the sample to its final output. Subsequently, the inferred basic facts and logical clauses are harnessed for training the machine learning model and preserved as symbolic knowledge, respectively. The process involves optimizing consistency to maximize the alignment between the conjectures and the background knowledge. To tackle this intricately complex challenge, we reframe it as a task that involves searching for a function that can make educated guesses regarding potentially erroneous primitive facts.

A baseline ML/LLM is used for learning the perception model LLM. Given an input instance x, LLM can predict the pseudo-labels LLM(x) as groundings of possible primitive concepts in x. These concepts can potentially be used in recommendation. When the pseudo-labels contain mistakes, the LLM needs to be fine-tuned, where the labels are the revised pseudo-labels $r(x)$ returned from logical abduction.

The task of abductive learning can be formalized as follows. The input of abductive learning consists of a set of labelled training data $D = \{<\boldsymbol{x}_1; y_1>; \ldots; <\boldsymbol{x}_n; y_n>\}$

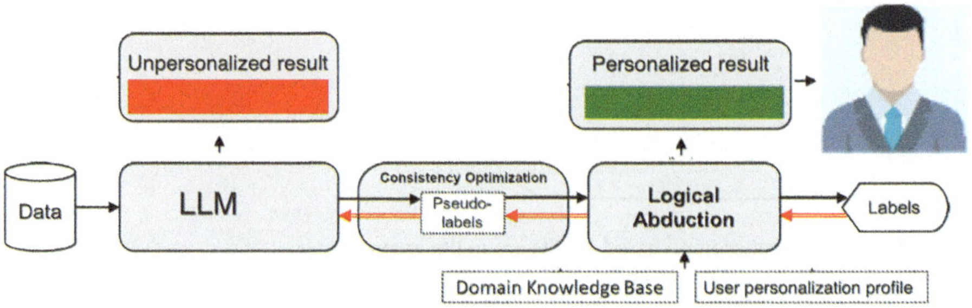

Fig. 18 Personalized and unpersonalized results with LLM and logical abduction

about a target concept C and a domain knowledge base B, where $\boldsymbol{xi} \in X$ is the input data, $yi \in \{0; 1\}$ is the label for $\boldsymbol{x}_i$ of target concept C, and B is a set of first-order logical clauses. The target concept C is defined with unknown relationships among a set of primitive concepts symbols $P = \{p_1; \ldots; p_r\}$ in the domain, where each p_k is a defined symbol in B. The target of abductive learning is to output a hypothesis model $H = p \cup \Delta_C$, in which:

1. p: $X \rightarrow P$ is a mapping from the feature space to primitive symbols, i.e., it is a perception model formulated as a conventional machine learning model
2. Δ_C is a set of first-order logical clauses that define the target concept C with B, which is called knowledge model

The hypothesis model should satisfy:

$$\forall x;y \in D\left(B \cup \Delta_C \cup P(x) \models y\right): \tag{1}$$

'|=' is a logical entailment.

Target concept C is viewed as a purpose of the prompt: once it is run, we expect to obtain C.

The perception model LLM is not personalized. LLM model and the knowledge model Δ_C are mutually dependent:

1. To learn Δ_C, the perception results $p(x)$ (the set of groundings of the primitive concepts in x) is required.
2. To obtain p, we need to get the ground-truth labels $p(x)$ for training, which can only be logically derived from $B \cup \Delta_C$ and y.

In an un-personalized LLM, the perceived primitive symbols $p(x)$ is most likely incorrect; therefore, it should be called pseudo-groundings or pseudo-labels. As a consequence, the inference of Δ_C based on Eq. (1) would be inconsistent.

When the personalization knowledge model Δ_C is inaccurate, the logically derived pseudo-labels LLM(x) might also be wrong, which harms the fine-tuning of LLM. In either way, they will interrupt the learning process.

Formally, an abductive logic program can be defined as follows: An abductive logic program is a triplet <B; A; IC>, where B is background knowledge, A is a set of predicates obtained by abduction, and IC is the integrity constraints. Given some observed facts O, the program outputs a set of ground abduction results of A, such that:

$B \cup \Delta \models O$
$B \cup \Delta \mid = IC$
$B \cup \Delta$ is consistent

Δ serves as a hypothesis that explains how an observation O could hold according to the background knowledge B and the constraint IC. Δ can be obtained in NL form if we request LLM to derive O from B and the constraint IC via chain of thoughts.

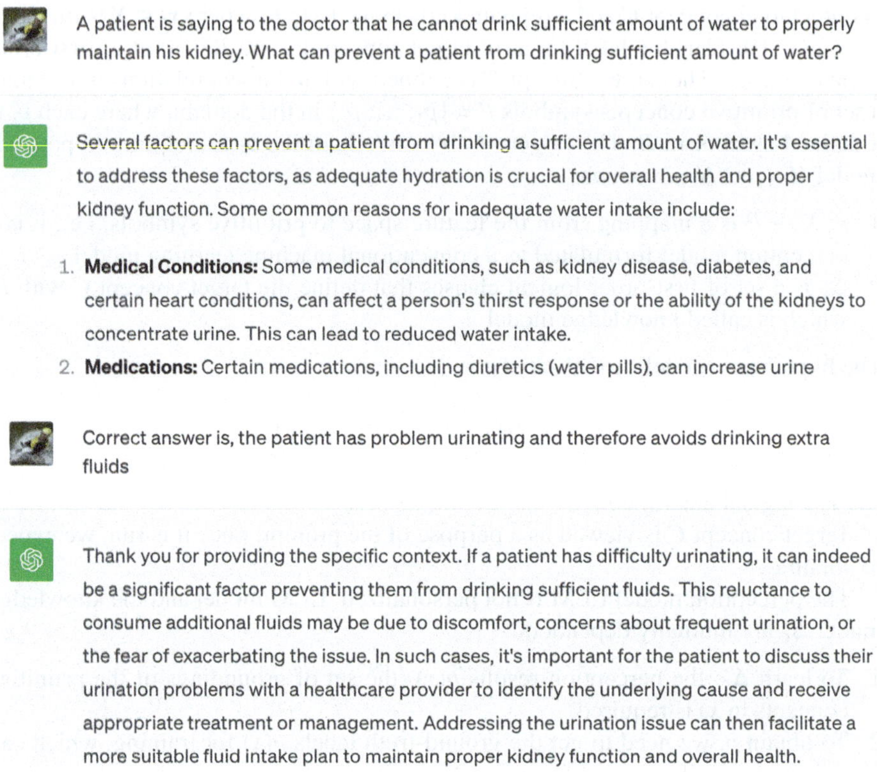

Fig. 19 Encouraging LLM to perform abduction

Abductive reasoning takes observed facts about data instances and also takes the combined LLM + reasoning about prompt prediction model $H = p \cup \Delta_C$. Given a fixed Δ_C, the system produces explanation by abduction $p(X)$ according to B and Y. However, when the learning model p has been obtained, the system builds an improved explanation Δ_C according to $B \cup Y \cup P(\boldsymbol{x})$.

$p(X) = \bigcup_{i=1}^{n}\{p(x_i)\}$ for coarse labels for all instances $X = \bigcup_{i=1}^{n}\{x_i\}$, and $Y = \bigcup_{i=1}^{n}\{y_i\}$ are the improved labels (correctly personalized results) for X.

An LLM can also be requested to perform abduction independently (Fig. 19).

7 Evaluation

We draw the comparison between our personalization algorithm and competitive LLM personalization approach in the shopping domain and then apply the optimized algorithm in the health domain. Traditional personalization systems commonly use task-specific metrics like ranking-oriented metrics, NDCG, AUC, and recall for assessing model performance. However, the integration of LLMs into recommender systems brings about significant changes in evaluation tools. Conventional metrics may not adequately capture the novel capabilities and altered recommendation generation processes, necessitating the development of new evaluation tools. In LLM-powered systems, evaluating user preferences is crucial, demanding a user-centric approach. Metrics such as user satisfaction, engagement, and overall experience become pivotal considerations, with user satisfaction surveys and feedback questionnaires serving as valuable options. It is also important to consider factors like diversity, novelty, serendipity, user retention rates, interpretability, and fairness of personalized recommendations. We will do that for a health recommendation domain.

The dataset is randomly split into training, validation, and testing sets using 4:2:1 ratio. Negative training dataset is created using random negative sampling. We follow personalization strategy of Wei et al. (2019) for evaluation. For the validation and testing sets, we pair each observed user-item interaction with 1000 items that the user has not previously observed. We use the following metrics: *Precision@K*, *Recall@K*, and *NDCG@K*.

We adopt the leave-one-out approach to evaluate the performance of each method, a commonly employed technique in many related studies. In this approach, for each user, we designate the last interacted item as the test data, the item just before the last interaction as validation data, and the remaining items for training. The evaluation of each method is conducted on the complete item set without any form of sampling.

For the validation and testing sets, we pair each observed user-item interaction with 1000 items that the user has not previously interacted with. It is crucial to emphasize that there is no overlap between the negative samples in the training set and the unobserved user-item pairs in the validation and testing sets, ensuring the independence of the evaluation data.

To assess the performance of top-K recommendations, we utilize widely used metrics such as Precision@K, Recall@K, and NDCG@K. In our case, we set $K = 10$, indicating that we consider the top 10 recommendations. We present the average scores across five different splits of the testing sets, providing a comprehensive evaluation of the recommendation performance.

We compare Multilayered Perceptron (MLP, Zhou et al. 2022), PALR (Chen et al. 2023), LLM-Rec (Lyu et al. 2023), and FMLP-Rec (Zhou et al. 2022) in the four datasets of sequential item-based recommendations: Movielens, Recipe, Yelp, and Sports. Movielens dataset includes 25 million ratings and one million tag applications applied to 62,000 movies by 162,000 users (https://grouplens.org/datasets/movielens/). Yelp is a dataset for business recommendation.

The baseline model employs a Multilayered Perceptron (MLP). The augmented text embeddings and the original content description embeddings are combined through concatenation, and the resulting vector is fed into a two-layer MLP. The first MLP layer's output dimension, along with the input/output dimensions of the second MLP layer, is set to 128. The first MLP layer incorporates a ReLU activation function and a dropout layer. Following this, the dot product of the latent embeddings of the user and the item is computed, and the resulting value is then processed through a Sigmoid function. This Sigmoid function transforms the dot product into a final relevance score, representing the relationship between the user and the item.

We show three types of performances of our system on a popular dataset (Table 3):

1. Ours baseline with LLM finetuning. Its implementation in health domain is available at Galitsky-github (2024).
2. Baseline extended with symbolic abduction.
3. Baseline extended with meta prompts.
4. Baseline extended with meta prompts and neural abduction.
5. Meta prompt + neuro-symbolic abduction.

The rightmost column shows the averaged improvement over our baseline (grayed row 7).

One can observe that our baseline performance is below all the competitive systems. We were not able to match FMLP-Rec on Sports database in all three of our architectures but improved its performance in Yelp domain. The best performance with a stand-alone component is achieved:

Table 3 Results of personalized recommendation

	Movielens			Recipe			Yelp		Sports		Improvement over baseline, %
	Precision@10	Recall@10	NDCG@10	Precision@10	Recall@10	NDCG@10	HR@5	NDCG@10	HR@5	NDCG@10	
MLP	0.291	0.244	0.363	0.032	0.068	0.058					
LLM-Rec	0.315	0.277	0.395	0.039	0.084	0.070					
FMLP-Rec							0.273	0.502	**0.380**	0.316	
PALR		0.072*	0.044*								
Ours baseline (LLM finetuning)	0.307	0.210	0.320	0.029	0.065	0.043	0.28	0.460	0.355	0.261	0
Ours with symbolic abduction	0.321	**0.254**	0.390	0.041	0.062	0.075	**0.291**	0.489	0.371	0.332	10.5
Ours with meta prompts	0.304	0.231	0.403	0.0361	**0.072**	0.068	0.273	0.465	0.376	0.309	4.0
Ours with learned (neural) abduction	0.312	0.226	0.384	0.046	0.063	0.074	0.254	0.460	0.389	**0.342**	9.1
Ours with integrated neuro-symbolic abduction and	**0.325**	0.239	**0.412**	**0.050**	0.066	**0.078**	0.290	**0.502**	**0.397**	0.336	15.5

- 10.5 using symbolic abduction
- 9.1 neural abduction
- 4.0—using meta prompts, in comparison with our baseline

The hybrid system boosts individual components by 5–10%. In comparison with LLM-Rec, FMLP-Rec, and PALR, our performance of stand-alone components is comparable, but our hybrid system outperforms FMLP-Rec and PALR.

Symbolic abduction is the winner in the Movielens dataset, outside of NDCG measure, and also in the Recipe dataset, outside of Recall@10 measure. We observe that both symbolic and neural abductions and also meta-prompts give a substantial boost to personalization performance. Hence, these are promising directions to improve overall recommendation accuracy as LLMs are being further developed.

8 Deployment of LLM Personalization

8.1 Service-Based Architecture for Web 3.0 Personalization

Figure 20 shows an overview of the framework, including three major components:

1. Interface
2. Orchestrator
3. External sources

The interface serves as a connection point between users and agents, offering interactive tools accessible via mobile, desktop, or Web applications. It incorporates multimodal communication channels, encompassing text and audio. Queries from users are received by the interface and then relayed to the Orchestrator. It is

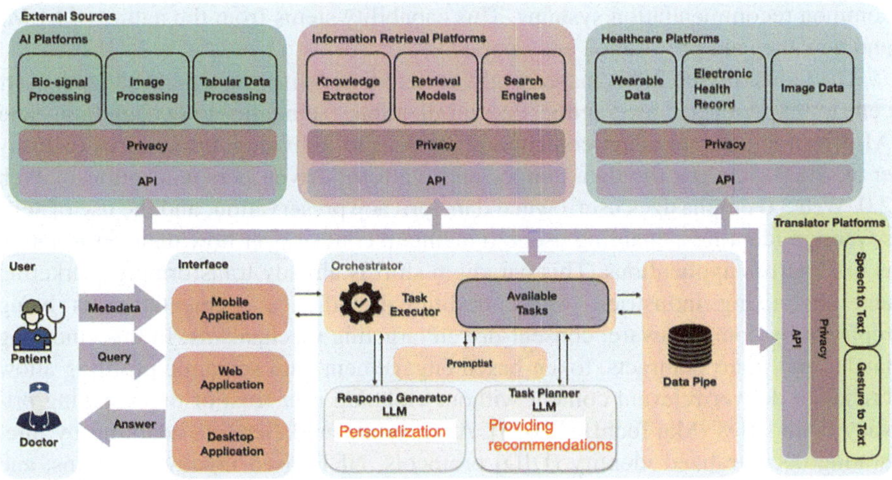

Fig. 20 A service-based personalization architecture

important to note that questions can be presented through various modes of human communication, such as text, speech, or gestures. Within this framework, users have the capability to provide metadata alongside their queries, which may include images, audio, gestures, and more. As an example, a user might capture an image of their meal and inquire about its nutritional values or calorie content, with the image serving as metadata.

The *Orchestrator* assumes the responsibility of problem-solving and decision-making to generate suitable responses based on user queries. It incorporates Neisser's (1976) Perceptual Cycle Model, enabling the Orchestrator to perceive, transform, and analyze input queries and metadata for response generation. In this process, input data is aggregated, transformed into structured data, and analyzed to plan and execute actions. The Orchestrator engages with external sources to obtain necessary information, conduct data integration and analysis, and extract insights, among other functions. Here, we delineate five key components of the Orchestrator.

The Task Executor, functioning within the Orchestrator, executes actions by adhering to the decision-making process established by the Task Planner and invoking modules upon request. The Task Executor has dual responsibilities. Firstly, it serves as a data converter, translating input multimedia data into prompts. When the Task Executor receives user queries, including text, audio, and more, it calls on relevant external platforms to convert them into text.

9 Discussion and Conclusions

He et al. (2023) observe that even without fine-tuning, LLMs can outperform existing fine-tuned conversational recommendation models in terms of relevance. Surprisingly, LLMs can function as stand-alone recommendation systems without requiring additional feature engineering or manual processes typically involved in common recommendation systems. This capability stems from the nature of LLMs and how they are pre-trained and operate.

The focus of Web 3.0 is increasingly centered on delivering a personalized, user-centric experience, leveraging decentralization, semantic understanding, and AI-driven insights to adapt content and services to individual preferences (Murray et al. 2023). Unlike the data-hungry and platform-driven personalization of Web 2.0, Web 3.0 emphasizes user-owned data, privacy preservation, and the use of self-sovereign identities, enabling users to maintain control over how their information is used across applications. This paradigm shift is already transforming marketing and advertising industries, where traditional audience segmentation is being replaced by context-aware, consent-driven targeting mechanisms. In this emerging landscape, smart contracts, token-based engagement, and semantic profiling allow brands to deliver relevant content without violating user trust or overreaching privacy boundaries (MarTechBot 2025). As a result, marketers are adapting by integrating decentralized identity (DID) protocols, NFT-based loyalty programs, and blockchain analytics into their strategies, paving the way for more transparent and personalized digital interactions.

The following are the advantages of LLM-based personalization in comparison with baseline LLM:

1. Engages with user via interactions
2. Resents outputs in a verbalized and explainable way
3. Receives feedback from the user
4. Makes adjustment on top of the feedback from passively making search
5. Has recommendations that help proactively figure out user's need and seek for user's preferred items

Also, LLMs may go beyond simply making personalized search and recommendation:

1. Take notes of users' important information within their memory
2. Make personalized plans based on memorized information when new demands are raised
3. Execute plans by leveraging tools like search engines and recommendation systems

As to the difficulties in LLM personalization, they are as follows:

1. Personalization requires an understanding of user preference, which is a domain-specific knowledge rather than the commonsense knowledge learned by LLMs.
2. Unclear how to effectively and efficiently adapt of LLMs for personalized services.
3. LLMs could memorize user's confidential information while providing personalized services.
4. There is a risk that most of the recommendations are similarity based (i.e., items that are semantically similar to the items purchased so far), though I do think with extensive training data of user purchase history, we can ameliorate this issue via the "collaborative filtering" approach that this method could mimic.
5. Also, since there is no weighting assigned to various actions or events a user undertakes, from exploration to the eventual purchase of an item, our reliance is solely on the LLM's prediction of the most probable next token(s) for recommendations. We are unable to consider factors such as items bookmarked, items observed for a duration, items added to the cart, and similar user interactions.

It is noted that incorporating abduction and meta-prompts significantly enhances personalization performance. Consequently, these avenues hold promise for enhancing overall recommendation accuracy as LLMs continue to undergo further development.

A personalized agent designed to represent users in decision-making processes involving the consideration of mental attributes of both human and automated agents was introduced in (Galitsky 2002). The proposed framework included the Natural Language Multiagent Mental Simulator (NL_MAMS) as a component for reasoning and user profile specification. This simulator had the capability to predict the entire range of possible mental states a few steps ahead based on the current mental state of the involved agents.

In their study, Zhou et al. (2022) discovered that user behavior data logged often contains noisy interactions. Through empirical analysis, they demonstrated the efficacy of filtering algorithms from the digital signal processing community in mitigating the impact of noise on recommendation models based on LLMs. Building upon this insight, the authors introduced the FMLP-Rec model, incorporating learnable filters for enhancing sequential recommendation tasks.

Galitsky (2016) introduced a framework aimed at extracting user interests from social network profiles, such as those on Facebook, in order to tailor recommendations for products and services. Traditional personalization systems, relying on keyword matching of user interests and product attributes, demonstrated low recall. This limitation necessitated a shift toward a more generalized, category-based framework. The challenge lies in the need for substantial reasoning about products and their categories, aligning the taxonomy of a user's expressed public profile with that of products and services. To address inconsistencies between these taxonomies, Galitsky proposed a mapping expressed as a Defeasible Logic program. In this program, a potential mapping can be overridden by others if additional relevant information becomes available. The study introduces a Facebook app and a mobile app with similar functionality, designed for recommending activities. The accuracy and usability of these apps are evaluated. Additionally, a content management system supporting personalized recommendations is outlined as part of the proposed framework.

Galitsky (2021) proposed an innovative approach to conversational recommendation that deviates from traditional methods of questioning users about their preferences. Instead, our recommender system monitors the user's interactions with others, including customer support agents (CSAs), and intervenes in the conversation only when there is a pertinent recommendation to make, and the timing is appropriate. The Recommender Joining Conversation (RJC) model utilizes information extraction, discourse and argumentation analyses, as well as dialogue management techniques to generate recommendations for products and services based on the user's needs inferred from the ongoing conversation. We specifically address scenarios where customers discuss their issues with CSAs, aiming to resolve problems and receive recommendations for products tailored to address these issues. Evaluations conducted in various human-human and human-chatbot dialogues demonstrate that RJC is an effective and unintrusive method, providing highly relevant and persuasive recommendations.

Recent research has demonstrated the successful application of personalized word embeddings in addressing ambiguity in various downstream tasks (Welch et al. 2022). Garimella et al. (2017) explored the impact of location and gender on associations with words such as "health" and other stimuli like "stack"—Does it evoke thoughts of books or pancakes? Welch et al. (2020) delved into different associations, such as the interpretation of "embodying" an idea, which may vary depending on religious or economic beliefs. Similarly, the word "wicked" may denote "evil" or serve as an intensifier, contingent on geographical location (Bamman et al.

2014). These instances illustrate how personalized representations can enhance distinctions in meaning. However, static representations have their limitations. A diversity of meaning for various combinations of horses, men, and heads is shown in Fig. 21.

We concur with Wiesner and Pfeifer (2014) in asserting that collaborative filtering is unsuitable for an HRS. Collaborative filtering involves examining user profiles across users, posing a significant security risk, especially considering the stringent confidentiality requirements of EHR Record systems. While it might be argued that the similarity measuring across user profiles is conducted by the recommendation engine (a machine) rather than humans, assuring users that their profiles remain private proves challenging. In practice, explaining this concept to an average user, concerned about data confidentiality, becomes complex. Additionally, any technical implementation must be meticulously designed to prevent security vulnerabilities. Processing data from different users in the same computational units simultaneously (i.e., server-side sessions) could potentially expose security holes, making it easier for hackers to exploit and access another user's profile data.

We now demonstrate that LLMs are good at resolving ambiguities; this property is naturally leveraged by recommenders. Let us now look at comparison of "testing" and "sanitation." Observations reveal that in certain contexts, "testing" may refer to assessing the functionality of a device, and "sanitation" may relate to a pest control issue. In another context, both terms may signify conditions related to the COVID-19

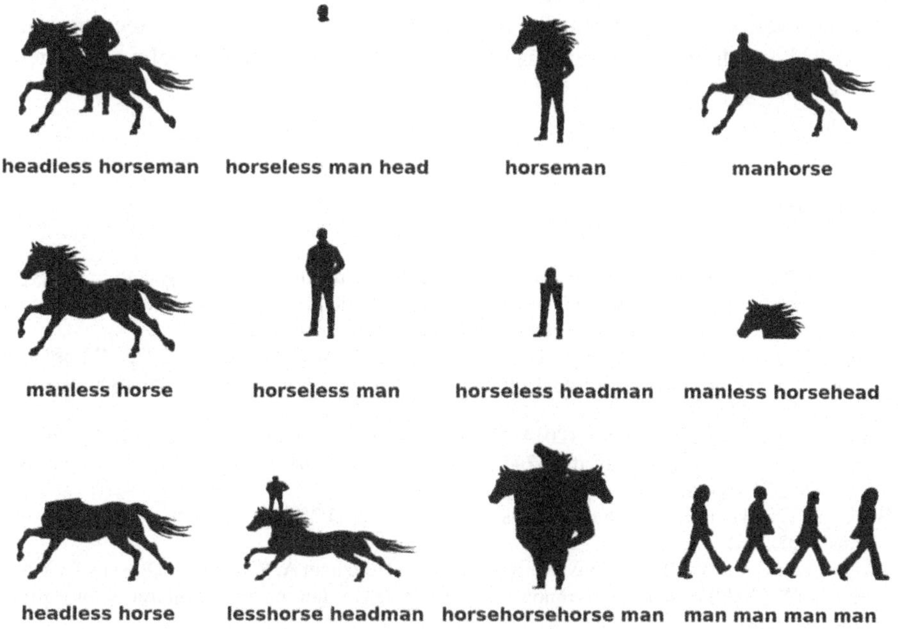

Fig. 21 Visualization of superpositions of horse-man-head

pandemic. Personalized LLMs, designed to better predict individual expressions, could effectively address these nuances, as they learn dynamic encodings of words.

The evaluation results presented in this study highlight both the challenges and opportunities in leveraging advanced reasoning techniques—such as symbolic abduction, neural abduction, and meta-prompting—to enhance recommendation systems. While our baseline system underperformed compared to existing competitive models like FMLP-Rec, LLM-Rec, and PALR, the integration of these novel components significantly improved performance across multiple domains.

In particular, our architectures demonstrated domain-specific strengths: while we did not match FMLP-Rec on the Sports dataset, we surpassed it in the Yelp domain. The stand-alone implementations of symbolic abduction, neural abduction, and meta-prompts achieved notable improvements of 10.5%, 9.1%, and 4.0%, respectively, over the baseline. More importantly, the hybrid approach combining these components yielded an additional 5–10% performance boost, outperforming FMLP-Rec and PALR overall.

Symbolic abduction emerged as the top performer on the Movielens and Recipe datasets in most evaluation metrics, demonstrating its potential for structured and knowledge-rich domains. Neural abduction showed flexibility and adaptability, particularly in less structured environments, while meta-prompts provided a lightweight but effective means of personalization enhancement.

These findings suggest that incorporating reasoning-based components (chapter "Adversarial Integration of LLM and Logic Program") into traditional recommendation frameworks can lead to meaningful gains in accuracy and personalization. As large language models continue to evolve, integrating such strategies will likely become even more impactful, paving the way for next-generation recommender systems that combine the strengths of machine learning with symbolic and generative reasoning.

References

Aboufoul M (2023) Using large language models as recommendation systems. A review of recent research and a custom implementation. https://towardsdatascience.com/using-large-language-models-as-recommendationsystems-49e8aeeff29b

Balepur N, Padmakumar V, Yang F, Feng S, Rudinger R, Boyd-Graber JL (2025) Whose boat does it float? Improving personalization in preference tuning via inferred user personas. arXiv:2501.11549

Bamman D, Dyer C, Smith NA (2014) Distributed representations of geographically situated language. In: Proceedings of the 52nd annual meeting of the Association for Computational Linguistics, Volume 2: Short papers

Benet J (2014) IPFS—content addressed, versioned, P2P file system. https://arxiv.org/abs/1407.3561

Bhagavatula C, Le Bras R, Malaviya C, Sakaguchi K, Holtzman A, Rashkin H, Downey D, Yih W t, Choi Y (2020) Abductive commonsense reasoning. In: International conference on learning representations

Buterin V (2015) Ethereum White Paper: a next-generation smart contract and decentralized application platform. https://ethereum.org/en/whitepaper/

Chan X, Wang X, Yu D, Mi H, Yu D (2024) Scaling Synthetic Data Creation with 1,000,000,000 Personas. arXiv preprint arXiv:2406.20094.

Chen D, Xie LJ, Kim B, Wang L, Hong CS, Wang L-C, Han Z (2020) Federated learning based mobile edge computing for augmented reality applications. In: 2020 International conference on computing, networking and communications (ICNC), pp 767–773

Chen J, Liu Z, Huang X, Wu C, Liu Q, Jiang G, Pu Y, Lei Y, Chen X, Wang X, Lian D, Chen E (2023) When large language models meet personalization: perspectives of challenges and opportunities. arXiv:abs/2307.16376

Dai W, Xu Q, Yu Y, Zhou Z (2019) Bridging machine learning and logical reasoning by abductive learning. Neural Information Processing Systems

Deb B, Awadallah AH, Zheng G q (2022) Boosting natural language generation from instructions with meta-learning. In: Proceedings of the 2022 conference on empirical meth ods in natural language processing. Association for Computational Linguistics, Abu Dhabi, United Arab Emirates, pp 6792–6808

Denecker M, Kakas AC (2002) Abduction in logic programming. In: Kakas AC, Sadri F (eds) Computational logic: logic programming and beyond: essays in honour of Robert A. Kowalski, Lecture notes in computer science, vol 2407. Springer, pp 402–437

Di Palma D, Biancofiore GM, Anelli VW, Narducci F, Di Noia T, Di Sciascio E (2023) Evaluating ChatGPT as a recommender system: a rigorous approach. https://arxiv.org/abs/2309.03613

Ficler J, Goldberg Y (2017) Controlling linguistic style aspects in neural language generation. arXiv preprint arXiv:1707.02633

Flek L (2020) Returning the N to NLP: towards contextually personalized classification models. In: Proceedings of the 58th annual meeting of the Association for Computational Linguistics, pp 7828–7838

Friedman L, Ahuja S, Allen D, Tan T, Sidahmed H, Long C, Xie J, Schubiner G, Patel A, Lara H et al (2023) Leveraging large language models in conversational recommender systems. arXiv preprint arXiv:2305.07961

Galitsky B (2002) Designing the personalized agent for the virtual mental world. In: AAAI FSS-2002 symposium on personalized agents, pp 21–29

Galitsky B (2016) Providing personalized recommendation for attending events based on individual interest profiles. Artif Intell Res 5(1):1–13

Galitsky B (2021) Recommendation by joining a human conversation. In: Artificial intelligence for customer relationship management: solving customer problems. Springer, Cham

Galitsky B (2025) LLM-based personalized recommendations in health. In: Heath applications of neuro-symbolic AI. Elsevier

Galitsky B, Kovalerchuk B (2014) Improving web search relevance with learning structure of domain concepts. In: Clusters, orders, and trees: methods and applications: in honor of Boris Mirkin's 70th birthday. Springer, New York, pp 341–376

Galitsky-github (2024) Treatment recommendation personalization system. https://github.com/bgalitsky/LLM-personalization/blob/main/Treatment_Recommendation_Personalization_System.ipynb

Gao T, Fisch A, Chen D (2021) Making pre-trained language models better few-shot learners. In: Proc. of ACL-IJCNLP, pp 3816–3830

Gao Y, Sheng T, Xiang Y, Xiong Y, Wang H, Zhang J (2023) Chat-rec: towards interactive and explainable LLMs-augmented recommender system. arXiv preprint arXiv:2303.14524

Garimella A, Banea C, Mihalcea R (2017) Demographic-aware word associations. In: Proceedings of the 2017 conference on empirical methods in natural language processing

Geng S, Liu S, Fu Z, Ge Y, Zhang Y (2022) Recommendation as language processing (RLP): a unified pretrain, personalized prompt & predict paradigm (P5). In: Proceedings of the 16th ACM conference on recommender systems (RecSys '22). Association for Computing Machinery, New York, pp 299–315. https://doi.org/10.1145/3523227.3546767

Greyling C (2023) The Evolution Of Prompt Engineering. https://cobusgreyling.substack.com/p/the-evolution-of-prompt-engineering. Last downloaded March 3 2026

Goodman N (2023) Meta-prompt: a simple self-improving language agent. https://noahgoodman.substack.com/p/meta-prompt-a-simple-self-improving

He Z, Xie Z, Jha R, Steck H, Liang D, Feng Y, Majumder BP, Kallus N, McAuley J (2023) Large language models as zero-shot conversational recommenders. In: Proceedings of the 32nd ACM international conference on information and knowledge management (CIKM '23), October 21–25, 2023, Birmingham, United Kingdom

Hou Y, Dong H, Wang X, Li B, Che W (2022) MetaPrompting: learning to learn better prompts. In: Proceedings of the 29th international conference on computational linguistics, pp 3251–3262

Janthong N, Brissaud D, Butdee S (2010) Combining axiomatic design and case-based reasoning in an innovative design methodology of mechatronics products. CIRP J Manuf Sci Technol 2(4):226–239

Kirk HR, Vidgen B, Röttger P, Hale SA (2023) Personalisation within bounds: a risk taxonomy and policy framework for the alignment of large language models with personalised feedback. arXiv preprint arXiv:2303.05453

Li XL, Liang P (2021) Prefix-tuning: optimizing continuous prompts for generation. In: Proc. of ACL-IJCNLP, pp 4582–4597

Lu SCY, Liu A (2012) Abductive reasoning for design synthesis. CIRP Ann 61(1):143–146

Lyu H, Jiang S, Zeng H, Wang Q, Zhang S, Chen R, Leung C, Tang J, Xia Y, Luo J (2023) LLM-Rec: personalized recommendation via prompting large language models. arXiv2307.15780

MarTechBot (2025) Web3 technologies focus on decentralization and user control over digital assets. Are they of any use to marketers? https://martechorg/what-are-the-marketing-opportunities-in-web3-technologies/. Accessed 25 May 2025

Michael J, Mahdi S, Rein D, Petty J, Dirani J, Padmakumar V, Bowman SR (2023) Debate helps supervise unreliable experts. arXiv preprint arXiv:2311.08702

Mistry U, Thakker R, Gupta M, Obaidat S, Tanwar S, Kumar N, Rodrigues JJPC (2021) Medblock: an AI enabled and blockchain-driven medical healthcare system for covid-19. In: ICC 2021—IEEE international conference on communications, pp 1–6

Murray A, Kim D, Combs J (2023) The promise of a decentralized internet: what is Web3 and how can firms prepare? Bus Horiz 66(2):191–202

Narayanan A, Bonneau J, Felten E, Miller A, Goldfeder S (2016) Bitcoin and cryptocurrency technologies: a comprehensive introduction. Princeton University Press

Neisser U (1976) Cognition and reality. W. H. Freeman and Company, San Francisco, CA

Peirce CS (1974) Collected papers of Charles Sanders Peirce, vol 5. Harvard University Press

Rafailov R, Sharma A, Mitchell E, Ermon S, Manning CD, Finn C (2023) Direct preference optimization: your language model is secretly a reward model. In: Proceedings of the 37th International Conference on Neural Information Processing Systems (NIPS '23). Curran Associates Inc., Red Hook, NY, USA, Article 2338, 53728–53741

Shen M, Tan Z, Niyato D, Liu Y, Kang J, Xiong Z, Zhu L, Wang W, Shen X (2023) Artificial intelligence for Web 3.0: a comprehensive survey. arXiv:2309.09972

Syed B et al (2020) Adapting language models for non-parallel author-stylized rewriting. Proc AAAI Conf Artif Intell 34(05):9008

Talmor A, Herzig J, Lourie N, Berant J (2019) CommonsenseQA: a question answering challenge targeting commonsense knowledge. In: Proceedings of the NAACL-HLT 2019, Minneapolis, MN, USA, pp. 4149–4158

Tian Y, Liu A, Dai Y, Nagato K, Nakao M (2024) Systematic synthesis of design prompts for large language models in conceptual design. CIRP Ann 73(1):85–88

Tseng YM, Huang Y-C, Hsiao T-Y, Chen WL, Huang C-W, Meng Y, Chen YN (2024) Two tales of persona in LLMs: a survey of role-playing and personalization. In: Findings of the Association for Computational Linguistics: EMNLP2024. Association for Computational Linguistics, Miami, FL, pp 16612–16631

Wang A, Singh A, Michael J, Hill F, Levy O, Bowman S (2018) GLUE: a multi-task benchmark and analysis platform for natural language understanding. In: Proceedings of the 2018 EMNLP workshop BlackboxNLP: analyzing and interpreting neural networks for NLP. Association for Computational Linguistics, Brussels, Belgium, pp 353–355

Wang Y, Chu Z, Ouyang X, Wang S, Hao H, Shen Y, Gu J, Xue S, Zhang J, Cui Q, Li L, Zhou J, Li S (2024) LLMRG: improving recommendations through large language model reasoning graphs. Proc AAAI Conf Artif Intell 38(17):19189–19196. https://doi.org/10.1609/aaai.v38i17.29887

Wei Y, Wang X, Nie L, He X, Hong R, Chua T-S (2019) MMGCN: multi-modal graph convolution network for personalized recommendation of micro-video. In: ACM MM'19, NICE, France, Oct. 21–25. https://github.com/weiyinwei/MMGCN

Wei J, Wang X, Schuurmans D, Bosma M, Xia F, Chi E, Le QV, Zhou D et al (2022) Chain-of thought prompting elicits reasoning in large language models. Adv Neural Inf Process Syst 35:24824–24837

Weidinger L, Uesato J, Rauh M, Griffin C, Huang P-S, Mellor J, Glaese A, Cheng M, Balle B, Kasirzadeh A et al (2022) Taxonomy of risks posed by language models. In: Proceedings of the 2022 ACM conference on fairness, accountability, and transparency, pp 214–229

Welch C, Kummerfeld JK, Perez-Rosas V, Mihalcea R (2020) Compositional demographic word embeddings. In: Proceedings of the 2020 conference on empirical methods in natural language processing

Welch C, Gu C, Kummerfeld J, Pérez-Rosas V, Mihalcea R (2022) Leveraging similar users for personalized language modeling with limited data. ACL, pp 1742–1752

White J, Fu Q, Hays S, Sandborn M, Olea C, Gilbert H, Elnashar A, Spencer-Smith J, Schmidt DC (2023) A prompt pattern catalog to enhance prompt engineering with ChatGPT. Preprint arXiv:2302.11382. https://arxiv.org/abs/2302.11382

Wiesner M, Pfeifer D (2014) Health recommender systems: concepts, requirements, technical basics and challenges. Int J Environ Res Public Health 11(3):2580–2607

Xu X, Weber I, Staples M, Zhu L, Bosch J, Bass L, and Pautasso C (2016) A Taxonomy of Blockchain-Based Systems for Architecture and Design. IEEE International Conference on Software Architecture (ICSA).

Xue G-R, Han J, Yu Y, Yang Q (2009) User language model for collaborative personalized search. ACM Trans Inf Syst. 27. https://doi.org/10.1145/1462198.1462203

Yang D, Flek L (2021) Towards user-centric text-to-text generation: a survey. In: International conference on text, speech, and dialogue. Springer, Cham

Zhang F, Cecchetti E, Croman K, Juels A, Shi E (2016) Town crier: an authenticated data feed for smart contracts. In: Proceedings of the 2016 ACM SIGSAC conference on computer and communications security (CCS '16)

Zhao W, Chiu J, Hwang J, Brahman F, Hessel J, Choudhury S, Choi Y, Li X, Suhr A (2024) UNcommonsense reasoning: abductive reasoning about uncommon situations. In: Proceedings of the 2024 conference of the North American chapter of the Association for Computational Linguistics: human language technologies, volume 1: Long papers. Association for Computational Linguistics, Mexico City, Mexico, pp 8487–8505

Zhou K, Yu H, Zhao WX, Wen J-R (2022) Filter-enhanced MLP is all you need for sequential recommendation. In: Proceedings of the ACM web conference 2022 (WWW'22). Association for Computing Machinery, New York, pp 2388–2399

Zhuang Y, Liu Z, Qian P, Liu Q, Wang X, He Q (2021) Smart contract vulnerability detection using graph neural networks. In: IJCAI'20

Zyskind G, Nathan O, Pentland A (2015) Decentralizing privacy: using Blockchain to protect personal data. In: 2015 IEEE security and privacy workshops (SPW)

Creating Ad Sequences with Optimal Discourse Structure by LLM

1 Introduction

With the rapid advancements in artificial intelligence (AI) and large language models (LLMs) in recent years, their applications have expanded across various fields, significantly transforming industries such as advertising, media, e-commerce, education, and more (Gao and Huang 2021). The rise of AI has established a technical foundation for intelligent operations in the advertising industry (Lai 2021).

By analyzing consumer behavior, LLMs offer valuable insights that help advertisers craft more effective strategies, enhancing the efficiency of ad information processing and decision-making (Malthouse and Copulsky 2023). This represents a major breakthrough in the field. In the area of ad targeting, various machine learning techniques are employed to optimize online advertising, particularly by refining the scope of target audiences and improving user segmentation (Chandra et al. 2022; Dwivedi et al. 2021).

For advertisers, the primary appeal of online ads lies in their ability to strategically target users based on factors like demographics and online behavior. Targeted ads are widely believed to be more effective than non-targeted ones. Research supports this view, showing that users are more likely to engage with and click on targeted ads (Tucker 2014). As a result, ad targeting has become a key focus for many researchers (Rafieian and Yoganarasimhan 2021).

LLMs also assist creative teams by generating diverse and engaging advertising content through the analysis of large datasets (Wiredu 2023). For example, Toyota developed an "intuition-driven" ad scenario where content can be optimized in real-time based on location, timing, and customer profiles (Huang and Rust 2021). Using LLMs in advertising campaigns and content creation can elevate creative quality and increase ad effectiveness. Moreover, integrating LLMs with reinforcement learning (RL) allows for the optimization of ads to align more closely with users' actual needs, thereby boosting ad effectiveness and purchase conversion rates

B. Galitsky, *Applications of Neuro-Symbolic Artificial Intelligence*,
https://doi.org/10.1007/978-3-032-16524-4_8

(Muhlhoff and Willem 2023). For instance, eBay leverages LLMs to develop descriptive and predictive models, providing users with precise ad content tailored to their preferences, such as price points and other criteria (Kumar et al. 2019).

1.1 Ad Sequencing

Ad sequencing is a storytelling strategy in digital advertising that allows advertisers to deliver ads in a specific order to their target audience, based on previous interactions with the brand. This technique leverages dynamic creative optimization to fine-tune message delivery, ensuring that each interaction with the audience is personalized. In this chapter, we will explore the concept of ad sequencing. Ad sequencing enables businesses to deliver targeted and personalized ads in a deliberate sequence to potential customers, guided by their prior engagement with the brand. The goal is to create a cohesive and relevant experience for customers by presenting tailored messages at each stage of their interaction journey. Ad sequencing is crucial because consumers are constantly exposed to multiple ads from various brands across different channels, making it harder for marketers to capture their attention. By using ad sequencing, businesses can effectively communicate their stories by adapting their message delivery to the customer's engagement level with the brand.

The process of ad sequencing involves analyzing customer data and behavior to anticipate their interests, preferences, and intent. It breaks down long-form content into smaller, bite-sized pieces delivered sequentially across different channels, tailored to each stage of the user's journey. This approach helps increase conversion rates by providing personalized messaging aligned with the customer's position in the sales funnel. Key elements for successful ad sequencing are shown in Fig. 1.

Existing approach, dynamic creative optimization (DCO) is an ML technique that selects the best combination of creative elements for ad delivery based on user data. In ad sequencing, DCO is vital for crafting personalized messages using various ad formats, visuals, and copy, all tailored to user behavior.

Optimizing message sequencing is essential in ad sequencing to ensure that messages are delivered in a specific order, creating a seamless experience for the customer. To optimize message sequencing, advertisers should:

1. Develop a clear narrative that aligns with the brand's overall message
2. Personalize messaging based on user behavior
3. Start with a compelling hook to grab attention immediately
4. Use high-quality visuals and copy to tell the story effectively
5. Continuously test and optimize for improvements

The benefits of ad sequencing are shown in Fig. 2.

Ad sequencing is a powerful strategy for delivering personalized messages that resonate with the target audience while also telling the brand's story in a compelling and engaging way (Sugarman 2009). By following best practices, you can create an effective ad sequencing campaign that drives results.

Fig. 1 Components of ad sequencing

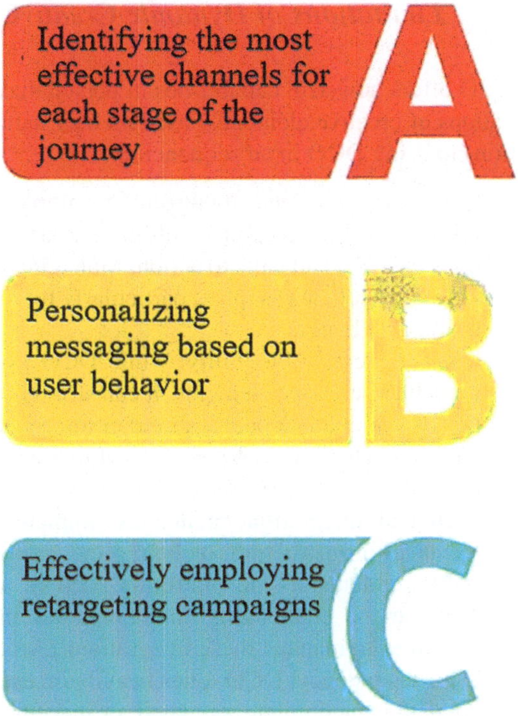

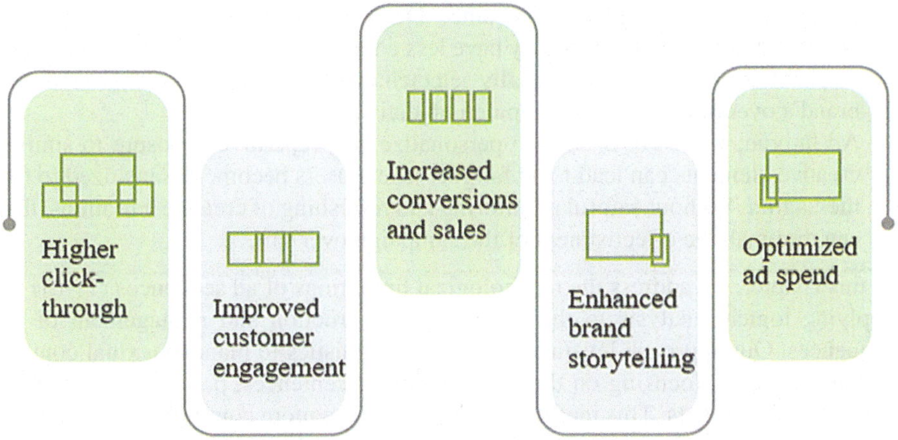

Fig. 2 Benefits of ad sequencing

1.2 Limitations of Dynamic Creative Optimization

DCO offers powerful capabilities in personalizing ads by selecting the best combinations of creative elements based on user data. However, there are several limitations to using DCO in ad sequencing:

1. Complex setup and management: implementing DCO requires significant upfront work in creating multiple variations of ad components, such as headlines, visuals, and calls to action. Managing and optimizing these variations over time can become complex, particularly for campaigns with multiple touchpoints or large audiences.
2. Limited creative flexibility: while DCO excels at optimizing combinations of predefined creative elements, it can struggle with maintaining the broader narrative or emotional appeal required in some storytelling campaigns. The automated nature of DCO may prioritize performance metrics over creativity, potentially leading to generic or fragmented messaging.
3. Technical integration challenges: implementing DCO requires sophisticated technical infrastructure, including integration with data management platforms (DMPs), demand-side platforms (DSPs), and creative management systems. This can be challenging for companies that lack the technical expertise or resources to manage these integrations (Affstaff 2024).
4. Data dependency: DCO relies heavily on user data to optimize ads effectively. If the data is incomplete, outdated, or inaccurate, the optimization may be flawed, leading to suboptimal ad performance.
5. Limited control for advertisers: since DCO automates much of the creative decision-making, marketers may have less control over the final output. This can be problematic if the automatically selected combinations do not align with the brand's overall message or campaign objectives.
6. Ad fatigue: while DCO aims to personalize ads, repetitive exposure to similar creative elements can lead to ad fatigue, where users become desensitized to the messaging. Without careful monitoring and refreshing of creative variations, this can diminish the effectiveness of the campaign over time.

In this chapter, we address the technological limitations of ad sequences (1)–(3) by applying logical analysis to the automated construction and management of ad sequences. Our approach leverages discourse linguistics to manage textual content at a higher level, focusing on the organization of sentences, paragraphs, sections, and entire documents. This method contrasts with the more commonly used semantic and syntactic analyses in marketing. By integrating discourse analysis, we aim to elevate the effectiveness of ad sequencing beyond the conventional methods.

The logical analysis of ad sequences will be implemented through an LLM, which will not only facilitate discourse analysis but also support other stages of ad sequence delivery. This allows for a more structured and coherent narrative flow throughout the ad campaign, enhancing the overall user experience.

Discourse linguistics examines the general organization of text and the thought structure of its authors (Blakemore 1988). This makes it particularly promising for the logical simulation of ad sequences, as it enables the system to understand and simulate how ideas and messages are structured and delivered across different stages of a campaign. By focusing on the macro-structure of content, discourse analysis helps ensure that the sequence of ads maintains a consistent and logical flow, engaging the audience at a deeper level.

This approach offers a significant advantage over traditional methods, which often focus on optimizing individual elements (e.g., keywords or phrases) without considering the broader narrative context. By analyzing the discourse, we can create ad sequences that are not only personalized but also coherent and engaging, leading to more effective storytelling and better audience retention. Ultimately, this methodology aims to bridge the gap between technological automation and the need for sophisticated, narrative-driven ad campaigns.

1.3 Main Components of Ad Management System

The relationship among the four key elements of AI-driven advertising—Targeting, Personalization, Content Creation, and Ad Optimization—is deeply interconnected, with each component reinforcing and depending on the others. Grounded in computational advertising, these elements form a cohesive framework that maximizes the effectiveness and efficiency of advertising campaigns (Fig. 3).

Targeting and *personalization* (Sect. 2) are two closely related aspects of AI advertising, each addressing a different part of the advertising process. Targeting focuses on identifying the audience most likely to engage with an advertisement by analyzing demographic data, behavioral patterns, preferences, and other relevant factors. Essentially, it answers the question, "Who should see the ad?" Personalization, on the other hand, takes the insights gained from targeting and tailors the content of the ad to match the preferences and behaviors of individual users or user segments. It answers the question, "What type of ad should they see?" By delivering content that resonates on a personal level, personalization enhances the likelihood of user engagement. Together, targeting and personalization ensure that the right users see the most relevant ads, optimizing the chances for a positive response.

Content creation plays a vital role in executing effective personalization strategies. Leveraging advancements in LLMs and other AI tools, content creation involves generating advertising content that aligns with the targeted users' preferences and needs. This content can range from text to multimedia formats such as images, audio, and video. The creative appeal and relevance of the content are critical factors in the success of personalized advertising. In this relationship, content creation supports personalization by providing the necessary materials that make the ads engaging and effective. High-quality, personalized content enhances the user experience and increases the likelihood of a successful advertising campaign.

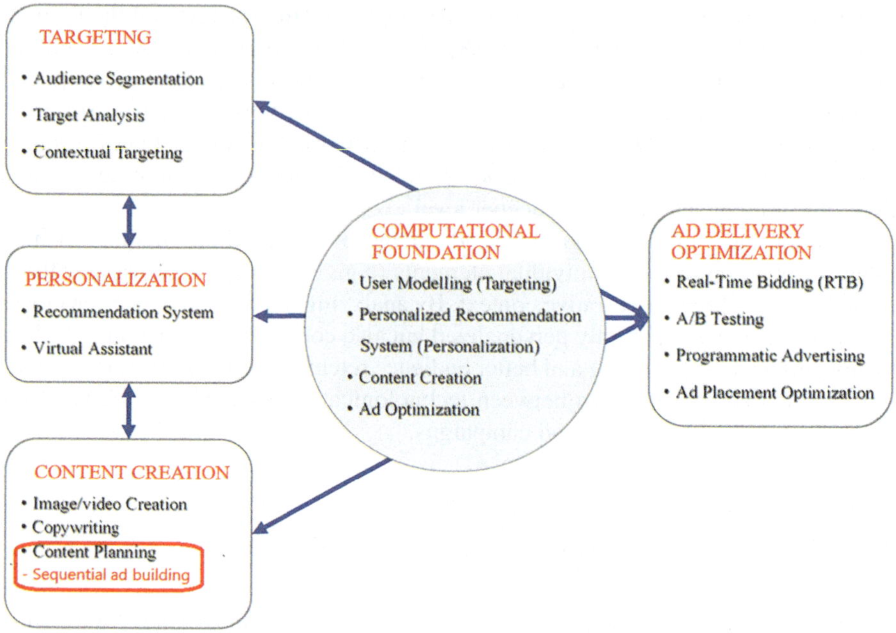

Fig. 3 Interaction between the four main components of ad management system. The focus of this chapter is depicted in the bottom-right corner

Ad optimization is the final piece of the puzzle, integrating the efforts of targeting, personalization, and content creation into a cohesive strategy that maximizes advertising effectiveness and return on investment (ROI). Ad optimization involves data-driven strategies that adjust ad delivery parameters, such as frequency, timing, and placement, to ensure the ads reach the right audience at the right time in the right context. Drawing on the insights from targeting, ad optimization ensures that the ads are served to the most responsive user groups. Personalization and content creation provide the creative elements that drive engagement, and ad optimization fine-tunes the delivery to maximize impact. This process often involves continuous testing and refinement, using performance data to adjust strategies dynamically. By constantly analyzing the effectiveness of different ad variations and placements, ad optimization enhances the overall success of advertising efforts.

Together, these four elements—*targeting*, *personalization*, *content creation*, and *ad optimization*—create a feedback loop that enhances the overall advertising process. Targeting identifies the audience, personalization ensures that the content resonates with them, content creation provides the materials that make this connection possible, and ad optimization fine-tunes the entire process to maximize results. Each component relies on data generated by the others, creating a cycle of continuous improvement that drives more effective and efficient advertising strategies.

This integrated approach enables advertisers to move beyond traditional methods, leveraging AI to deliver highly personalized and optimized campaigns that can adapt in real time, ultimately leading to more successful and impactful advertising outcomes—possible return on investment in ads.

1.4 Contribution

We present **LLM and Discourse Ad Generation and Targeting (LDAGT)**, a novel framework that leverages the generative capabilities of large language models (LLMs) to create high-quality advertising content while orchestrating ad delivery through discourse-informed narrative structures. Unlike conventional advertising systems, which often treat content creation, targeting, and sequencing as separate processes, LDAGT unifies these tasks into a single adaptive pipeline.

At the core of LDAGT is the integration of **discourse tree modeling** with **keyword-driven targeting**, enabling the system to construct ad sequences that follow a coherent rhetorical progression. By mapping ad components to nodes in a discourse tree, the framework ensures that each stage of the campaign—whether introducing brand awareness, elaborating on product benefits, or delivering a persuasive call to action—flows logically and persuasively into the next. This structural consistency enhances narrative impact and reinforces brand messaging across multiple touchpoints.

The targeting process is further optimized through **adaptive LLM prompt engineering** and **reinforcement learning–based feedback loops**, which dynamically refine both content and keyword selection in response to user engagement data. These feedback mechanisms enable LDAGT to continuously adapt to audience behavior, ensuring high relevance and improved click-through performance over time. By aligning keyword targeting, ad sequencing, and narrative structure within a single system, LDAGT achieves a strong correspondence between generated ad text and its intended search or display context.

Experimental evaluations demonstrate that LDAGT delivers substantial improvements in key advertising metrics, including engagement rates, conversion rates, and return on ad spend. The results indicate that discourse-informed generation, when coupled with data-driven targeting, provides a scalable and effective strategy for producing personalized, persuasive, and context-aware ad campaigns.

2 Targeting Dimensions

Intuitively, targeting should increase an advertiser's profit by allowing the advertiser to stop wasting money on users who are not interested in the advertiser's products (Skiera 2022). However, it is difficult to ascertain the impact of tar getting on an advertiser's profit because targeting affects profit in three ways:

1. Targeting often comes with extra data costs for the advertiser (or an increased price of the ad impression), which negatively affects the advertiser's profit.
2. Targeting reduces the number of reachable users, which may decrease the total number of conversions and, in turn, the advertiser's profit.
3. Despite the above, targeting should improve the performance of advertising campaigns via an increase in at least one of the following metrics: the probability of a click, the probability of a conversion, and the (long term) margin per conversion (Beales 2021; Ahmadi et al. 2024).

2.1 IP-Based Targeting

Ad targeting based on a user's IP address is a method commonly used in digital marketing to deliver relevant ads based on the user's geographic location or network-related information. Here is how IP-based ad targeting typically works.

IP address to location mapping: IP addresses can be mapped to approximate geographic locations (country, region, city, etc.) using IP geolocation databases. These databases are updated regularly to reflect the latest allocations and reassignments of IP addresses. Advertisers can deliver ads to users in specific geographic regions based on their IP address. For example, businesses can target local ads to people in a particular city or region. Applications include local business ads, region-specific offers and language-specific content.

By identifying the Internet Service Provider (ISP) associated with the IP address, advertisers can target users based on their network provider. This might be useful for marketing specific Internet services or devices. Some IP addresses can indicate that a user is part of a corporate network. Advertisers can use this information to target ads specific to certain industries or companies.

Advertisers can whitelist or blacklist certain IP ranges. For example, they can exclude IP addresses associated with non-targeted regions or specific competitors from seeing their ads. Also, if a company is hosting an event or conference, they can target ads to specific IP ranges that belong to the venue's network to ensure that event attendees see their ads.

Limitations of IP-based targeting are as follows:

1. Accuracy: IP-based geolocation isn't always precise, especially with dynamic IP addresses, VPNs, or mobile networks. Users may appear to be in a different location than they actually are.
2. Broad targeting: Since IP-based targeting is often limited to geographic or network information, it doesn't offer the same level of precision as other forms of targeting, such as behavioral or demographic targeting.

Tools and platforms include Google Ads or Facebook Ads. Major ad platforms allow for geo-targeting based on IP address. Advertisers can set up campaigns that focus on users from specific locations (O'Brien 2018). IP Geolocation services tools like MaxMind, IP2Location, or GeoIP services from providers like Google Cloud or AWS can provide accurate IP-to-location mapping.

By leveraging IP addresses, advertisers can reach users based on where they are or what network they are using, making their ad campaigns more relevant and effective. However, it's important to combine IP-based targeting with other methods for a more holistic and accurate approach.

2.2 Enhanced LLM-Based Reasoning for Behavioral Targeting

Large language models (LLMs) can significantly enhance behavioral targeting by improving the reasoning behind ad placements, user segmentation, and personalized messaging. By integrating LLMs into behavioral targeting systems, advertisers can better understand user behavior, preferences, and intent, leading to more effective and personalized advertising strategies. Below are several use cases where LLMs can support better reasoning for behavioral targeting:

1. Contextual understanding of user intent analyzes search queries, social media posts, and content consumption to infer deeper user intent. An LLM can analyze the nuances of a user's recent search queries and social media activity to determine if they are casually browsing for information or if they have a strong purchase intent. For example, if a user searches for "best electric cars" and engages with content about sustainability, the LLM can infer an environmentally conscious purchase intent and suggest targeted ads for eco-friendly products. LLMs excel at understanding context and can distinguish between different levels of intent, improving the precision of ad targeting.
2. Crafting personalized ad copy based on a user's online behavior and preferences. By analyzing a user's browsing history and social media interactions, an LLM can generate personalized ad copy that resonates with the user's preferences. If a user frequently engages with content about healthy living, the LLM can tailor an ad to emphasize the health benefits of a product, making the messaging more compelling.
3. Segmenting audiences based on psychological traits, values, attitudes, and lifestyles inferred from text analysis. An LLM can analyze user-generated content such as reviews, social media posts, and blogs to categorize users into psychographic segments (e.g., trendsetters, bargain hunters, health-conscious individuals). Advertisers can then target each segment with ads that align with their values and preferences. LLMs can detect subtle patterns in language that indicate underlying psychological traits, enabling more refined and meaningful audience segmentation.
4. Continuously adapting ad strategies based on real-time analysis of changing user behavior. An LLM can monitor a user's interaction with ads, Web sites, and content over time to detect shifts in behavior or preferences. For instance, if a user's

browsing patterns suggest a sudden interest in a new hobby, the LLM can recommend shifting ad campaigns to target that new interest. LLMs can reason about changes in user behavior and adjust targeting strategies dynamically, leading to more relevant ad placements.

5. Detecting user sentiment and emotions from text interactions to adjust ad content accordingly. An LLM can analyze a user's tone in emails, chat messages, or social media posts to determine their emotional state. If the user expresses frustration or dissatisfaction, the system can avoid pushing aggressive sales ads and instead offer supportive or problem-solving content. LLMs provide deeper emotional and sentiment analysis, allowing for more empathetic and context-aware ad targeting.
6. Mapping the user's journey across different touchpoints to provide cohesive and logical ad sequencing. An LLM can analyze the stages of a user's journey (e.g., awareness, consideration, decision) by reviewing their interactions with various content and ads. The LLM can then recommend ad content that fits the user's current stage in their journey. For example, if a user is in the consideration phase for purchasing a car, the LLM can suggest comparison ads instead of awareness-building ads. LLMs can reason about the user's position in their decision-making process, ensuring ads are delivered at the right time and with the right message.
7. Understanding user behavior across different platforms (e.g., social media, e-commerce sites, email) to create a unified ad strategy. An LLM can integrate data from multiple sources, such as a user's social media activity, purchase history, and email interactions, to create a holistic profile. This allows for consistent and contextually relevant ads across channels, ensuring that users receive a cohesive brand experience. LLMs can synthesize information from various sources, providing a more complete picture of user behavior and enabling better-targeted cross-channel campaigns.
8. Predicting future behavior and preferences based on past interactions. An LLM can analyze patterns in a user's past behavior to predict future interests. For instance, if a user frequently engages with tech-related content and makes gadget purchases, the LLM can predict that they are likely to be interested in upcoming tech releases and target them with pre-launch ads. LLMs can enhance the predictive capabilities of behavioral targeting by identifying subtle trends and patterns in user behavior, leading to proactive and anticipatory ad strategies.
9. Providing recommendations based on the context of user behavior, such as location, time, or activity. An LLM can consider contextual factors, such as a user's current location or the time of day, to suggest relevant ads. For example, if a user is searching for lunch options during midday, the LLM can prioritize ads for nearby restaurants or food delivery services.

Current research on behavioral targeting shows that behavioral information may increase ad effectiveness (Aziz and Telang 2016; Rafieian and Yoganarasimhan 2021). Studies discover the positive effects of behavioral targeting on purchase intent and sales, with the former increasing by about 2/4 (Goldfarb and Tucker 2011) and the latter by 1/30 (Johnson et al. 2020). Aziz and Telang (2016) find that users with high baseline purchase probabilities responded positively to ads, increasing their purchase probability by up 1/35 percentage points. Moreover, Rafieian and Yoganarasimhan (2021) demonstrate with their targeting strategy based on behavioral and contextual information that improved CTR by 2/3 over the baseline system. These gains mainly stemmed from behavioral rather than contextual information.

By integrating LLMs into behavioral targeting strategies, advertisers can gain a deeper understanding of user behavior, improve personalization, and create more engaging and effective ad campaigns. LLMs enable advanced reasoning capabilities that go beyond simple data analysis, allowing for more thoughtful and context-aware targeting.

2.3 *Enhanced LLM-Based Reasoning for Demographic Targeting*

Companies such as Google, Facebook/Instagram, Twitter, and Spotify have built digital advertising ecosystems that provide advertisers with self-service platforms to purchase and test hundreds of audience segments. Also known as "audience lists," "user lists," or "data segments," these user groups share specific attributes such as demographics, income, region, interests, or behaviors (Cutura 2017). For instance, advertisers can currently target their Google ads using:

1. Affinity segments (based on people's interests and habits)
2. In-market segments (based on recent purchase intent)
3. Similar segments (based on interests similar to those of the advertiser's Web site visitors or existing customers)
4. Detailed demographic segments (based on long-term life facts)
5. Life-event segments (people who are amid important life milestones; see, e.g., Google (2023))

Facebook's core audiences similarly cover a wide range of location-, demographic-, interest-, and behavior-based segments for targeting (Meta 2023), resulting in more than 600 segments (Fig. 4).

LLMs can significantly enhance demographic targeting by providing deeper insights into demographic profiles, understanding the nuances of specific groups, and helping advertisers create more relevant and personalized campaigns. Below are several use cases where LLMs can support better reasoning for demographic targeting:

1. Understanding cultural nuances: Tailoring ads to different cultural backgrounds by understanding language nuances, traditions, and cultural sensitivities. An LLM can analyze demographic data (e.g., ethnicity, language, cultural practices) to tailor ad content that resonates with specific cultural groups. For example, an ad campaign targeting Hispanic audiences in the United States can incorporate

Behavioral Targeting	Contextual Targeting
Leverages cookies, and may be affected by evolving privacy regulations concerning the use of cookies in browsers.	Not subject to privacy regulations because it does not use or collect cookies.
Relies on hard, historical data of users' search history and what sites they spend the most time on.	Uses context of a user's search "journey" to show them the right ads.
Limits control over the website on which the ads appear.	Ads align with what a user is searching as well as with their values, which allows for advertising without sacrificing the reputation.
Historical data or past behavior may not always be relevant because it may not align with the user's current needs or buying patterns.	Contextual ads are always relevant because they're based on what a user is actually searching, watching, or reading

Fig. 4 Behavioral and contextual targeting (Shannon 2021)

culturally relevant themes, holidays, and language variations such as Spanglish. Reasoning enhancement is important as well: LLMs can reason about cultural references, ensuring that ads are contextually appropriate and resonate with the target demographic.

2. Age-specific content personalization manages ad content that is age appropriate and tailored to different generational groups (e.g., Gen Z, millennials, boomers). For example, an LLM can analyze text and behavioral data to segment audiences by age group and create ad content that appeals to each group's preferences. Furthermore, it can generate playful, meme-based content for Gen Z while producing more professional, informative content for baby boomers. LLMs can understand generational differences in language, humor, and values, allowing for more precise targeting and better engagement.
3. Gender-sensitive marketing crafts gender-sensitive ads that avoid stereotypes and appeal to modern gender identities. For instance, an LLM can analyze demographic data to understand the gender identities and preferences of an audience. It can help create gender-inclusive marketing campaigns that resonate with diverse gender groups, avoiding traditional stereotypes and promoting inclusivity.

LLMs can reason about gender-specific language, preferences, and sensitivities, ensuring that ads are respectful, relevant, and inclusive (Deiss and Henneberry 2020).

4. Geo-specific targeting customizes ad campaigns based on the geographic location of users, considering regional preferences, climates, and lifestyles. In particular, an LLM can analyze the geographic location of users and tailor ad content accordingly. An ad for winter clothing would be targeted at users in colder regions, while users in tropical climates might see ads for summer apparel. LLMs can reason about regional variations in climate, culture, and lifestyle, helping advertisers create more geographically relevant campaigns.
5. Income-based targeting creates ads that cater to different income levels, adjusting the messaging, product positioning, and offers accordingly. An LLM can analyze data related to a user's income level (inferred from purchase history, location, and other indicators) and create ads that are relevant to their financial situation. For example, luxury products would be marketed to higher-income segments, while value-oriented products would be highlighted for lower-income groups. LLMs can reason about economic indicators and tailor marketing messages that are sensitive to the financial realities of different demographics.
6. Educational background and professional targeting tailors content based on the educational background and professional experience of the target audience. An LLM can analyze users' educational and professional backgrounds to create ads that resonate with their knowledge level and career stage. For example, a tech product might be marketed differently to an engineer with a PhD than to a college student studying computer science. LLMs can reason about the educational and professional context of users, ensuring that ads are appropriately targeted based on knowledge and experience.
7. Family and life stage targeting adjusts ad content based on family status and life stage (e.g., single, married, parents, empty nesters). For instance, an LLM can analyze a user's demographic data and life stage to suggest relevant products and services. For example, new parents might see ads for baby products, while empty nesters might receive ads for travel and leisure activities. LLMs can reason about the implications of different life stages, helping advertisers create campaigns that resonate with users' current needs and priorities.
8. Health and wellness targeting customizes ads based on demographic health data, such as age-related health concerns or regional health trends. An LLM can analyze demographic health data to create ads that address specific health concerns, for example, targeting an older demographic with ads for joint supplements or wellness programs tailored to age-related health needs. LLMs can reason about health trends and demographic-specific health concerns, improving the relevance and impact of health-related marketing campaigns.
9. Event-specific targeting tailors ad content to demographic groups based on relevant events (e.g., festivals, sports events, national holidays). An LLM can identify upcoming events relevant to certain demographic groups and create targeted campaigns around those events. For example, targeting sports fans during major tournaments or creating holiday-specific campaigns for different cultural

celebrations. LLMs can reason about the timing and relevance of events to specific demographic groups, optimizing ad placements and engagement.

By leveraging LLMs for demographic targeting, advertisers can go beyond basic demographic data to understand the deeper context of users' lives, preferences, and needs. LLMs offer advanced reasoning capabilities that allow for more nuanced and effective targeting, resulting in more relevant and personalized ad experiences for diverse demographic groups.

2.4 Introduction to the Discourse of Targeted Advertisement

The discourse of targeted advertisement refers to the language, communication strategies, and underlying ideologies that shape the way advertisements are created, delivered, and perceived by specific audiences. It involves the use of personalized messaging to engage particular demographic groups, psychographic segments, or behavioral clusters. The discourse of targeted advertising includes nine components:

1. Language of personalization: targeted ads often use language that feels personal and relevant to the individual. This might include addressing the user by name, referencing specific behaviors, or tailoring content based on their preferences or location. Discourse emphasizes creating a connection with the user by delivering messages that resonate with their identity, interests, and needs. The language used is often direct, relatable, and sometimes informal, aiming to establish trust and familiarity.
2. Data as a narrative: discourse of targeted advertising increasingly revolves around data-driven insights. Advertisers analyze data to craft messages that seem more tailored to individual users or groups. This can be reflected in the choice of words, images, and offers that speak directly to the user's previous online behaviors. Behind the scenes, the discourse involves discussions around the role of algorithms, machine learning, and artificial intelligence in shaping advertising. These technologies allow advertisers to create and refine messages based on vast amounts of data.
3. Language of segmentation: targeted advertising divides the audience into segments based on various factors like age, gender, location, interests, and behaviors. The discourse focuses on how different segments respond to specific messaging, and the language is crafted accordingly to appeal to these segmented identities.
4. Representation and stereotyping: ad discourse also deals with the representation of different demographic groups. Targeted ads may reinforce certain stereotypes or challenge them, depending on how the advertiser chooses to address the segment. This includes the use of culturally specific references, images, and language that resonate with the target group.
5. Behavioral cues and persuasion include the language of persuasion: targeted ads often use persuasive language tailored to the psychological triggers of the

audience. This includes urgency (e.g., "limited-time offer"), social proof (e.g., "people like you also bought this"), or emotional appeals (e.g., "feel confident every day"). Behavioral influence is important as well: ad discourse explores how certain types of language and imagery influence consumer behavior, nudging them toward specific actions such as making a purchase, signing up for a service, or sharing content.

6. Cross-platform communication and platform-specific language are reflected in discourse managing targeted advertising that varies across different platforms (e.g., social media, search engines, video platforms). The language used in targeted ads is adapted to the norms of each platform, whether it's a concise call-to-action on Twitter, a visually driven narrative on Instagram, or a conversational tone in a chatbot. Ad sequence discourse also involves ensuring consistency across various channels. Advertisers aim to create a seamless experience where the targeted message feels consistent and coherent, regardless of whether the user encounters it on a Web site, app, or social media.
7. Ad sequence discourse described the engagement-driven language. Targeted ads often encourage user interaction and engagement, using language that invites clicks, likes, shares, or comments. This part of the discourse involves crafting messages that prompt the user to take action. Discourse also involves analyzing how users respond to ads, including their feedback, reviews, and interactions. This feedback loop helps refine future targeted messaging.
8. Discourse facilitates storytelling: targeted ad sequences increasingly employ storytelling techniques to build a narrative around the product or service. The discourse of storytelling in ads is designed to connect emotionally with the target audience by framing the product as part of a larger, meaningful story that resonates with their lives.
9. Discourse is also a tool for framing. How an ad is framed depends on the target demographic. For example, an ad targeting eco-conscious consumers might frame a product as environmentally friendly, emphasizing sustainability through both language and imagery.

The discourse of targeted advertising is a complex interplay of language, personalization, and modern LLM technology. It is about crafting messages that resonate with specific audiences, leveraging data and algorithms to predict what will be most effective, while navigating the balance between persuasion and ethical considerations. This discourse is continually evolving as technologies like AI and machine learning advance, making targeted ads more sophisticated and personalized.

3 Computing Ad Sequencing

Adaptive ad sequencing introduces a forward-looking approach to the advertiser's ad allocation challenge. Unlike traditional methods that only consider immediate user engagement based on current data, adaptive sequencing accounts for future user behavior and exposure to ads. This forward-thinking approach is illustrated by distinguishing between information from past user interactions and the anticipated data that future engagements might provide.

However, many advertising platforms shy away from forward-looking models due to the added complexity they introduce. The predominant approach in the industry relies on supervised learning and contextual bandit algorithms, which focus solely on optimizing ad delivery based on real-time data, disregarding potential future exposures and engagements (Theocharous et al. 2015).

The reluctance to adopt forward-looking models stems from several factors. Firstly, the computational complexity of integrating future projections into ad allocation models can be daunting. Secondly, the potential returns from such an approach remain uncertain, making it difficult for advertisers to justify the investment in a more dynamic framework.

Ultimately, an advertiser's decision to embrace a forward-looking model depends on whether incorporating future user behavior data can deliver better outcomes. Advertisers must weigh the benefits of predictive modeling against the increased complexity and uncertainty, considering whether the potential for improved engagement and conversion justifies the shift from traditional methods. If successful, adaptive ad sequencing could revolutionize the way advertisers allocate their budgets, enhancing long-term effectiveness by anticipating and influencing future user actions rather than merely reacting to present circumstances.

A chart for the ad sequencing decision based on ad exposure discourse tree is shown in Fig. 5.

The user is at the fourth exposure in the session, and the advertiser needs to decide which ad to show to this user. The advertiser is enabled with ad forecasting possibility to accounts for the futures exposures when making the decision.

The first ad A just introduces a generic product. Then ad B introduces a more specific version of a product from a family. After that, ad C introduces a

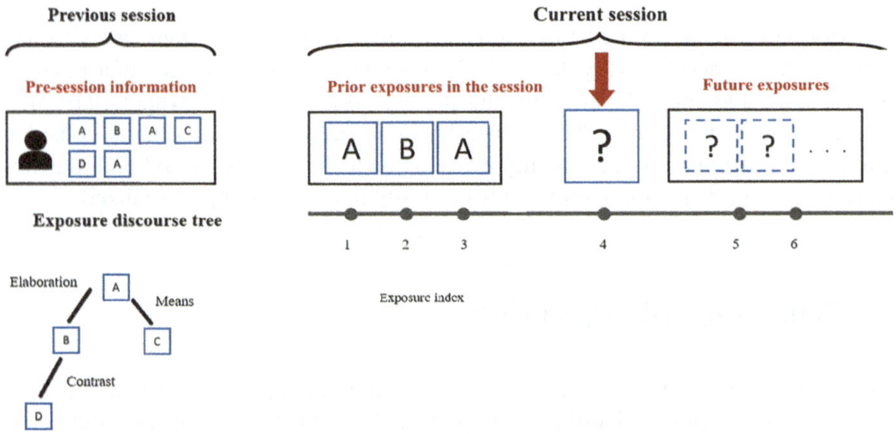

Fig. 5 A chart for the ad sequencing decision based on ad exposure discourse tree

complementary product that is needed to enable A. Finally, ad D is introduced for a product from a competitive family, so that the user choses if she prefers A family of products or D family of products (Fig. 6).

3.1 Ad Network as an Auction

The ad network designs an auction to sell ad slots. In our setting, the ad network runs a quasi-proportional auction with a cost-per-click payment scheme. As such, for a given ad slot and a set of participating ads A with a bidding profile $<b_1, b_2, b_{|A|}>$, the ad slot is allocated to ad a with the following probability:

$$\pi_0(b;m) = \frac{b_a m_a}{\sum_{j \in A} b_i m_i}$$

where m_a is ad a's quality score, which is an expression for the profitability of ad a. The ad network does not customize quality scores across auctions. The subscript$_0$ in π_0 expresses the baseline allocation policy. The payment scheme is cost-per-click, similar to Google's sponsored search auctions. That is, ads are first ranked based on

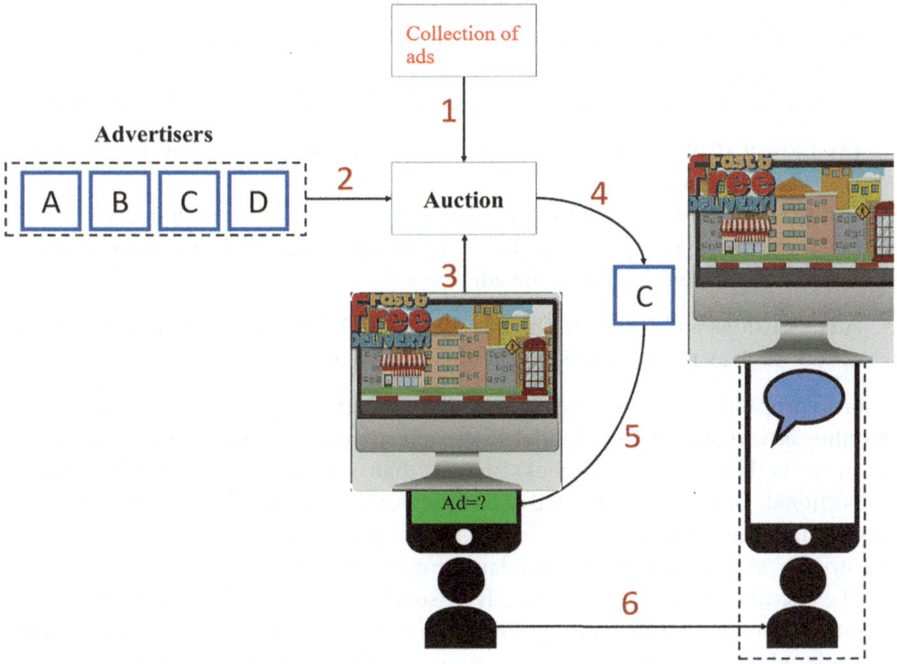

Fig. 6 An auction for sequential ads

their product of bid and quality score, and the winning ad pays the minimum amount that guarantees their rank if a click happens on their ad.

Google's sponsored search auctions are the mechanism by which Google determines which ads appear at the top of search results, and in what order, when users enter search queries. This process involves advertisers bidding to have their ads shown to users who search for relevant terms. Sponsored search refers to the ads that appear at the top or bottom of the search engine results page (SERP) when users enter specific search queries. These ads are labeled as "sponsored" or "ad" to differentiate them from organic search results. Every time a user performs a search, an auction is triggered instantly. The auction determines which ads appear, their ranking, and how much each advertiser pays if a user clicks on their ad.

Advertisers bid on keywords that they think are relevant to their business. For example, a shoe retailer might bid on keywords like "buy running shoes." When a user enters a query, Google matches the query with the relevant ads based on the keywords chosen by advertisers. An auction is initiated for all the ads that are eligible to appear based on the keyword match. Advertisers are not guaranteed placement; instead, their ads compete in the auction.

The position of an ad in the auction is determined by its Ad Rank. Ad Rank is calculated as

$$AdRank = BidAmount \times QualityScore.$$

based on several factors:

1. Bid amount: The maximum amount an advertiser is willing to pay per click (CPC bid).
2. Quality score: This measures the relevance and quality of the ad and includes:
   (a) Expected click-through rate (CTR): How likely the ad is to be clicked when shown.
   (b) Ad relevance: How closely the ad matches the user's search query.
   (c) Landing page experience: The relevance and quality of the page users are taken to when they click the ad.
3. Ad extensions and formats: The impact of additional information provided with the ad, such as phone numbers, site links, or location information.

A quasi-proportional auction with a cost-per-click (CPC) payment scheme is a type of online advertising auction model where advertisers bid for ad placement, and the payment is based on user clicks rather than impressions. The term "quasi-proportional" refers to the method used to allocate ad slots to advertisers, where the probability of an ad being shown is roughly proportional to the advertiser's bid, but not strictly proportional as in a standard proportional auction.

The auction setup is as follows: advertisers bid for ad slots, typically indicating the maximum amount they are willing to pay for each click on their ad (CPC bid). Multiple advertisers compete for the same ad placement on a Web site or search engine results page. The term "quasi-proportional" implies that the allocation of ad slots is influenced by the advertisers' bids, but not in a strictly proportional way. Unlike a fully proportional auction, where the probability of an ad being shown is directly proportional to the bid amount, a quasi-proportional auction adjusts the allocation in a more complex way. This could involve factors such as ad quality, relevance, or user engagement history, in addition to the bid amount.

The goal is to balance maximizing revenue for the platform (by giving higher-bid ads a better chance) with ensuring that users see relevant and high-quality ads. Cost-per-click (CPC) payment works as follows: advertisers only pay when their ad is clicked, rather than paying based on the number of times the ad is shown (impressions). The amount is usually determined by a second-price auction mechanism, where the winning advertiser pays slightly more than the next highest bidder's price, rather than their own maximum bid. As an example, let us consider three advertisers (A, B, C) bidding for ad slots with the following CPC bids: advertiser *A* with $2.00 per click, advertiser B with $1.50 per click, and advertiser C with $1.00 per click.

In a quasi-proportional auction, the probability that each advertiser's ad will be shown is roughly proportional to their bid but also adjusted based on factors like ad quality. If advertiser A has a slightly lower ad quality than B and C, their bid might be de-emphasized, and advertiser B could have a better chance of winning the auction even though they bid less. Ultimately, the ad that gets the click will trigger the CPC payment. This approach balances the need for effective advertising with ensuring that ad placements are not solely determined by the highest bid, providing a more equitable and performance-based auction system.

3.2 Ad Allocation Process

The ad allocation process is implemented as follows:

1. The ad network designs an auction to sell ad slots.
2. Advertisers participating in the auction make the following choices:
    (a) Design their banner.
    (b) Decide which impressions they want to target.
    (c) Decide how much to bid.
3. Whenever a user starts a new session in an app, a new impression is being recognized, and a request is sent to the publisher to run an auction.
4. The auction takes all the participating ads into account and selects the ad probabilistically based on the weights. Note that all the participating ads have the chance to win the ad slot. This is in contrast with more widely used deterministic mechanisms like second-price auctions, where the ad with the highest product of bid and quality score always wins the ad slot.
5. The selected ad is delivered.
6. Each ad exposure lasts a certain period of time, like a minute. During this time, the user makes two key decisions:

(a) Whether to click on the ad.
(b) Whether to stay in the app or leave the app and end the session.

If the user clicks on the ad, the corresponding advertiser has to pay the amount determined by the auction. After a certain time period, if the user continues using the app, the ad network treats the continued exposure as a new impression and repeats steps 3 to 6 until the user leaves the app.

In the fiercely competitive world of digital advertising, securing prime ad placements is like playing a high-stakes game of strategy, where every move is carefully planned to win the attention of potential customers. Ad placement is influenced by a delicate balance of factors, from the advertiser's budget and audience behavior to the constantly shifting algorithms of ad platforms. Success in this landscape requires advertisers to navigate these elements with precision, adapting to changes while staying ahead of the competition (Fig. 7). The factors are as follows:

1. Budget allocation: advertisers must judiciously allocate their budgets to bid for premium ad spaces. For instance, a company might decide to spend more on placements during peak shopping seasons to maximize visibility.
2. Audience targeting: understanding the demographics and interests of the audience is crucial. A travel agency might target ads toward users who have recently searched for vacation destinations.
3. Ad quality and relevance: the quality and relevance of the ad itself can affect its placement. A well-designed ad with high relevance to the user is more likely to win a favorable spot.
4. Bidding strategies: advertisers employ various bidding strategies to outmaneuver competitors. A common strategy is to set a higher bid for placements on popular Web sites or during primetime slots.
5. **Discourse-based ad sequence planning**, which is a focus of this chapter (Fig. 8).
6. Real-time adjustments: the ability to make real-time adjustments to bids based on analytics can give advertisers an edge. For example, if an ad is performing

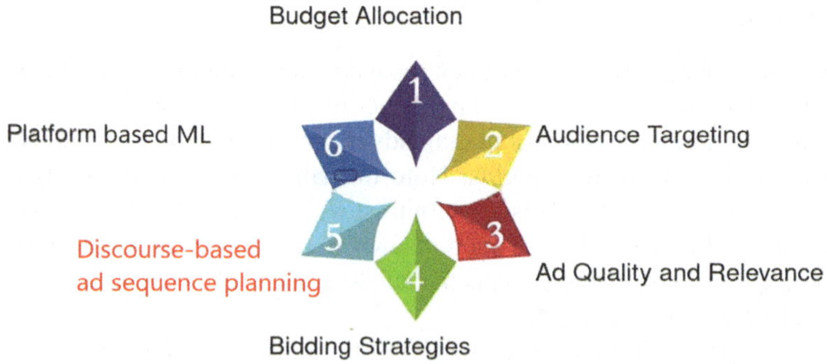

Fig. 7 Prime ad placement components

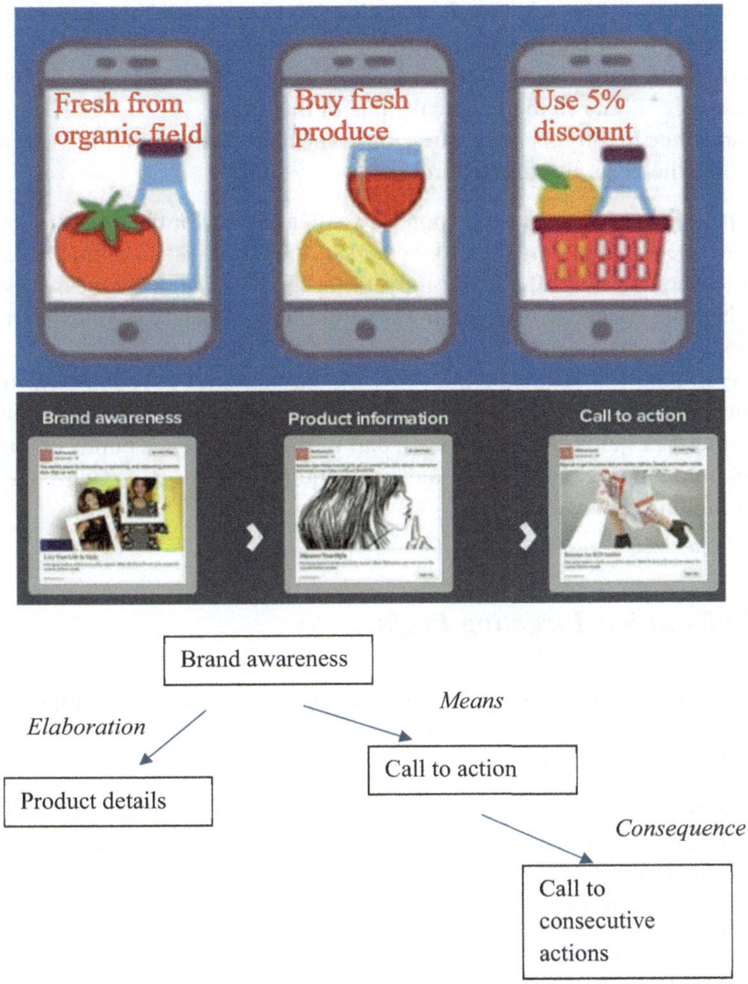

Fig. 8 A discourse chain of ads

well on a particular site, the advertiser might increase the bid for that site to secure more placements.

7. Platform algorithms: each advertising platform has its own set of algorithms that determine ad placement. Advertisers need to stay informed about these algorithms to tailor their strategies accordingly.

Figure 8 illustrates a simple ad sequence along with its corresponding discourse tree. The discourse tree visually represents the logical structure of the ad sequence, guiding the user through a carefully crafted persuasive journey. This structure is organized using rhetorical relations such as *Elaboration*, *Means*, and *Consequence*.

- *Elaboration* provides additional details, enriching the initial message by explaining the features or benefits of the product or service.
- *Means* shows how the product can be used or how it solves a specific problem, connecting the user to the practical value of the offer.
- *Consequence* highlights the positive outcomes or rewards of taking action, driving home the necessity of the purchase.

Each step of the ad sequence corresponds to a mental step for the user, progressively convincing them that buying the product or service is not just an option but a necessity. The discourse tree effectively mirrors the user's thought process, systematically addressing concerns, highlighting benefits, and ultimately leading them toward a decision to purchase.

By organizing the sequence this way, the ad taps into psychological triggers, increasing engagement and maximizing the likelihood of conversion. This discourse-based framework helps ensure that each part of the ad sequence builds on the previous one, maintaining coherence and momentum and aligning with the user's decision-making process.

3.3 Calculating Targeting Profit

To calculate the revenue from the targeted campaign in comparison with untargeted, we define revenue functions for each case.

The advertiser's revenue (p) is a function of:

1. The number of users who purchase the advertiser's product (Q)
2. The (long-term) margin per conversion (m, which can represent customer lifetime value (CLV))
3. The cost per conversion

No-targeting case will have a subscription "-", and a targeting audience segment is denoted by i. The totality of segment is denoted by S, $i \in S$.

In case where an exact profit margins data on ad conversions is unavailable, advertisers can measure advertisement results such as registrations, leads, and certain Web site activities including booking the service or demanding a quote.

The advertiser's no-targeting revenue is as follows:

$$p_{-} = N_{-} CTR_{-} CR_{-} \left(m_{-} - \frac{CPM_{-}}{1000 CTR_{-} CR_{-}} \right),$$

where N_{-} is the number of ad impressions, CTR_{-} is click-through rate, CR_{-} is a conversion rate, m_{-} is the absolute margin per conversion, and CPM_{-} is the advertising cost in CPM. This is what an advertiser pays for 1000 non-targeted ad impressions.

The targeting revenue is computed analogously

$$p_i = N_i CTR_i CR_i \left(m_i - \frac{CPM_i + P_i}{1000 CTR_i CR_i} \right),$$

where P_i is the data cost in CPM. This is an extra expense that an advertiser pays to target the respective audience segment *i*.

It is obvious that advertiser have a boost in revenue when targeting users who are more likely to be interested in the advertiser's products. At the same time, targeting may decrease the number of ad impressions by reducing the number of available users because it excludes those who do not meet the targeting criteria. Also, targeting is associated with additional expense for the advertiser due to increased number of impressions for targeted users.

Four coefficients need to be introduced to representing the changes in CTR, CR, margin, and reach, respectively, through targeting audience segment *i*. As a result, $CTR_i = \alpha_i CTR\_$, $CR_i = \beta_i CR\_$, $m_i = \delta_i m\_$, and $N_i = \gamma_i N\_$. There should be an increase in CTR, CR, and margin and a decrease in reach (i.e., $0 < \delta_i < 1$).

Data cost can be defined as a function of reach, $P_i = 1/\delta_i^{\hat{\gamma}}$. $\hat{\gamma} > 0$, an estimated parameter expressing the observation that a segment is more expensive with a larger $\hat{\gamma}$. It can be transferred from previous ad campaigns.

Now we have for targeting revenue:

$$p_i = \delta_i N\_CTR\_CR\_ \left(\alpha_i \beta_i \gamma_i m\_ - \frac{CPM_i + 1/\delta_i^{\hat{\gamma}}}{1000 CTR\_CR\_} \right).$$

Breakeven performance of a segment *i* can now be expressed as

$$\left(\alpha_i \beta_i \gamma_i\right)^* = \frac{1}{\delta_i} - \frac{CPM\_ - \delta_i \left(CPM_i + 1/\delta_i^{\hat{\gamma}}\right)}{1000 \delta_i CTR\_CR\_m\_}.$$

This equation expresses a breakeven performance related to targeting audience segment *i* that makes targeting as profitable as no-targeting (with the reach δ_i). Advertisers need to evaluate whether their targeting is actually improving their revenue and targeting makes sense.

4 A Discourse Chain of Ads

Research has revealed that individuals exposed to a strategically sequenced series of advertisements are significantly more likely to convert, with conversion rates reported to be 14–17 times higher than those who were exposed only to stand-alone ads (Visioneerit 2024). This dramatic increase underscores the power of sequential advertising in building a narrative, reinforcing brand messaging, and gradually guiding the audience through a well-structured journey from awareness to action.

By delivering ads in a specific order that aligns with the consumer's decision-making process, brands can foster deeper engagement, enhance recall, and ultimately drive more successful conversions. This finding highlights the importance of moving beyond isolated, one-off ads and adopting a more cohesive, story-driven approach to advertising.

We introduce a novel approach called LLM and Discourse Ad Generation and Targeting (LDAGT), which harnesses the power of LLMs to generate ad content and structure ad delivery using discourse trees. LDAGT optimizes keyword targeting while utilizing adaptive LLM prompt engineering and reinforcement learning-based feedback loops to enhance ad relevance and effectiveness. By integrating ad content generation with ad sequence delivery and keyword targeting, LDAGT ensures a cohesive discourse structure across a sequence of ads, along with a strong alignment between the generated ad text and targeted keywords. This leads to significantly improved ad performance.

4.1 Ad Sequence Templates

We now consider more complex discourse structures of ad sequences. Figure 9 shows a discourse representation of an ad sequence where a proposed product is presented as an opposite entity to another product, which is not recommended because of certain negative characteristics. This is a well-known pattern to strengthen the benefits of one product by explaining how it is better than the other product.

To strengthen an advantage of a given product, an efficient way is to show that skillful users can use this advantage and unskillful cannot (Fig. 10).

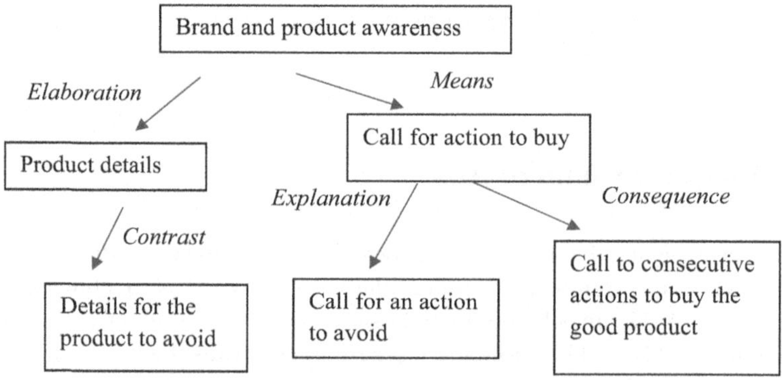

Fig. 9 A DT for opposing a good product to a product to avoid

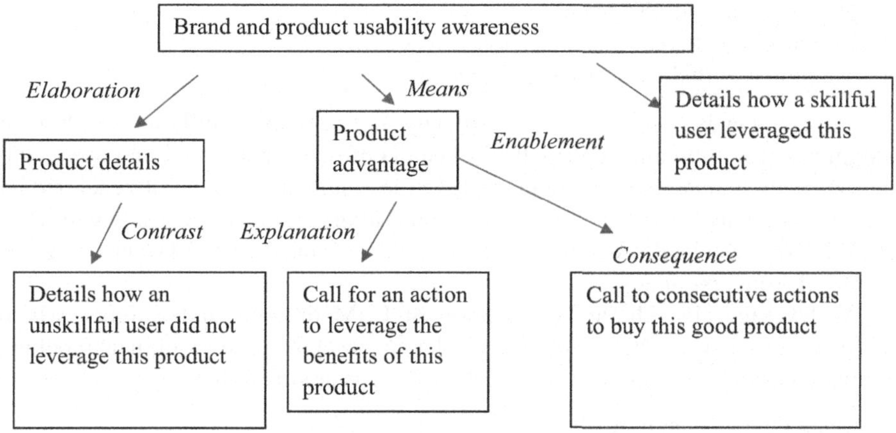

Fig. 10 DT for opposing a skillful user to unskillful

4.2 LLM Support

Kumar and Khanna (2024) proposed adaptive prompt engineering and reinforcement learning-based feedback loops to enhance the relevance and effectiveness of ads. The adaptive prompt engineering technique dynamically adjusts the prompts based on user behavior and ad performance, while the reinforcement learning-based feedback loop enables continuous optimization of the ad generation and targeting process.

To construct the prompts and optimize them, we rely on a combination of contextual information, such as user demographics, browsing history, and search queries. Incorporating user demographics and browsing history into prompt engineering via algorithmic optimization can improve the relevance of generated ad content by about 1/5 having the fixed to generic prompts as a baseline (Yan et al. 2021). Domain-specific knowledge, such as product attributes and brand guidelines, is also valuable to make sure the delivered ad content is informative and convincing. More than 2/3 of prospective buyers and users are more likely to engage with ads that are relevant to their interests and preferences (Interactive Advertising Bureau 2020). The prompts are adaptively updated based on real-time user interactions and feedback. For example, if a user clicks on an ad, the corresponding prompt is reinforced to generate similar ad content in the future. Conversely, if an ad receives low engagement, the prompt is adjusted to explore alternative ad content strategies. Chen et al. (2021) showed that dynamic prompt adjustment based on user feedback can improve the click-through rate (CTR) of generated ads by 12% compared to static prompts, served as inspiration for this adaptive approach.

Given a prompt and a set of discourse trees templates for sequential ads, the LLM generates multiple candidate ad variations for each sequence element and for each template. These variations are assessed as candidate ad sequence elements based on:

1. Their relevance to the prompt
2. Discourse coherence
3. Alignment with the advertising objectives

The assessment is done based on perplexity and semantic similarity and human judgments from domain experts. It has been confirmed that combining automatic metrics with human judgments can improve the quality assessment of generated ad content by about 1/4 compared to using stand-alone automatic measurement (Liu et al. 2020). Obtained sequence of ads is also associated with relevant keywords extracted from text of ads.

We leverage algorithmic optimization of LLM prompts and weights, which improves efficiency when LMs are used one or more times within a pipeline. Without algorithmic optimization, the default steps are the following:

1. Break the problem down into steps.
2. Prompt the LLM well until each step works well in isolation.
3. Tweak the steps to work well together.
4. Generate synthetic examples to tune each step.
5. Use these examples to fine-tune smaller LMs to cut costs.

Every time advertisement pipeline is updated, all prompts (or fine-tuning steps) may need to be re-done.

To make the advertising pipeline more systematic and powerful, algorithmic optimization employs several key strategies:

1. Separation of workflow and parameters: It decouples the program's flow (modules) from the parameters (language model prompts and weights) at each step, allowing for better control and flexibility.
2. Introduction of optimizers: These are LM-driven algorithms designed to tune both prompts and weights based on specific metrics, optimizing the performance of the system to achieve desired outcomes.

The algorithmic optimization system, DSPy, enhances the reliability of powerful models like GPT and local models such as T5-base or Llama (Opsahl-Ong et al. 2024). It achieves this by improving output quality and identifying and mitigating specific failure patterns. DSPy translates the same program into different instructions, few-shot prompts, or weight updates (fine-tuning) for each language model. This represents a new paradigm where language models and their prompts become optimizable components within a larger system that learns from data, driving more efficient and effective ad delivery.

4.3 Meta-Learning

We extend DSPy approach to prompting with meta-learning (Fig. 11).

Is it necessary to train an ML model for ad sequence to evaluate its potential effectiveness? Or can the performance of ML model for the prediction of ad performance be estimated based solely on its architecture and training parameters? These

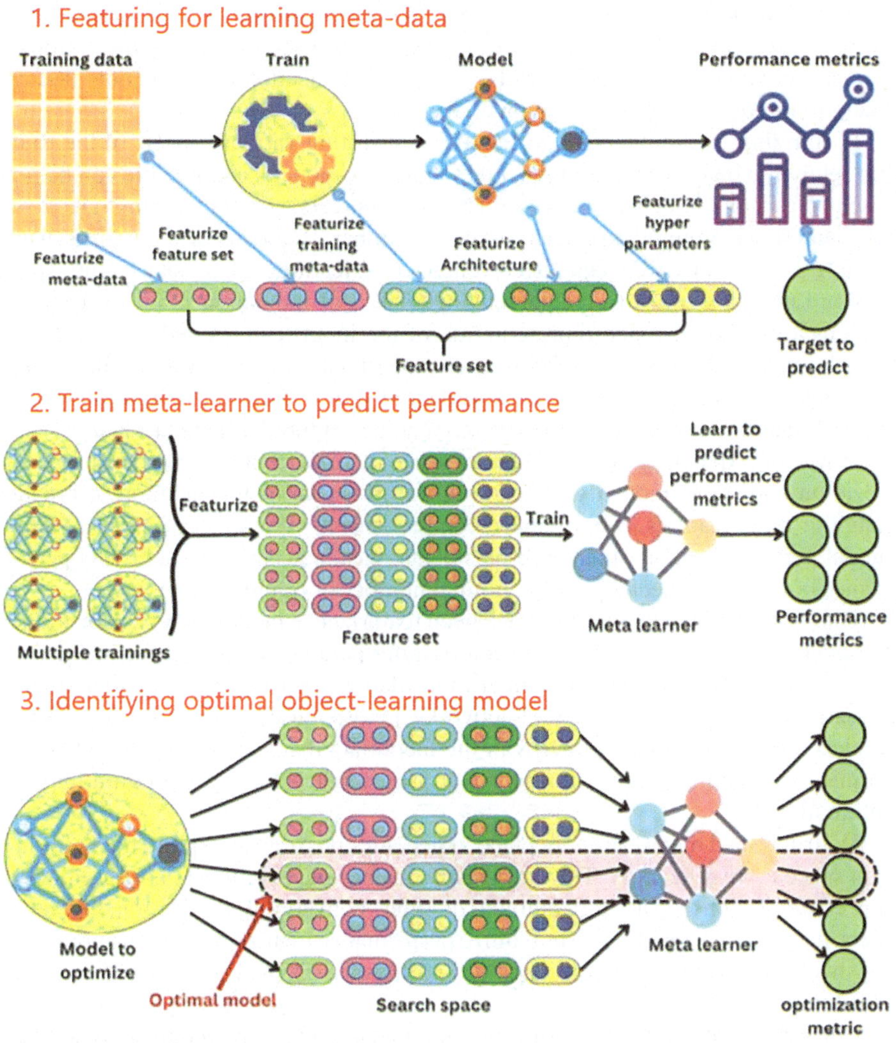

Fig. 11 Meta-learning for model optimization

questions are central to meta-learning, an approach that seeks to identify the patterns that make one model outperform another for a particular learning task (Benveniste 2024).

Meta-learning focuses on optimizing the learning process itself by leveraging prior knowledge about model performance. The basic idea involves transforming information about the training configuration into a set of features and then training a separate model, known as a meta-learner, to predict performance metrics based on those features. Once trained, this meta-learner can be used to navigate the optimization space when tuning other models, guiding the selection of parameters that are likely to improve performance (Finn et al. 2017).

Featurizing learning meta-data involves converting various aspects of the training setup into quantifiable features. For example, a model's architecture can be represented as a one-hot encoded vector. Hyperparameters, such as learning rate, batch size, and other training parameters like the number of epochs or the type of hardware (e.g., CPU or GPU), can also be encoded as features. This meta-feature space can be further extended to include characteristics of the dataset used for training, such as a one-hot encoded representation of the input features and the number of samples involved. This allows for the inclusion of data-related variables like feature selection in the optimization process. Essentially, any factor that could influence a model's learning behavior or performance can be captured as a meta-feature.

However, expanding the meta-feature space increases both the potential for optimization and the complexity of learning the target variable. Including a broader range of meta-features offers a more comprehensive view of the factors affecting model performance, but it also presents challenges in accurately predicting performance outcomes, especially with a limited number of training examples.

Once the training experiments have been featurized, a meta-learner can be trained to understand the relationship between training parameters and performance metrics. Given the typically small number of available samples, simple models such as linear regression or shallow neural networks are often used as meta-learners. These models attempt to capture patterns in the meta-data that correlate with higher performance, enabling more informed decisions when selecting configurations for future models.

With a trained meta-learner in place, it becomes possible to explore the optimization space more efficiently. The meta-learner can evaluate billions of different training configurations in a short amount of time, converging quickly on the set of meta-features that are likely to maximize performance metrics. Common approaches to this optimization process include reinforcement learning and supervised fine-tuning.

Fine-tuning, in particular, can refine the meta-learner's predictions. When specific training data is available or when the focus shifts to a subset of the optimization space, training a few new models and feeding their performance metrics back into the meta-learner allows for better-targeted optimization. This iterative process enhances the meta-learner's ability to guide the search for optimal training configurations, ultimately improving the efficiency and effectiveness of the model development process.

By leveraging meta-learning, it becomes possible to automate and accelerate the search for high-performing models, reducing the reliance on trial-and-error experimentation and paving the way for more systematic model optimization.

4.4 *Reinforcement Learning–Based Feedback Loop for A/B Ad Testing*

Loop continuously optimizes the ad generation and targeting process based on real-time user interactions and feedback. The optimization problem is formulated as a Markov decision process (MDP), where the state is the currently delivered ads in a sequence, current position in a discourse tree, and the user's behavior. MDP action is the created ad content and the targeted keywords. The reward is based on how well the ad performed, expressed via the click-through rate (CTR) and conversion rate (CR), assisting with learning the optimal policy for all steps of ad sequence creation and delivery.

A deep reinforcement learning (RL) algorithm, such Deep Q-Network (DQN) (Wang et al. 2018), can be used to learn the optimal policy. The DQN agent interacts with the environment by selecting an ad sequence template, generating ad as each element of this sequence, observing user reactions, and receiving rewards. The agent aims to maximize the cumulative reward over time, leading to improved ad performance. Employing DQN for ad selection can increase the CTR by a quarter and the CR by 1/6 compared to traditional rule-based methods (Zhao et al. 2018).

The feedback loop allows LDAGT to continuously adapt and refine its ad composition targeting, retargeting, and reengagement strategies based on real-time user interactions. This enables LDAGT to quickly identify effective ad content and keyword combinations and adjust its policies and strategies accordingly. By incorporating reinforcement learning, LDAGT can learn from its actions and optimize its performance over time, as shown in the work of where a RL-based approach improved the overall ad revenue by almost a third compared to a baseline without RL (Zhao et al. 2018).

4.5 *Retargeting and Reengagement*

Aziz and Telang (2016) study the effect of using more cookie information on accuracy of purchase predictions in the course of retargeting, leveraging impression and bid-level data. The authors use regression analysis and counterfactual simulations and observe up to 35% increase in purchase intent. Example ad here are where discourse structure controls re-engagement as a part of retargeting campaign.

Retargeting and *reengagement* are two crucial strategies in digital advertising that focus on reconnecting or keeping in a loop with users who have already interacted with a brand but may need a nudge to complete a desired action or to stay engaged with the brand. Retargeting reaches out to potential customers who have previously visited a Web site, engaged with an ad, or shown interest in a product but have not completed a specific action, such as making a purchase. The idea is to "retarget" these users with ads as they browse other Web sites or social media platforms, reminding them of what they were interested in and encouraging them to return and complete the action (Austin 2020).

A sequence of Facebook ads takes users through the marketing funnel versus a set of standard individual call-to-action ads (Fig. 12).

Instead of focusing on getting the immediate conversion from key prospects, a social media ad campaign can serve a set of creatives sequentially to tell the brand story and lead viewers through the marketing funnel before going for the hard sell. This story is structured in the form of discourse tree with rhetorical relations *Elaboration*, *Cause*, *Means* and *Explanation*. It turns out that a sequential storytelling social media ad campaign can outperform a typical campaign in which each ad is designed to generate an immediate conversion.

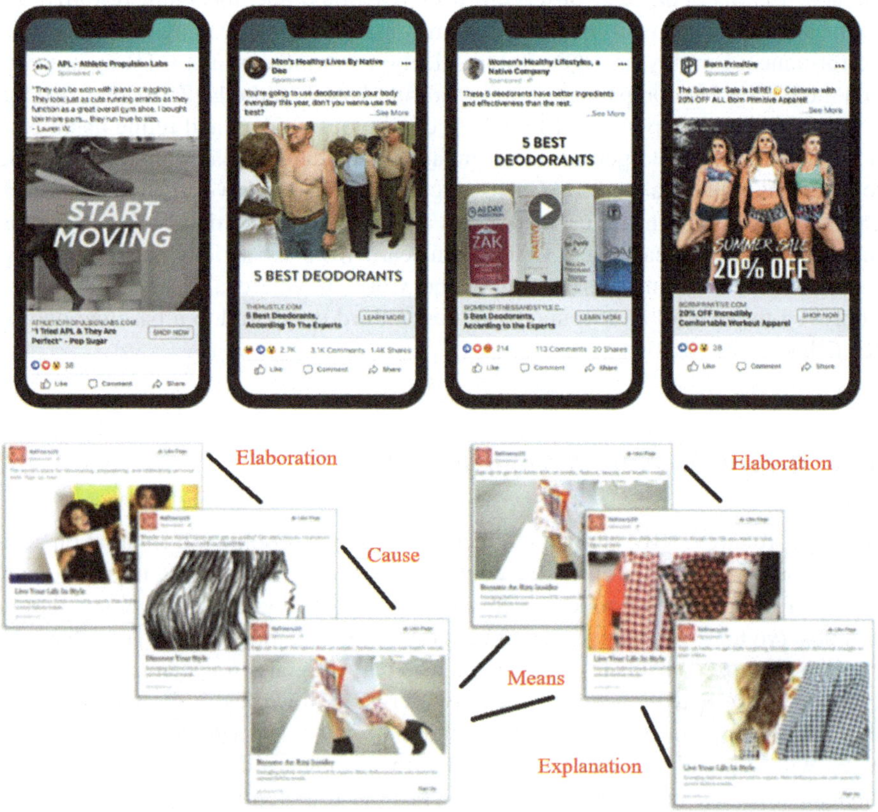

Fig. 12 Another example of a sequence of ads and its discourse representation

4.6 *Benefits of Sequential Advertising with Management at Discourse Level*

Sequential advertising offers numerous advantages, with one of the most significant being its effectiveness in storytelling. Research shows that ads that combine storytelling with a call to action (CTA) are more successful than those focused solely on the CTA. By presenting ads in a narrative format, brands can create more engaging, immersive, and personalized experiences for their audiences. In addition to storytelling, sequential advertising provides other key benefits:

1. Versatility: sequential advertising is a flexible strategy that can be applied across virtually any platform and channel, including social media giants like Facebook. This adaptability allows brands to reach their audience wherever they are.
2. Reduced ad fatigue: repeatedly showing the same ad can quickly lead to ad fatigue, causing the audience to lose interest. Sequential advertising allows for varied content, keeping the audience engaged and targeting different segments more effectively.
3. Focused targeting: with sequential advertising, you can tailor the ads to specific buyer personas or target users at different stages of their journey. This precision targeting ensures that the content resonates with the right people at the right time.
4. Maximized return on ad spend: by honing in on the most relevant audience with sequential messaging, brands can make a greater impact, leading to more efficient use of their ad budget and higher returns.
5. Increased viewership: sequential advertising combats ad fatigue by delivering varied and engaging content over time. This keeps the audience interested, which correlates with higher viewership and attention rates.
6. Higher conversions: data shows that sequential advertising can lead to 14×–17× higher conversion rates when two or more ads are used in sequence. This is because the strategic sequencing of creatives allows for a more focused delivery, driving higher engagement and turning quality leads into paying customers.
7. Sequential advertising with management at the discourse level is a powerful tool that not only tells a compelling brand story but also delivers targeted, engaging content that drives better results across the board.

5 System Architecture

The LDAGT framework consists of four main components:

1. LLM-based single ad generation
2. LLM-based discourse management system for ad sequence
3. Targeting, retargeting, and reengagement ad delivery system
4. Reinforcement learning–based ad feedback

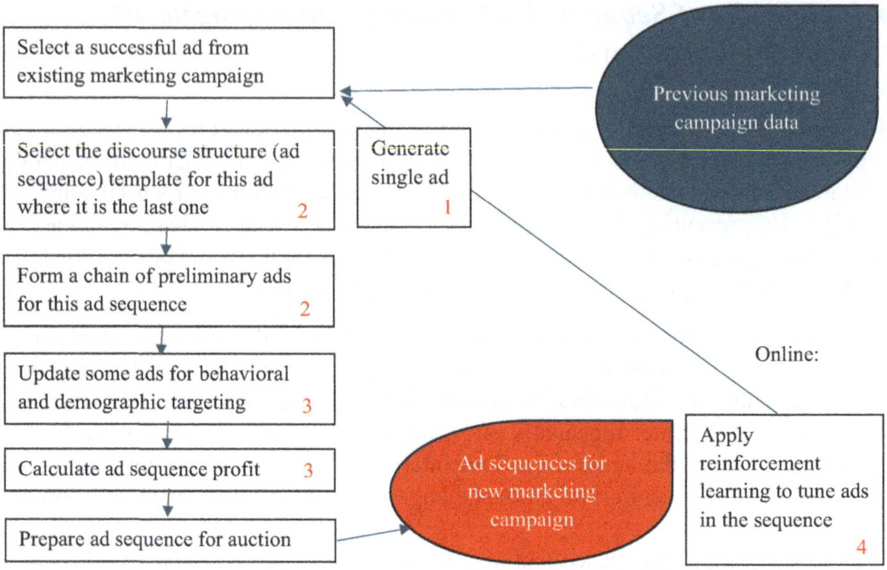

Fig. 13 System architecture

These components work together to generate relevant and effective sequential ad content while continuously optimizing the targeting strategy based on user feedback.

The chart in Fig. 13 shows how an LLM can be used to create and optimize ad sequences using past marketing data, discourse structures, and reinforcement learning. It starts with data from previous marketing campaigns, which include examples of ads and their performance. From there, the process either generates a new ad with the LLM or selects a successful ad from the past. That ad becomes the final step in a larger ad sequence, and earlier ads are created to lead up to it in a coherent narrative.

The sequence is then adjusted for specific audience targeting and projected profitability. Once the ads are optimized, they are prepared for auction on advertising platforms and launched as part of a new campaign. After launch, reinforcement learning is used to update and improve the ads in real time based on how well they perform. Data from this new campaign is added to the historical campaign data so the system can keep improving over time.

6 Evaluation

We compare LDAGT with the following five baseline approaches:

1. Template-based ad generation: This method generates ad content using predefined templates and fills in the slots with relevant product information.

Template-based methods have been widely used in the industry due to their simplicity and efficiency (Mita et al. 2024).

2. Rule-based keyword targeting: This method selects targeted keywords based on a set of predefined rules, such as term frequency and relevance to the product category. Rule-based methods have been shown to be effective for keyword targeting in various domains (Chen and Gunawan 2023).
3. LLM-based ad generation: This method generates ad content using an LLM but does not perform keyword targeting. LLM-based ad generation has gained attention in recent years due to the success of large language models in text generation tasks (Meguellati et al. 2024).
4. LLM-based keyword targeting: This method selects targeted keywords using an LLM but does not generate ad content. LLM-based keyword targeting leverages the semantic understanding capabilities of LLMs to identify relevant keywords (Kathiriya et al. 2022).
5. LLM-based single ad generation and targeting (Kumar and Khanna 2024).

Instead of relying on a high volume of randomized controlled trials (RCTs) to test different targeting strategies and their combinations, we utilize the model developed by Ahmadi et al. (2024) to simulate and systematically compare various targeting approaches. While RCTs are the gold standard for assessing the causal impact of ads and determining the profitability of a limited number of targeting strategies, they can be both costly and time-consuming. Running RCTs on hundreds of audience segments and their combinations poses significant challenges. By leveraging the simulation model, we can narrow down the most promising strategies that should be tested in real-world scenarios.

The simulator analyzes different advertising strategies that various advertisers might implement during their campaigns. Click-through rates (CTRs), conversion rates (CRs), and profit margins show significant variation depending on the product, advertisement network, and advertiser. The simulation adjusts baseline CTRs, CRs, and margins under no-targeting conditions, exploring four distinct factor levels for each: low, mid-low, mid-high, and high.

The targeting study covers 12 different scenarios, incorporating data from hundreds of audience segments, including those from major platforms like Facebook, as documented in studies by Yan et al. (2021), Kathiriya et al. (2022), and Kumar and Khanna (2024). These audience segments are categorized by factors such as age, gender, device type (mobile/desktop), user preferences, and online behavior. This comprehensive simulation provides valuable insights, allowing advertisers to focus on the most effective strategies before investing in large-scale, real-world testing.

Table 1 illustrates the improvement in ad performance compared to a non-targeting baseline, as well as baseline ad creation and targeting components. For each row, a specific parameter range is set (such as mid-low for CTR), while other parameters like CR and margin (m) are averaged.

The study evaluates performance across 12 different ad delivery rates and cost settings, with the remaining parameters averaged. In all baseline evaluations, as well as in the LDAGT case, higher CR values consistently result in better

Table 1 Ad performance rate for various advertising parameters and ad sequence techniques

Targeting parameter	Fuzzy value	Range		Template-based ad generation	Rule-based keyword targeting	LLM-based ad generation	LLM-based keyword targeting	LLM-based **single ad** generation and targeting	LLM-based **ad sequence** generation and targeting
		From	To						
CTR_	Low	0.5	0.75	1.07	1.12	1.15	1.22	**1.28**	1.26
	Mid-low	0.75	1	1.12	1.14	1.15	1.24	1.25	**1.27**
	Mid-high	1	1.25	1.13	1.17	1.21	1.28	1.32	**1.37**
	High	1.25	1.5	1.13	1.16	1.20	1.28	1.28	**1.36**
CR_	Low	1.5	1.75	1.21	1.30	1.29	1.35	1.40	**1.45**
	Mid-low	1.75	2	1.23	1.28	1.35	**1.45**	1.43	1.43
	Mid-high	2	2.25	1.27	1.25	1.34	1.42	1.47	**1.52**
	High	2.25	2.5	1.26	1.30	1.29	1.40	1.43	**1.50**
$ (m_)	Low	150	250	1.19	1.27	1.36	1.40	**1.49**	1.46
	Mid-low	250	500	1.23	1.30	1.32	1.38	**1.52**	1.50
	Mid-high	500	750	1.20	1.31	1.34	1.41	1.47	**1.52**
	High	750	1000	1.18	1.26	1.33	1.37	**1.50**	1.49

performance than higher CTR values. Due to the approximate averaging of parameters, the sum of performance for the top, middle, and bottom four rows is not identical.

The rightmost column presents the performance of LDAGT compared against five different baselines for ad sequence or individual ad construction. LDAGT demonstrates superior performance in more than half of the cases (7 out of 12), outperforming the "LLM-based single ad generation and targeting" approach, which wins in 5 cases. Additionally, LDAGT significantly outperforms other baselines (in columns 4–7).

The results indicate that the application of discourse analysis for automatic ad sequence construction leads to a substantial improvement in ad performance. This supports the conclusion that discourse-based ad sequence management significantly outperforms baseline methods that rely on semi-automatic ad sequence generation or fixed sequence structures.

7 Analysis of Advertisement Discourse

Discourse analysis emphasizes language as a tool that constructs texts and talk. It does not only analyze the text itself but also the processes that govern its production and reception, from producers to the target audience. Rather than using language to imply the presence of underlying psychological constructs, it focuses on how people use language to express their inner self and state of mind. The discourse analysts are more concerned with studying what people are doing while talking than what they are saying in their talk.

By using discourse analysis as a means of examining the social processes that create ads, advertising techniques, and discursive strategies, an in-depth analysis of the advertising discourse can be conducted (Baig 2013). The study of social processes of a discourse is dealt under a contemporary approach to discourse analysis, which is called Critical Discourse Analysis (CDA).

In CDA, we trace three components of Speech Act theory:

1. Locution involves all the linguistic elements used in this advertisement. These linguistic items are the utterances that contain a meaningful effect in their production.
2. Illocution, in the discourse under discussion, is the communicative intent of the advertiser that is to persuade the consumers in such a way that they would make Standard Chartered their choice.
3. Perlocution is the degree of influence that the advertiser could have upon the viewers.

If the meaningful utterances (locutionary act) in an ad generate a strong effect upon the listeners/viewers controlling their minds and actions, then the perlocutionary aim of the advertiser is achieved.

As speech act is a "thing" to be done through "words," the speech act being performed in this piece of discourse is that of the speaker's/advertiser's communicative intent, the illocutionary act, which is persuasion.

According to Searle (1975)'s classification of speech acts, this discourse would falls under the class of commissives. As commissives are the utterances that commit the speaker to some future action, the claims made in the advertisement are a kind of commissive speech act being performed by the advertiser. Such as claiming:

Classifying the speech act performed in the chosen discourse in terms of struc-

To be here for Customers
Here for Progress in Usability
Here for the Long Run!
Here for Good!

ture, it could be called an indirect speech act, which exhibits an indirect relationship between the linguistic form and the function of the utterance.

The structure or the linguistic form of this advertisement is that of interrogative utterances. As demonstrated below: *Can a Bank really stand for something? Can it balance its ambition with its Conscience to do what it must not what it can? Can it not only look at the profit it makes but that how it makes it makes that profit and stand beside people not above them? Where every solution depends on each person, simply by doing good, can a bank in fact be great?*

Communicative function: The communicative function being performed by "text and talk" in the advert is that of persuasion, as the advertiser aims at influencing the opinions and ideologies of the consumers. Thus, the linguistic structure and communicative function do not coincide in the chosen discourse, for the interrogation does not require a response from the viewers; rather it aims at questioning the nature and performance of a "bank" in general.

8 Conclusions

In this chapter, we demonstrated that incorporating discourse analysis significantly enhances ad sequence creation and improves ad delivery performance. We identified a set of discourse tree structures that are particularly effective for ad sequences. Applying this proposed discourse analysis technology on a large scale in practice will further refine the selection of optimal discourse structures for sequential ads.

There are four dimensions of application of AI in the advertising domain (Gao et al. 2023, Fig. 14):

1. Targeting primarily relies on ML to precisely identify and target the audience group most likely to respond positively to the advertisement.

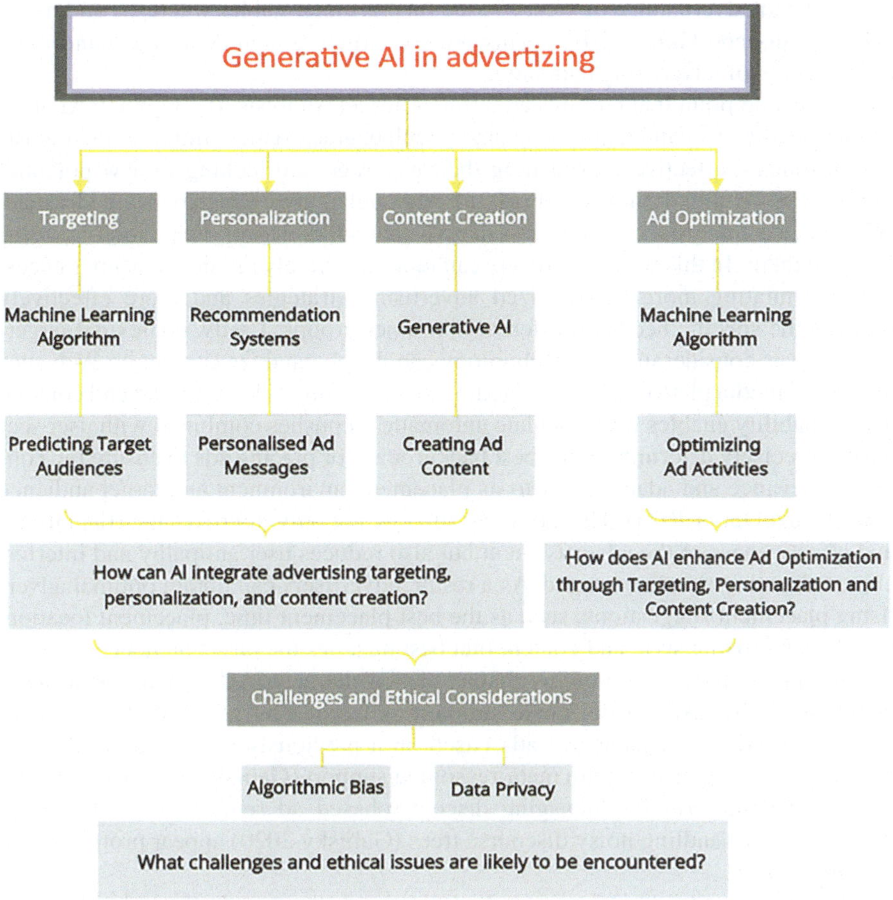

Fig. 14 Four pillars of AI in advertisement

2. Personalization mainly uses technologies like the Recommendation System and Virtual Assistant to tailor the most relevant and appealing advertising content for each user content creation utilizes Generative AI and NLP technology to generate creative content that can pique users' interest.
3. Ad optimization leverages LLMs and RL techniques to adjust advertising strategies dynamically, maximizing advertising effectiveness and return on ad investment.

Audience segmentation algorithms can analyze consumer data and target specific user subsets through very refined descriptive targeting criteria (Bateni et al. 2017), identifying different customer groups (Chandra et al. 2022). Based on this, it is possible to identify each group's unique needs and preferences, predicting which ads are most likely to succeed, thus achieving scientific targeting. By segmenting customer data, advertisers can more accurately target their audiences and tailor a personalized advertising experience based on consumer habits, interests, and needs (Theodoridis and Gkikas 2019). This precise targeting significantly enhances the effectiveness of advertising campaigns.

The second point that requires attention is Target Analysis. By applying AI algorithms, advertisers can identify consumers with characteristics similar to their existing customers, effectively expanding their coverage and locking in new potential user groups (Muhlhoff and Willem 2023). Although no two consumers are identical, AI's precise analysis can identify shared characteristics or behavioral patterns between them. In this way, advertisers can more accurately predict consumer needs, thus formulating more personalized advertising strategies and more effectively meeting the specific needs of different consumer groups. Lastly, contextual targeting deserves consideration. AI technology can deeply analyze content on Web sites and social media platforms, understanding and grasping its background and context. This capability enables it to calculate automatic ad pushes combined with user scenarios, precisely determining the best background for placing ads to ensure the content's relevance and adapt ability to its placement environment and target audience (Bansal and Gupta 2023). This targeting strategy not only improves the adaptability and effectiveness of the advertisement but also reduces user antipathy and interference, enhancing user acceptance. As a result, advertisers can obtain optimal advertising placement suggestions, such as the best placement time, placement location, and the advertising style and content that best matches the target audience.

The approach described in this chapter can be characterized as **neuro-symbolic**, as the symbolic, explicit discourse structure is learned by an LLM through fine-tuning to predict a sequence of ads based on a previously successful seed ad. A neuro-symbolic approach with meta-reasoning support (Galitsky et al. 2023) would be an ideal platform for managing discourse-based ad sequences. Additionally, techniques for handling noisy discourse trees (Galitsky 2020) appear promising for this application.

The author's interest in automated ad creation began nearly three decades ago. The Scenario Synthesis system proposed in Galitsky (1998) aimed to automate the creation of advertisement scenarios, enhancing their appeal, making them more natural and memorable, and improving composition performance. This system was based on a model of formal logical anecdotes derived from natural language, with deductive tools developed for their representation. This tool, using meta-language reasoning about action, spatial-temporal relationships, knowledge, and belief, was capable of understanding anecdotes and synthesizing formal scenarios. The Scenario Synthesis prototype was designed to generate multiple scenarios for Internet advertisements.

The scenario synthesis project later evolved into ad persuasiveness technology, presented in Galitsky and Kuznetsov (2008). We conducted a comparative analysis of two sources of argumentation-related information to evaluate the validity of scenarios in agent interactions. The first source is the overall structure of a scenario, which includes communicative actions in addition to attack relations and is learned from prior multi-agent interactions. Our earlier studies introduced a concept-based learning technique for this source, representing scenarios as directed graphs with labeled vertices (for communicative actions) and arcs (for temporal and attack relations). The second source involves traditional methods of handling argumentative structures in dialogues, assessing the validity of individual claims. We developed a system where data from both sources are visually specified to assess the validity of customer complaints. Evaluating the contribution of each source shows that both sources of argumentation-related information are essential for building persuasive scenarios.

We observed that the LLM and Discourse Ad Generation and Targeting (LDAGT) system, which leverages the capabilities of LLMs to generate and optimize ad content, offers a significant enhancement in the overall ad creation process. By structuring the ad delivery using discourse trees, the LDAGT system organizes advertisements in a logical and persuasive manner that aligns with the user's cognitive journey from initial exposure to final conversion.

The discourse trees guide the flow of each ad sequence by incorporating rhetorical relations, such as Elaboration, Justification, and Consequence, ensuring that the narrative structure of the ads is cohesive and compelling. This systematic organization allows the ad sequence to progressively address potential user concerns, highlight product benefits, and deliver persuasive arguments that culminate in a strong call to action.

In addition to improving the quality of the ad content, the LDAGT system significantly enhances ad targeting by dynamically adjusting to user data, such as previous interactions, click-through rates, and conversion rates. By continuously refining the ad structure based on real-time feedback and leveraging past performance data, LDAGT ensures that each ad is not only relevant but also strategically optimized for maximum engagement.

The combined impact of discourse-driven content generation and data-informed targeting leads to a notable increase in ad efficiency. This approach not only boosts user engagement and conversion rates but also reduces the time and resources required for ad creation, making it a powerful tool for marketers seeking to optimize their advertising strategies.

References

AffStuff (2024) Ad sequencing: better targeting and ad relevance (affstuff.com). https://affstuff.com/glossary/ad-sequencing/

Ahmadi I, Nabout NA, Skiera B, Maleki E, Fladenhofer J (2024) Overwhelming targeting options: selecting audience segments for online advertising. Int J Res Mark 41(1):24–40

Austin A (2020) Facebook retargeting ads for ecommerce: how to address the full funnel & drive more remarketing conversions. https://commonthreadco.com/blogs/coachs-corner/facebook-ads-retargeting

Aziz A, Telang R (2016) What is a digital cookie worth? SSRN working paper series. https://papers.ssrn.com/sol3/papers.cfm?abstract_id=2757325

Baig M (2013) Analyzing the advertising discourse—a journey from sight to mind. Int J Appl Linguist English Literat 2(1)

Bansal S, Gupta M (2023) Towards using artificial intelligence in neuromarketing. In: Gupta M, Jindal P, Bansal S (eds) Promoting consumer engagement through emotional branding and sensory marketing, IGI Global, pp. 16–23

Bateni M, Esfandiary H, Mirrokni V, Seddighin S (2017) A study of compact reserve pricing languages. In: Proceedings of the AAAI conference on artificial intelligence, San Francisco, CA, USA, vol 31(1), pp 363–368

Beales H (2021) The value of behavioral targeting. http://delivery.acm.org/10.1145/2960000/2956586/a4-hummel.pdf?ip=137.205.11.221&id=2956586&acc=OA&key=4D4702B0C3E38B35%2E4D4702B0C3E38B35%2E4D4702B0C3E38B35%2E5945DC2EABF3343C&__acm__=1557226973_29f1fdbf6ce5cb39a17ede7a256e138b

Benveniste D (2024) Meta-learning for model optimization. https://www.facebook.com/photo/?fbid=10163627781329338

Blakemore D (1988) The organization of discourse. Cambridge University Press

Chandra S, Verma S, Lim WM, Kumar S, Donthu N (2022) Personalization in personalized marketing: trends and ways forward. Psychol Mark 39(8):1529–1562

Chen AH-L, Gunawan S (2023) Enhancing retail transactions: a data-driven recommendation using modified RFM analysis and association rules mining. Appl Sci 13(18):10057. https://doi.org/10.3390/app131810057

Chen Y, Meng Z, Huang J (2021) Dynamic prompt adjustment for ad generation based on user feedback. In: Proceedings of the 14th ACM international conference on web search and data mining, pp 93–101. https://doi.org/10.1145/3437963.3441758

Cutura K (2017) Advertising on Google: the high performance cookbook. Packt Publishing

Deiss R, Henneberry R (2020) Digital marketing for dummies. Wiley

Dwivedi, Y. K., Hughes, L., Ismagilova, E., Aarts, G., Coombs, C., Crick, T., Duan, Y., Dwivedi, R., Edwards, J., Eirug, A., Galanos, V., Vigneswara Ilavarasan, P., Janssen, M., Jones, P., Kumar Kar, A., Kizgin, H., Kronemann, B., Lal, B., Lucini, B., & .Williams, M. D. (2021). Artificial intelligence (AI): multidisciplinary perspectives on emerging challenges, opportunities, and agenda for research, practice and policy. Int J Inf Manage, 57, 1–47

Finn C, Abbeel P, Levine S (2017) Model-agnostic meta-learning for fast adaptation of deep networks. In: 34th international conference on machine learning, Sydney, Australia, PMLR 70, p 2

Galitsky B (1998) Scenario synthesizer for the internet advertisement. Proc J Conf Infol Sci Duke Univ 3:197–200

Galitsky B (2020) Utilizing discourse structure of noisy user-generated content for chatbot learning. US Patent 10599885

Galitsky B, Kuznetsov SO (2008) Scenario argument structure vs individual claim defeasibility: what is more important for validity assessment? In: International conference on conceptual structure

Galitsky B, Ilvovsky D, Goldberg S (2023) Shaped-charge learning architecture for the human–machine teams. Entropy 25(6):924

Gao B, Huang L (2021) Toward a theory of smart media usage: the moderating role of smart media market development. Math Biosci Eng 18(6):7218–7238

Gao B, Wang Y, Xie H, Hu Y, Hu Y (2023) Artificial intelligence in advertising: advancements, challenges, and ethical considerations in targeting, personalization, content creation, and ad optimization. SAGE Open 13(4)

Goldfarb A, Tucker C (2011) Online display advertising: targeting and obtrusiveness. Mark Sci 30(3):389–404. https://doi.org/10.1287/mksc.1100.0583

Google (2023) About audience reporting. https://support.google.com/google-ads/answer/10566573

Huang MH, Rust RT (2021) A strategic framework for artificial intelligence in marketing. J Acad Mark Sci 49:30–50

Interactive Advertising Bureau (IAB) (2020) Consumer insights: relevance and personalization in advertising. https://www.iab.com/insights/consumer-insights-relevance-personalization-advertising/

Johnson GA, Shriver SK, Du S (2020) Consumer privacy choice in online advertising: who opts out and at what cost to industry? Marketing Science, 39(1):33–51.

Kathiriya S, Mullapudi M, Karangara R (2022) Optimizing ecommerce listing: LLM based description and keyword generation from multimodal data. Int J Sci Res

Kumar L, Khanna K (2024) Optimizing contextual advertising with large language models: a unified approach to ad content generation and targeting. J Adv Res Eng Technol 3(1):103–116

Kumar V, Rajan B, Venkatesan R, Lecinski J (2019) Understanding the role of artificial intelligence in personalized engagement marketing. Calif Manag Rev 61(4):135–155

Lai Z (2021) Research on advertising core business reformation driven by artificial intelligence. J Phys Conf Ser 1757(1):012018

Liu Y, Zhang M, Ma S (2020) Combining automatic metrics with human judgments for quality assessment of generated ad content. In: Proceedings of the 29th ACM international conference on information and knowledge management

Malthouse E, Copulsky J (2023) Artificial intelligence eco systems for marketing communications. Int J Advertising 42(1):128–140

Meguellati E, Han L, Bernstein A, Sadiq S, Demartini G (2024) How good are LLMs in generating personalized advertisements? In: Companion proceedings of the ACM on web conference 2024, pp 826–829

Meta (2023) Core Audience Targeting. https://www.facebook.com/business/news/Core-Audiences. Last downloaded Feb 4, 2026

Mita M, Murakami S, Kato A, Zhang P (2024) Striking gold in advertising: standardization and exploration of ad text generation. In: Proceedings of the 62nd annual meeting of the Association for Computational Linguistics (volume 1: long papers). Association for Computational Linguistics, Bangkok, Thailand, pp 955–972

Muhlhoff R, Willem T (2023) Social media advertising for clinical studies: ethical and data protection implications of online targeting. Big Data Soc 10(1):1–15

O'Brien (2018) Facebook ads & Facebook marketing mastery guide. https://www.udemy.com/course/facebook-ads-facebook-marketing-mastery-guide/

Opsahl-Ong K, Ryan MJ, Purtell J, Broman D, Potts C, Zaharia M, Khattab O (2024) Optimizing instructions and demonstrations for multi-stage language model programs. arXiv:2406.11695

Rafieian O, Yoganarasimhan H (2021) Targeting and privacy in mobile advertising. Mark Sci 40(2):193–218. https://doi.org/10.1287/mksc.2020.1235

Searle JR (1975) A taxonomy of illocutionary acts. Language, mind and knowledge. Minn Stud Philos Sci:344–369

Shannon M (2021) Embracing contextual vs. behavioral targeting. https://www.stackadapt.com/resources/blog/behavioural-vs-contextual/#compare

Skiera B (2022) Challenges of Marketing Automation: Linking MarTech & SalesTech. NIM Marketing Intelligence Review, 14(2), 11–17.

Sugarman J (2009) Advertising secrets of the written word: the ultimate resource on how to write powerful advertising copy. DelStar, Las Vegas, NV

Theocharous G, Thomas PS, Ghavamzadeh M (2015) Personalized ad recommendation systems for life-time value optimization with guarantees. In: Proceedings of the 24th International Joint Conference on Artificial Intelligence (IJCAI), pp. 1806–1812

Theodoridis PK, Gkikas DC (2019) How artificial intelligence affects digital marketing. In: Kavoura A, Kefallonitis E, Giovanis A (eds) Strategic innovative marketing and tourism, Springer Proceedings in Business and Economics, pp 1319–1327

Tucker C (2014) Social networks, personalized advertising, and privacy controls. J Mark Res 51(5):546–562. https://doi.org/10.1509/jmr.10.0355

Visioneerit (2024) Understanding social media analytics. https://www.visioneerit.com/blog/understanding-social-media-analytics

Wang J, Gou L, Shen HW and Yang H (2018) DQNViz: A Visual Analytics Approach to Understand Deep Q-Networks, in IEEE Transactions on Visualization and Computer Graphics, 25(1): 288–298. https://doi.org/10.1109/TVCG.2018.2864504

Wiredu J (2023) An investigation on the characteristics, abilities, constraints, and functions of artificial intelligence (ai): the age of ChatGPT as an essential. Inf Manag 108(3):62614–62620

Yan R, Nie J, Li X (2021) Incorporating user demographics and browsing history into prompt engineering for ad generation. In: Proceedings of the 44th international ACM SIGIR conference on research and development in information retrieval, pp 2078–2086

Zhao J, Qiu G, Guan Z, Zhao W, He X (2018) Deep reinforcement learning for sponsored search real-time bidding. In: Proceedings of the 24th ACM SIGKDD International Conference on Knowledge Discovery & Data Mining, pp. 1021–1030

A Computational Framework for Analyzing and Supporting Invention Dialogues

Boris Galitsky and Alexander Rybalov

1 Introduction

Innovation and invention have long been regarded as processes driven by creativity, problem-solving, and collaboration. At the heart of these processes lies the dialogue between inventors, which serves as a medium for exchanging ideas, defining goals, identifying constraints, and iteratively refining solutions. Such dialogues are inherently structured yet dynamic, reflecting the interplay between constraints imposed by the problem space and the conceptual flexibility required to generate novel solutions. This chapter proposes a formal model for inventor dialogues, framing invention as a process that can be analyzed through the lens of category theory and constraint satisfaction problems (CSPs).

Representing inventions in the form of dialogue has significant pedagogical value because it engages learners actively, enhances comprehension, and stimulates critical thinking. Dialogues allow students to follow a conversation between characters, making it easier to explore ideas step by step. Also, dialogues simulate a classroom discussion where learners can anticipate questions and mentally respond, reinforcing understanding. Dialogue introduces contrasting viewpoints, pushing learners to evaluate and synthesize ideas. Moreover, it encourages learners to think hypothetically—how an invention might impact society or solve specific problems.

An example scenario on 3D Bioprinting for Organ Transplants can look as follows:

There is indeed a noticeable scarcity of documented records capturing actual dialogues between famous inventors. While it is widely acknowledged that many

The original version of the chapter has been revised. A correction to this chapter can be found at https://doi.org/10.1007/978-3-032-16524-4_13

B. Galitsky, *Applications of Neuro-Symbolic Artificial Intelligence*, https://doi.org/10.1007/978-3-032-16524-4_9

Student A: "Why can't we just grow organs in labs?"
Mentor: "That's what 3D bioprinting tries to solve. Instead of waiting for donors, it prints tissues layer by layer."
Student B: "But what materials do they use?"
Mentor: "Special bio-inks made from living cells. For example, they use stem cells to build custom tissue structures."

groundbreaking inventions likely emerged from discussions, debates, and collaborative exchanges of ideas, concrete examples of such inventive dialogues remain elusive. Conversations between inventors and their peers may have played a pivotal role in refining concepts, addressing technical challenges, and sparking creative insights. However, the absence of preserved transcripts or detailed accounts has left this aspect of the inventive process largely speculative. Despite thorough searches, the authors were unable to uncover any verifiable examples of dialogues that directly contributed to specific inventions through online sources. This gap highlights the need for further archival research and exploration into personal correspondences, laboratory notebooks, and memoirs that might shed light on the conversational dynamics behind some of history's most transformative innovations.

One famous dialogue that led to a well-known invention occurred between Alexander Graham Bell and his assistant, Thomas Watson, during the development of the telephone in 1876 (Fig. 1).

While experimenting with the device, Bell famously said, "Mr. Watson, come here. I want to see you."

These were the first words transmitted via a telephone, marking a groundbreaking moment in communication history. This exchange demonstrated the practical

Fig. 1 Alexander Bell and his invented telephone

application of voice transmission over electric wires, revolutionizing global communication.

Another example is the Manhattan Project discussions involving Albert Einstein and Leo Szilard in the 1930s. Their letters to President Franklin D. Roosevelt, warning of Nazi Germany's potential to build atomic weapons, led to the development of the atomic bomb during World War II (Fig. 2).

Oppenheimer: "If we can sustain a chain reaction, what yields are we realistically expecting?"

Enrico Fermi: "In theory, with enough uranium-235 or plutonium, we could produce an explosive energy equivalent to thousands of tons of TNT."

Edward Teller: "And if the reaction runs away, the atmosphere—do we risk igniting it?"

Oppenheimer (pausing): "That's precisely what we must calculate. Safety first. But if the calculations hold..."

Hans Bethe: "Then the energy release could be controlled—directed—enough to end the war."

Oppenheimer (nodding): "But at what cost? Still, we have to proceed. We cannot let the Nazis get there first."

Traditional approaches to modeling inventions often emphasize brainstorming techniques, heuristic optimization, or rule-based reasoning systems. While these methods capture aspects of creativity, they frequently overlook the formalizable structure of invention dialogues, which integrate logical reasoning, constraint

Fig. 2 The hypothetical dialog around the Manhattan project

handling, and conceptual blending. This chapter addresses this gap by providing a mathematical framework that not only represents the informational flow within inventor dialogues but also models invention as a form of constraint satisfaction that balances feasibility with novelty.

We conceptualize invention as a dialogue-driven process where inventors collaboratively explore and refine a design space through constraints and transformations. In this framework:

- Category theory provides the algebraic tools to formalize relationships between concepts, capturing transformations and mappings between ideas.
- Constraint satisfaction problems define the boundaries within which invention operates, modeling constraints as logical and computational entities to ensure solutions are both feasible and innovative.

By integrating these two formal systems, we bridge the linguistic and mathematical dimensions of invention, enabling precise modeling of dialogue dynamics while preserving the exploratory nature of human creativity.

1.1 Contribution

This chapter makes the following contributions (Fig. 3):

1. Dialogue Formalism for Invention—It develops a formal model for inventor dialogues, highlighting their structure, patterns, and transitions.
2. Category-Theoretic Representation—It applies category theory to describe transformations in idea development, capturing abstraction, refinement, and integration processes.

Fig. 3 Visualization of contribution of this chapter

3. Constraint Satisfaction Problem Framework—It frames invention as a CSP, modeling constraints and dependencies within the design process (Galitsky 2025).
4. Unified Model for Creative Reasoning—It integrates the above methods to provide a unified computational framework for analyzing and supporting invention dialogues.
5. Developing pedagogical value for invention dialogues.

This research is motivated by the need for structured frameworks that support invention in domains ranging from engineering and design to AI-driven innovation platforms. The proposed model has implications for computational creativity, knowledge representation, and collaborative AI systems, offering tools to simulate and enhance human-machine co-creation.

2 Categorial Theory Formalization

Category theory provides a universal language for defining interactions among objects in all of mathematics, physics, computer science, and mathematical logic. Category theory has recently seen increasing use in machine learning, including dimensionality reduction (McInnes et al. 2018) and clustering. One unique aspect of defining machine learning is extending functors in a category (Mahadevan 2024).

2.1 First Example of an Invention Dialogue

We introduce the full dialogue and highlight its entities and components for categorial analysis (Fig. 4).

We identified three components of an invention dialogue highlighted in three ways:

1. Problem formulation
2. Problem resolution attempt
3. Successful break-through

Handling components and entities is important as we design prompts for LLMs to auto build invention dialogues. In prompt constructions, we need to outline the variable and constant parts of prompts. Domains in a dialogue come via the certain sequence, and entities could be varied to obtain a wide spectrum of dialogues.

Scene: Dr. Maxwell's workshop, filled with mechanical blueprints and prototypes of various inventions. Alex, a young and curious engineer, approaches Dr. Maxwell with an idea.

Alex: (excited) Dr. Maxwell, I've been thinking about how to make airplanes faster. Propellers seem to have reached their limit. I mean, we've improved engines, but the propeller design just can't push planes any faster through the air.

Dr. Maxwell: (leaning back, thoughtful) You're right, Alex. Propellers work by cutting through the air, but they encounter drag, especially at higher speeds. There's only so much power you can extract from that method. What do you have in mind?

Alex: What if... we could move the air faster than any propeller can spin? I've been reading about rocket propulsion—how rockets push gas backward to propel themselves forward. Couldn't we apply something like that to airplanes?

Dr. Maxwell: (raising an eyebrow) Rockets? They work in space because there's no atmosphere, but here, in the air... it's a different story. What you're describing sounds like pure thrust—without relying on an external propeller. You want to push the plane through the air by forcing air or gas backward, right?

Alex: Exactly! Imagine if we could compress the air coming in, ignite fuel with it, and then expel it at a high speed to create thrust. Instead of just spinning a propeller, we're pushing the plane forward directly by accelerating the air.

Dr. Maxwell: (grabs a pencil, sketching on a piece of paper) You're talking about continuous combustion in a confined space. We'd need to find a way to pull in air, compress it, mix it with fuel, ignite it, and push the exhaust out at much higher speeds. It would have to be an internal combustion engine, but without pistons, just airflow.

Alex: (pointing to the sketch) Exactly! A high-speed air intake, followed by compression. Then, we ignite the mixture and shoot it out the back, creating thrust. It's like a controlled explosion, propelling the plane forward. What if we could use turbines to compress the air before ignition?

Dr. Maxwell: (nodding, thinking out loud) A turbine, yes! If we install a compressor at the front to pull in air and compress it, then introduce fuel and ignite it, the high-speed exhaust could drive a turbine that helps keep the compression going... a self-sustaining cycle. The faster the exhaust, the faster the plane.

Alex: (excited) Exactly! And with no propeller to hold us back, the plane could go faster and higher without the same drag limitations. The exhaust gases would do all the work.

Dr. Maxwell: (grinning) You may have something here. This would be revolutionary—no need for a propeller. You'd be using the air itself as fuel for speed! The idea of using turbines to sustain compression and ignition—it could work. We would call this... a "jet engine."

Alex: (beaming) A jet engine... I like the sound of that. It's like we're not just flying anymore, we're truly propelling ourselves through the air.

Dr. Maxwell: (nodding) And once we perfect the design, we'll be able to achieve speeds no one thought possible with propeller engines. Let's start working on a prototype. We'll need to solve the challenges of heat, materials, and efficiency, but I believe we can do it. This could change aviation forever.

Alex: (smiling) Then let's get to work!

Fig. 4 An invention dialogue with highlighted entities and components

2.2 *Categorial Model*

To formalize a model of an invention dialogue using category theory, we start by defining key components such as objects, morphisms, functors, and natural transformations. Here's a breakdown of how to build the model step by step.

2.2.1 Objects in the Dialogue Model

In the categorical framework, **objects** in a dialogue represent participants, physical entities under discussion, and mental attributes associated with these entities. Specifically:

- Participants : People engaged in the dialogue (e.g., inventors, engineers)
- Physical Objects : The things being discussed, such as the engine or parts of the invention
- Mental Attributes : Intentions, ideas, or desires related to physical objects (e.g., "intent to improve an engine")

We will denote these objects as:

- P for a participant in the dialogue
- O for a physical object (e.g., the engine)
- A for a mental attribute (e.g., an intent to improve the engine)

An **epistemic category** models a state of knowledge or perspective of a dialogue participant. It contains both the physical objects of interest to a participant and the mental attributes associated with those objects.

For each participant P, an **epistemic category** E_P can be defined as:

- Objects in E_P: These are pairs (O, A), where O is a physical object and A is a mental attribute associated with O.
- Morphisms in E_P: These represent transitions between different mental states or perspectives regarding the same object or different objects. For instance, a morphism could represent the change in a participant's view about how to improve an engine or a shift in focus from one part of the engine to another.

An **epistemic functor** F maps one epistemic category (corresponding to an initial dialogue state) to another epistemic category (corresponding to a consecutive dialogue state). This captures the evolution of the dialogue, reflecting how a participant's knowledge or intentions change as the conversation progresses.

For each participant P, an epistemic functor $F{:}\ E_P^1 \rightarrow{:}\ E_P^2$ maps:

- Objects: (O, A) in $: E_P^1$ to new objects (O', A') in $: E_P^2$, where O' and A' represent updated knowledge or intentions about the physical object after some part of the dialogue.
- Morphisms: Any transition or change in mental state or knowledge about physical objects is mapped from one dialogue state to another.

The **category of all epistemic functors** consists of all possible functors between epistemic categories for each participant across different stages of the dialogue. This can be considered as representing the structure of the entire dialogue process.

Each participant's epistemic functor can be seen as a natural transformation between different epistemic categories. Specifically, a **natural transformation** $\eta{:}F \rightarrow G$ between two functors F, G would describe a transformation of dialogue states (and their associated changes in knowledge or mental attitudes) that respects the underlying structure of the participants' epistemic categories.

2.3 *Defining Natural Transformation*

A natural transformation is a mapping between two functors. The functors must have the same domain and codomain as each other. There is also the naturality square condition that must apply for this to be a natural transformation as we shall see below.

We take an object x in C, and we map it into a morphism in D (Fig. 5). For every object x in C, there is a morphism in D ($f\,x \rightarrow g\,x$) known as the components of α at x (written α_x).

So now we know what happens to an object in C, we now want to know what happens to structure in C, and for this we see what happens to a morphism in C. A morphism in C would map to two component morphisms in D. This diagram must commute for every morphism in C. So α_x is the component of the natural transformation at x, and α_y is the component of the natural transformation at y. For naturality we require that the square in D commutes (Fig. 6):

$$\alpha_y \cdot M = N \cdot \alpha_x$$

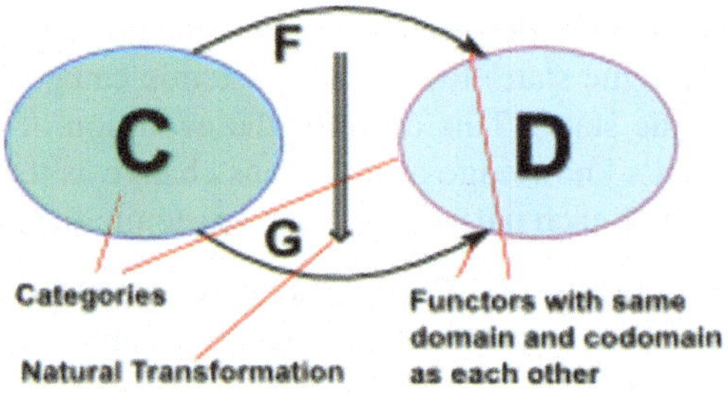

Fig. 5 Defining natural transformation

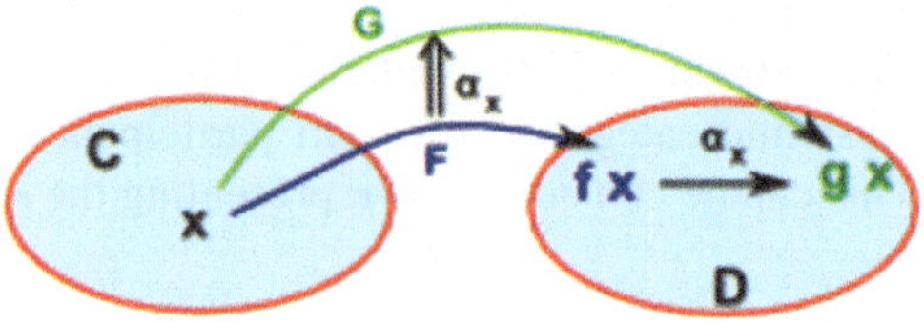

Fig. 6 Naturality square

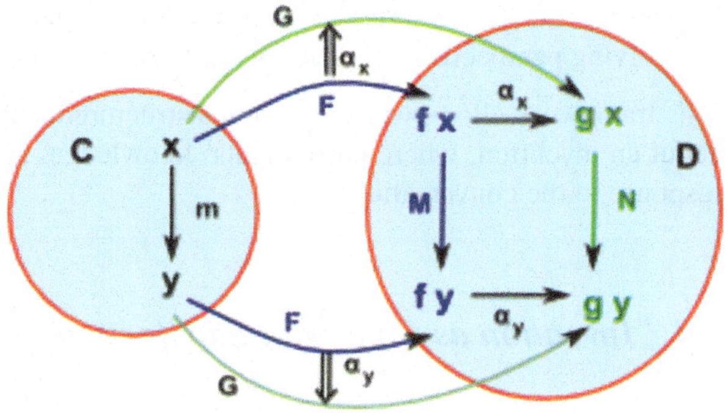

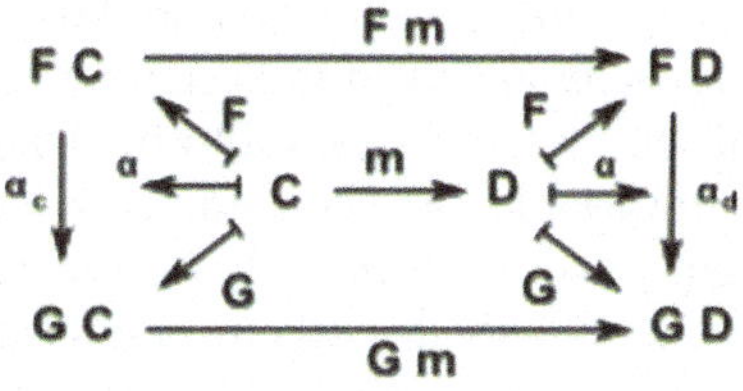

where: $M = F\,m$ and $N = G\,m$

$$\text{so}: \alpha_y \bullet F\,m = G\,m \bullet \alpha_x$$

2.3.1 Dialogue Model as Natural Transformation

The natural transformation formalizes the idea that different paths of reasoning, questioning, and discussing in the dialogue are systematically related. A dialogue model can now be viewed as a natural transformation between epistemic functors *F* and *G* for different participants *P*.

For example, if participant P_1 and participant P_2 are discussing how to improve an engine, and they start from different initial states (different epistemic categories $E^1_{P_1}$ and $E^1_{P_2}$), the dialogue serves as a natural transformation between the functors $F^1_{P_1}$ and $F^1_{P_2}$, representing the change in their perspectives.

In this model of invention dialogue:

- Objects represent participants, physical objects, and associated mental attributes.
- Epistemic categories capture the knowledge and intentions about physical objects for each participant.
- Epistemic functors map changes in knowledge and mental attributes over the course of the dialogue.

- Natural transformations between functors model the relationships between the evolving perspectives of different participants.

This framework allows us to represent a structured, categorical model of a dialogue about an invention, where participants' knowledge, ideas, and intentions evolve in response to the conversation.

2.4 Invention as Constrain Satisfaction

Formulating an invention as a CSP involves defining the problem in terms of variables, domains, and constraints. This method provides a structured way to identify and solve the core challenges involved in designing an invention. CSP consists of variables (elements of the system that need to be determined, domains (the set of possible values for each variable), and constraints (relationships or rules that define how the variables interact or limit their values). The goal is to assign values to the variables from their domains such that all constraints are satisfied.

The steps to formulate an invention via CSP are as follows:

1. Define the problem. Identify the key functional goals and requirements of the invention. Break them down into measurable components. For example, if invention is a solar-powered water purifier, then the goal is to purify water effectively, operate using solar power, and be portable and affordable.
2. Identify variables, determining the key design elements that influence the invention. These could be physical dimensions, performance characteristics, or system components. For the solar-powered water purifier, the variables are:

   - *Type of purification technology* (*x1*), *e.g., UV filtration, reverse osmosis*
   - *Solar panel size* (*x2*)
   - *Battery capacity* (*x3*)
   - *Material type* (*x4*)
   - *Weight* (*x5*)

3. Define domains, specifying the possible values each variable can take. Domains should reflect realistic choices or ranges.

   - *x1:* {*UV* filtration, *reverse osmosis, activated carbon*}
   - *x2:* {*50* cm^2 *to 500* cm^2}
   - *x3:* {*10 Wh to 50 Wh*}
   - *x4:* {*plastic, aluminum, stainless steel*}
   - *x5:* {*1 kg to 5 kg*}

4. Specify constraints, defining the relationships or rules that must hold between variables. These capture the functional requirements and practical limitations of the invention.

- The solar panel size (*x2*) must generate enough energy to power the purification system (*x1*).
- The battery capacity (*x3*) must store sufficient energy for nighttime operation.
- The total weight (*x5*) must be less than 5 kg to ensure portability.
- The material (*x4*) must be non-toxic and corrosion resistant.
- The cost of components must not exceed a certain budget.

5. Solve the CSP using a CSP solver to find values for the variables that satisfy all constraints. This step often involves optimization techniques to identify the best solution (e.g., minimize cost, maximize efficiency). In our example, the solver might determine:

   - *x1: UV filtration*
   - *x2: 200 cm*^{2} *solar panel*
   - *x3: 20 Wh battery*
   - *x4: plastic*
   - *x5: 3 kg*

2.5 From Categories to Constrain Satisfaction

From the standpoint of category theory, constraint propagation can be viewed as morphisms. Constraints, variables, and domains are treated as objects in a category. Propagation rules (e.g., arc consistency algorithms) are modeled as morphisms that map objects (CSP states or variable domains) to other objects. These morphisms encode how constraints reduce or refine possible variable assignments.

A functor is then a mapping between categories that lifts the propagation process to a higher level of abstraction. For instance, category 1 represents the CSP problem space (variables, domains, and constraints), and category 2 represents transformations or refined solutions (e.g., narrowed domains, more consistent variable assignments). The functor maps CSP objects and propagation morphisms in one category to another category where propagation has been applied systematically.

In an abstract CSP solvers, functors could represent the transition from the initial CSP state to a refined state after applying propagation algorithms:

1. Type systems for CSPs: in programming, functors might implement CSP propagation rules while preserving type safety and compositionality.
2. Optimization frameworks: using category theory, a functor could model the search process in constraint satisfaction, ensuring it respects certain mathematical properties (e.g., monotonicity or consistency).

If the reader is familiar with functional programming: a "constraint satisfaction propagation functor" could represent something like a monad or applicative functor that applies propagation rules to CSP states in a composable way. For instance, a functor might encapsulate the propagation process, allowing transformations like *propagate*: CSP → CSP to be composed while maintaining structure.

Hence, a CSP propagation morphism functor likely refers to a structured, compositional framework (inspired by category theory) for modeling constraint satisfaction and propagation. It highlights morphisms as transformations (like propagation rules) between CSP states and functors as mappings between categories that abstract the propagation process, enabling compositional and reusable algorithms.

3 Phases of Invention Dialogue

Invention dialogues share a common structure that drives the discovery process. Here is the breakdown of invention dialogue features (Fig. 7):

1. **Problem Identification**: Each dialogue begins with the recognition of a challenge or limitation in the current technology or approach. The characters often express frustration or curiosity about an issue that needs solving.

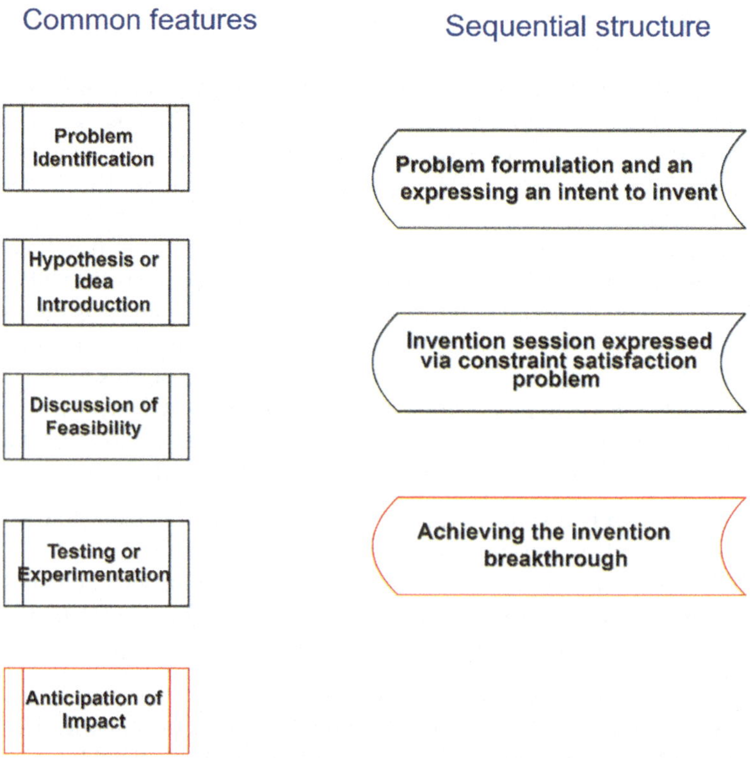

Fig. 7 Features and sequential structure of invention dialogue. Red frame indicates the components which are absent in unsuccessful inventions

2. **Hypothesis or Idea Introduction**: A new idea, hypothesis, or alternative approach is proposed, typically by a second character or as part of a collaborative brainstorming process. This is often based on an understanding of existing science or technology, suggesting a new path.
3. **Discussion of Feasibility**: The characters explore the feasibility of the proposed idea, often discussing its theoretical basis, challenges, or how it differs from current methods. There may be skepticism or clarification, but the characters ultimately show enthusiasm about testing the idea.
4. **Testing or Experimentation**: The conversation moves toward a plan for experimentation or testing the new idea. The characters outline a basic methodology to see if their theory holds.
5. **Anticipation of Impact**: The dialogue ends with excitement or realization of the potential impact of their innovation. The characters foresee how their discovery could revolutionize a field or solve a significant problem.

This structure mirrors real-world innovation processes, where problem-solving, hypothesis generation, collaborative brainstorming, and practical experimentation lead to groundbreaking discoveries.

In terms of its sequential structure, an invention dialogue includes three phases:

1. Defining the problem and setting the intent to invent. The process begins with clearly articulating the problem at hand. This involves understanding the challenge, identifying the constraints, and expressing a deliberate intent to discover a solution. The dialogue here focuses on framing the problem, setting objectives, and fostering a mindset geared toward innovation.
2. Collaborative problem-solving: the invention session. In this phase, the invention process is approached as solving a CSP. This involves systematically exploring potential solutions within the defined constraints, iterating through possibilities, and refining ideas. The dialogue revolves around brainstorming, analyzing trade-offs, and experimenting with creative strategies to overcome obstacles.
3. Achieving the invention breakthrough. The final phase is marked by the moment when a viable solution emerges, and the CSP is successfully resolved. The dialogue here shifts to celebrating the breakthrough, detailing the solution, and reflecting on the journey of discovery. This phase also sets the stage for translating the solution into actionable outcomes.

We visualize CSP as a state machine for jet engine invention dialogue (Sect. 2.1) in Fig. 8.

In category theory, we can define a *category of CSPs* where:

- Objects are CSP instances. Each CSP consists of variables, domains, and constraints.
- Morphisms between CSPs are constraint satisfaction morphisms, which map one CSP to another in a structure-preserving way (preserving the relationships defined by the constraints).

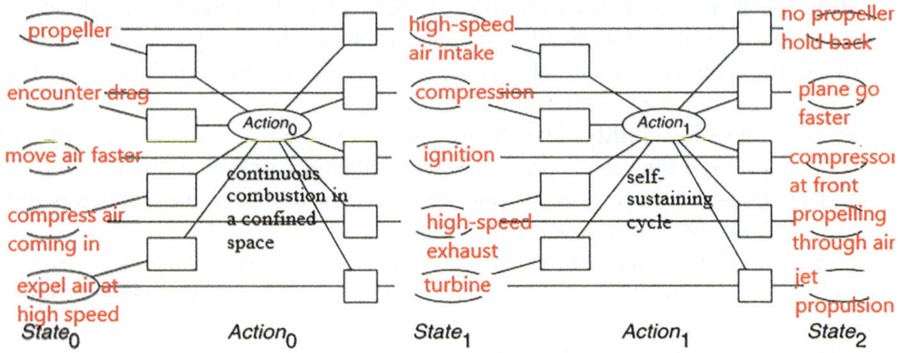

Fig. 8 Two actions in the course of invention of a jet engine. The first action is to use continuous combustion in confined space, and the second action is to achieve a self-sustain cycle in step 1

For two CSPs, *CSP1* and *CSP2*, a morphism $f{:}CSP1 \rightarrow CSP2$ might involve:

- A mapping of variables in *CSP1* to variables in *CSP2*
- A mapping of constraints in *CSP1* to constraints in *CSP2*, ensuring that solutions are preserved between *CSP1* and *CSP2*

A *functor* is a structure-preserving map between categories. In the context of CSPs, a *constraint satisfaction morphism functor* could map from the category of CSPs (let's call it *C_CSP*) to another category D, which could be, for example, a category representing solution spaces, sets, or algebraic structures).

Formally, a functor $F{:}C\_CSP \rightarrow D$ F: consists of:

- Object Mapping: For each CSP A in *C_CSP*, the functor maps *A* to an object *F*(*A*) in *D*. For instance, *F*(*A*) might represent the solution space of *A* as a set.
- Morphism Mapping: For each morphism $f{:}A \rightarrow B$ in *C_CSP*, the functor maps *f* to a morphism $F(f){:}F(A) \rightarrow F(B)$ in *D*. This morphism should preserve the structure induced by the CSP morphism *f*, meaning it should map solutions of *A* in a way that respects the constraints in *B*.

A few examples help illustrate different uses of constraint satisfaction morphism functors:

- Solution Space Functor: A functor from CSP to the category of sets *Set* might map each CSP A to the set of its solutions. A morphism between CSPs then maps solution sets, preserving the structure of possible assignments that satisfy the constraints.
- Algebraic CSP Functor: In algebraic CSPs, a functor might map CSPs to algebraic structures, such as groups or lattices, where solutions and constraints represent homomorphisms between different algebraic structures. This functor would help translate CSP problems into a form where algebraic tools can be applied.

- Graph Representation Functor: If we represent CSPs as graphs where variables are nodes and constraints are edges, a functor could map CSPs to graph categories, translating CSP morphisms into graph homomorphisms. This allows graph-theoretic techniques to be used to analyze and solve CSPs.

To qualify as a functor, a constraint satisfaction morphism functor must satisfy two properties:

1. Identity preservation: The identity morphism in CSP for each CSP A (which maps A to itself) must map to an identity morphism in D.
2. Composition preservation: For any two composable morphisms f and g in CSP, the functor must map the composition $g \circ fg$ to $F(g) \circ F(f) F(g)$.

Constraint satisfaction morphism functors offer a formal way to translate CSP problems into other categories where different methods of analysis and solution can be applied. This approach is particularly useful in theoretical computer science and artificial intelligence, where many real-world problems can be formulated as CSPs, and mapping these problems into different structures can provide insights into their solvability, complexity, and solution spaces.

3.1 Invention Failure

Invention attempts can be successful or unsuccessful. We will now provide prompts and examples of successful and unsuccessful invention dialogues on the same topic.

Prompt:

Compose a dialogue between two inventors inventing a frame for a backpack to move it on a mountain trail. The dialogue starts with problem formulation, followed by a discussion on satisfying all invention constraints one by one, and concluding with invention description

ChatGPT:

Dialogue: Inventing a Backpack Frame for Mountain Trails

Inventor 1 (Alex): Alright, we've got a challenge: how do we design a backpack frame that can make carrying heavy loads on a mountain trail easier? Traditional backpacks work, but they put a lot of strain on the back and shoulders.

Inventor 2 (Jordan): True. The key problem is balancing the weight efficiently while keeping mobility intact. What are the main constraints we need to address?

Alex: Let's break it down:

1. **Weight distribution**—the load needs to be evenly spread to avoid strain.
2. **Mobility**—it must allow smooth movement, especially on uneven trails.

3. **Stability**—it should keep the load secure, no swaying or shifting.
4. **Comfort**—it needs to be ergonomic, with padding and proper ventilation.
5. **Durability and weight**—strong but lightweight materials are essential.

Jordan: Great. For **weight distribution**, what if we create a frame that shifts the load to the hips instead of the shoulders?

Alex: Yes! A rigid external frame that transfers the load to the hips via a padded belt could work. Maybe with adjustable straps to fine-tune the center of gravity.

Jordan: Good. But we also need **mobility**. How do we ensure the wearer can bend, twist, and climb comfortably?

Alex: Let's make the frame articulated at key joints—around the shoulders and lower back—so it can flex with the body's movements, almost like an exoskeleton.

Jordan: Perfect! For **stability**, how do we keep the load from shifting as they move?

Alex: Compression straps along the sides of the frame. They'll hold the gear snugly and prevent any swaying. Maybe even some stabilizer fins that hug the body without restricting movement.

Jordan: Smart. Now, **comfort**. What can we do to reduce pressure points and improve ventilation?

Alex: Use mesh panels for airflow, combined with contoured foam padding on the hips and shoulders. And adjustable straps everywhere—custom fit is key.

Invention Description:

Alex: We've created a modular backpack frame designed specifically for mountain trails. It features a titanium external frame that transfers the load to the hips via a padded, adjustable belt. The frame is articulated at key joints, allowing full mobility while maintaining stability…

Prompt: now convert this dialogue into an unsuccessful invention, which run against something that cannot be done

Dialogue: Inventing a Backpack Frame for Mountain Trails (Unsuccessful Attempt)

Inventor 1 (Alex): We've got a tough challenge: designing a backpack frame that makes heavy loads easier to carry on a mountain trail. Traditional backpacks strain the back and shoulders—there has to be a better way.

Inventor 2 (Jordan): Agreed. Let's outline the main constraints:

1. Weight distribution—...

Alex: For **weight distribution**, what if we shift all the load to the legs instead of the shoulders?

Jordan: To the legs? Hmm. Maybe we could create a frame that extends down to the thighs, kind of like an exoskeleton, transferring the weight directly to the legs.

Alex: Exactly! The frame could bypass the upper body entirely and connect to some lightweight braces on the thighs. No more shoulder or back strain!

Jordan: That sounds revolutionary. But how do we ensure **mobility**? Wouldn't braces on the legs restrict movement?

Alex: Good point, but we could make them flexible. Maybe they could hinge at the knees and swivel at the hips. That way, hikers can still bend and climb easily.

Jordan: Interesting. But now for **stability**—how do we prevent the frame from shifting while the user moves?

Alex: We could attach the frame tightly to the waist and thighs, with locking mechanisms that adjust to the user's leg movements.

Jordan: I see the issue already. If the frame locks with the legs but still needs to move with them, we risk either limiting motion or causing chafing. And adding hinges increases complexity and failure points. Wouldn't this frame make the user feel like they're carrying a rigid cage?

Alex: True, but maybe ultralight materials can fix that. What if we use carbon fiber braces?

Conclusion: **Unsuccessful Invention**

Jordan: We aimed to develop a frame that shifts weight to the legs, bypassing the back and shoulders, but the constraints of mobility, stability, and user fatigue make this approach unworkable. The complexity of hinging and locking mechanisms, combined with the increased strain on leg muscles, renders the idea both impractical and unsafe.

4 Constraint Satisfaction for Invention Dialogues

4.1 Exploration-Exploitation

The "Exploitation vs. Exploration" discussion is simply a "new" fancy way of referring to a problem that is as old as making decisions about efficient resource allocation. It is a name that has been borrowed from probability theory, very likely via product management, data science, and financial services circles. Rocca (2021) provides a comprehensive definition for both:

1. Exploitation consists of taking the decision assumed to be optimal with respect to the data observed so far. This safe approach tries to avoid bad decisions as much as possible but also prevents from discovering potential better decisions.
2. Exploration consists of not taking the decision that seems to be optimal, betting on the fact that observed data are not sufficient to truly identify the best option. This more risky approach can sometimes lead to poor decisions but also makes it possible to discover better ones, if there exist any.

The problem of choosing between exploitation and exploration can be encountered in many situations where observations drive decisions and decisions lead to new observations.

The idea of a *constraint satisfaction exploration-exploitation functor* combines concepts from constraint satisfaction, exploration-exploitation trade-offs (common in optimization and reinforcement learning), and category theory. In this framework, we aim to map CSPs to strategies for balancing exploration and exploitation in solution spaces. Such a functor would formalize the relationship between CSP structures and the decision-making process for exploring new solutions versus exploiting known good ones.

Exploration CSP refers to searching or navigating through the solution space of a CSP to discover possible solutions. It is often associated with techniques like *brute-force search*, *backtracking*, or *random sampling*, where the focus is on exploring the entire or a large part of the solution space without necessarily favoring the best solutions. Exploration CSP includes searching through less-visited regions of the solution space to discover new solutions. In CSPs, this might involve testing variable assignments that have not been thoroughly examined, even if they have uncertain outcomes.

Exploitation CSP refers to using prior knowledge or feedback to focus on the most promising areas of the solution space. This is often associated with more targeted search methods like *hill climbing*, *simulated annealing*, or *greedy algorithms*, where the search is steered toward areas that are expected to yield good solutions based on past performance. Focusing on promising regions of the solution space that are known to produce feasible or optimal solutions, one refinines known assignments to maximize satisfaction or optimize a criterion.

CSP consists of variables with domains, constraints that restrict combinations of variable values, and solutions that are assignments of values to variables that satisfy all constraints. Categories of CSPs are as follows:

1. Objects are CSPs, each representing a different problem instance.
2. Morphisms are mappings between CSPs that preserve constraint structures, as in typical CSP morphisms. These could translate to transformations between problem instances where similar solutions or solving techniques can apply.

An *exploration-exploitation functor* is a mapping from the *category of CSPs* to a *category of decision processes* or *strategies* that captures both exploration and exploitation strategies for solving CSPs. Formally, a functor $F{:}CSP \rightarrow Strategy$ would map each CSP instance A to an object $F(A)$ in a category *Strategy*, where

F(*A*) represents a specific strategy or decision-making process for solving A using exploration-exploitation techniques. Similarly, a CSP morphism $f{:}A \rightarrow B$ maps to a morphism $F(f){:}F(A) \rightarrow F(B)$ in *Strategy*, which translates strategies between CSPs.

Components of the exploration-exploitation functor are:

1. Object Mapping: For each CSP *A*, the functor maps it to a specific exploration-exploitation strategy *F*(*A*) in *Strategy*. For example, *F*(*A*) could represent a probabilistic strategy that starts by exploring a broad range of assignments and then shifts toward exploitation as high-satisfaction solutions are identified.
2. Morphism Mapping: For each morphism $f{:}A \rightarrow B$ between CSPs, *F*(*f*) translates an exploration-exploitation strategy for A into a compatible strategy for *B*. This morphism mapping might involve adjustments in the balance between exploration and exploitation, based on similarities in the structure of *A* and *B*. For example, if *B* is known to be a stricter version of *A*, *F*(*f*) might favor exploitation strategies focused on tightening known solutions.

Example scenarios for the exploration-exploitation functor are:

1. Adaptive CSP Solvers: Suppose strategy *Strategy* represents a category of CSP solvers that adaptively balance exploration and exploitation. Given a CSP instance *A*, the exploration-exploitation functor could map it to a solver that initially explores diverse variable assignments to cover the search space and gradually converges on refined solutions.
2. Probabilistic Solution Methods: If *Strategy* includes probabilistic strategies, such as simulated annealing or genetic algorithms, the functor could map CSPs to such methods, tuning exploration and exploitation parameters based on the problem's complexity and constraint density. A morphism between CSPs could adjust these parameters to reflect the nature of constraints in the target problem.
3. Transfer Learning for CSPs: In machine learning contexts, this functor could formalize a transfer-learning approach for CSPs by mapping a known strategy for a source CSP *A* to a new CSP *B*. For instance, if *A* and *B* share structural similarities, *F*(*f*) could favor exploitation based on known solutions from *A* when applied to *B*, reducing redundant exploration.

The exploration-exploitation functor is useful for building generalized, adaptable solvers for CSPs. By formalizing strategies that dynamically balance exploration and exploitation, this functor can improve efficiency in finding solutions, especially for large or complex CSPs, enable the reuse of problem-solving strategies across related CSPs, and guide automated reasoning about when to explore new solutions versus refining known ones.

Hence, an exploration-exploitation functor in the context of CSPs leverages category theory to formalize a decision-making strategy for balancing between discovering new solutions and optimizing known ones, enabling more efficient and adaptive approaches to solving constraint satisfaction problems.

4.2 Functors Representing Exploration and Exploitation

A functor from the category of CSPs to another category can model different types of exploration and exploitation strategies, each representing a different way of mapping a CSP into another structure, possibly with the aim of solving it. Our first example is exploration functor for random walks. In this case, we might define a functor that models an *exploration* strategy through random walks or stochastic search techniques.

1. Category: $\{CSP\}$ (category of CSPs) → $\{Graph\}$ (category of graphs).
2. Object Mapping: Each CSP is mapped to a *search tree* or *state space graph*. Each variable in the CSP corresponds to a node, and the constraints define the edges between nodes (possible assignments to variables that respect constraints).
3. Morphisms: A CSP morphism f between two CSPs is mapped to a *graph traversal* or a *random walk* between the two graphs, where the search space is explored non-deterministically.

In this setting, the functor represents an exploration strategy where the solution space is explored via random decisions, and each step in the exploration corresponds to a state transition in the graph. In exploitation functor-greedy algorithm, a functor could represent an exploitation strategy, like a greedy algorithm, where the CSP is mapped to a solution space that is explored in a way that always seeks the most optimal solution at each step:

1. Category: $\{CSP\}$ → $\{Set\}$ (category of sets, where solutions are considered as sets).
2. Object Mapping: Each CSP is mapped to the *set of variables and their domains*, along with the *partial assignments* that are currently being considered. A CSP might be translated into a set of possible configurations or partial assignments, from which the best configuration is chosen at each step.
3. Morphisms: A morphism f between two CSPs might be mapped to a *greedy selection function*, which maps partial assignments or variable assignments to the most promising candidates according to some evaluation function (such as minimizing conflicts or cost).

In this case, the functor corresponds to an exploitation strategy where the solver chooses the "best" solution at each step based on the evaluation function, focusing the search on high-value areas of the solution space.

We proceed to an example of exploration-exploitation balance functor-multi-armed bandit. A more sophisticated functor could model the balance between exploration and exploitation, much like in the multi-armed bandit problem from reinforcement learning. The multi-armed bandit problem is a classic problem in probability theory and decision-making that captures the essence of balancing exploration and exploitation. This problem is named after the scenario of a gambler facing multiple slot machines (bandits) and needing to determine which machine to play to maximize their rewards. The problem has significant applications in various fields, including online advertising, clinical trials, adaptive routing in networks, and more. The goal is to explore new strategies (new parts of the solution space) while also exploiting known strategies that have worked well in the past.

1. Category: $\{CSP\} \rightarrow \{Probabilistic\}$ (category of probabilistic processes, e.g., Markov decision processes or reinforcement learning environments).
2. Object Mapping: Each CSP is mapped to a *probabilistic model* of solutions, where each possible solution or partial solution is associated with a probability or expected value. The CSP is seen as a decision problem where different solution paths are taken probabilistically, with some paths explored more thoroughly and others based on previous experience.
3. Morphisms: A morphism f between two CSPs could be represented as a probabilistic policy that decides whether to explore (search in a new area) or exploit (focus on previously successful solutions), potentially adjusting the balance over time.

In this case, the functor represents a strategy that adapts to the problem by both exploring new solutions and exploiting promising solutions from previous experience. This is similar to how algorithms like *ε-greedy* or *Upper Confidence Bound* work in the context of the multi-armed bandit problem.

Applications in CSP solvers include exploration-exploitation functors in search algorithm. These functors are useful in CSP solvers that combine global search (exploration) with local search (exploitation). For instance, algorithms like *Genetic Algorithms* or *Simulated Annealing* may explore the search space randomly at first (exploration) and then focus the search in promising regions as it converges to a solution (exploitation). A functor can formalize this shifting behavior within a category-theoretic framework.

Also, they can be applied to AI and reinforcement learning. In AI, especially reinforcement learning, exploration-exploitation trade-offs are crucial. Functors can model how an agent explores different solutions to a CSP and then exploits the best solutions as it gains more experience or feedback from previous actions.

In the context of CSPs, functors modeling exploration and exploitation strategies offer a way to formalize how CSP solvers navigate the solution space. Functors can map CSPs to different categories representing various search strategies (e.g., random walks for exploration, greedy algorithms for exploitation), and by doing so, they provide a structured framework for understanding the relationship between problem-solving techniques and their theoretical foundations.

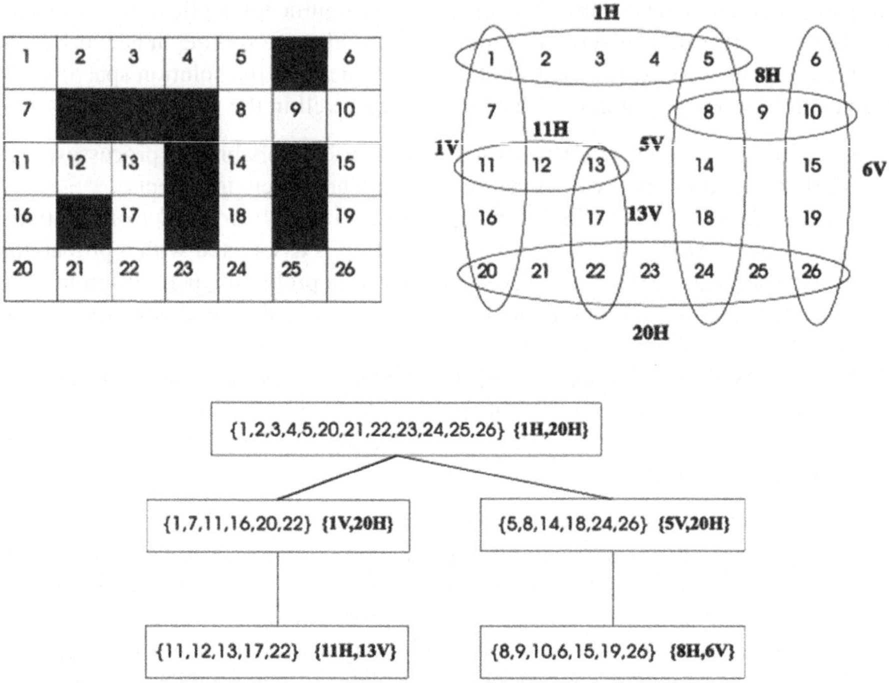

Fig. 9 Crossword puzzle and its associated hypergraph (on the top). A hypertree decomposition (on the bottom)

4.3 Hypergraph Representation for CSP

A crossword can be viewed as a partial case of invention. In the course of a dialogue, participants suggest solution words that might fit the crossword instantly, but rejected later, never fit or fit and remain. A crossword puzzle can be formulated as a CSP and then be represented as hypertree decomposition.

Figure 9 shows a combinatorial crossword puzzle. A set of English words is associated with each horizontal or vertical array of white boxes separated by black boxes. A successful solution to the puzzle problem, which we view as an invention, is an assignment of a character to each white box such that to each white array is assigned a word from its set of English words. This problem can be reduced to a CSP by assigning a CSP variable to each white box and by defining a constraint for each array of white boxes prescribing the existing English words that are mapped into it.

Solving a crossword in the course of a dialogue can be viewed as an invention process. A combinatorial crossword puzzle is a typical CSP (Dechter 1992). A set of legal words is associated to each horizontal or vertical array of white boxes delimited by black boxes. A solution to the puzzle is an assignment of a letter to each white box such that to each white array is assigned a word from its set of legal words. This problem is represented as follows.

There is a variable X_i for each white box and a constraint C for each array D of white boxes. (For simplicity, we just write the index i for variable X_i.) The scope of C is the list of variables corresponding to the white boxes of the sequence D; the relation of C contains the legal words for D.

Structural decomposition methods are considered in Gottlob et al. (2000), which identify tractable classes by exploiting the structure of constraint scopes as it can be formalized as a hypergraph whose nodes correspond to the variables and where each group of variables occurring in some constraint induce a hyperedge. A hypergraph can be defined as a generalization of a graph where edges, called hyperedges, can connect any number of nodes, not just two. Structural methods based on the notions of generalized hypertree width and tree width turn out to be efficient (Robertson and Seymour 1984). Gottlob et al. (2012) suggest a primal graph representation, where nodes correspond to variables and an edge between two variables indicates that they are related by some constraint.

The underlying idea is that solutions to CSP instances that are associated with acyclic (or nearly acyclic) structures can efficiently be computed via dynamic programming, by incrementally processing the structure according to some of its topological orderings.

An instance of a constraint satisfaction problem for an invention is a triple $I = <Var,U,C>$, where Var is a finite set of variables, U is a finite domain of values, and $C = \{C_1,C_2,\ldots,C_q\}$ is a finite set of constraints. Each constraint C_v, for $1 \leq v \leq q$, is a pair (S_v,r_v), where $S_v \subseteq Var$ is a set of variables called the invention scope and r_v is a set of substitutions (also called tuples) from variables in S_v to values in U, indicating the allowed combinations of simultaneous values for the variables in S_v. Any substitution from a set of variables $V \subseteq Var$ to U is extensively denoted as the set of pairs of the form X/u, where $u \in U$ is the value to which $X \in V$ is mapped. Then, a solution to I is a substitution θ: $Var \rightarrow U$ for which q-tuples $t_q \in r_1,\ldots,t_1 \in r_q$ exist such that $\theta = t_1 \cup \ldots \cup t_q$.

We now proceed to the example of representation for invention entities H (Figs. 10 and 11). In the hypergraph H shown on the left, which is associated with a

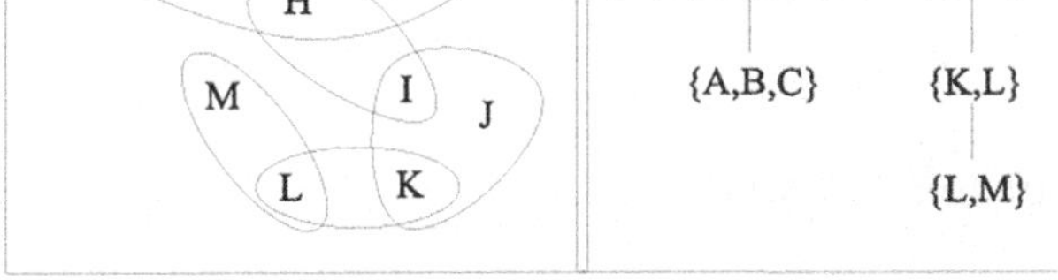

Fig. 10 A hypergraph H for invention entities, and a join tree $JT(H)$

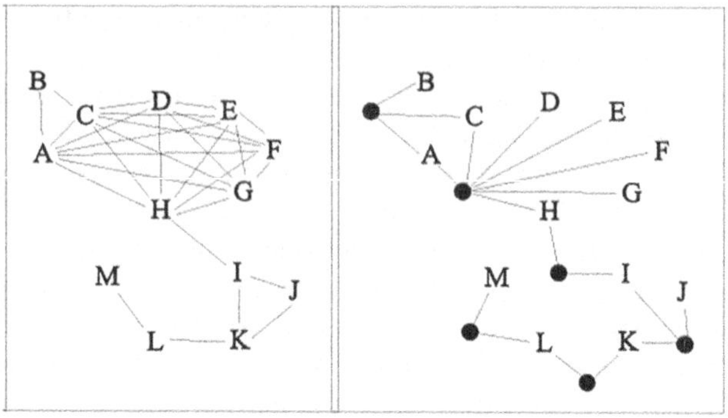

Fig. 11 The primal graph for invention entities *G*(*H*) and the incidence graph for them *inc*(*H*)

CSP formulation for an invention problem over the set of variables {*A*,*B*…}. A few constraints are defined over the instance whose scopes precisely correspond to the hyperedges in *E*(*H*); for instance, {*A*,*B*,*C*} is an example of constraint scope. *H* is acyclic: a join tree *JT*(*H*) for it is depicted on the right.

The structure of a CSP instance *I* is best represented by its associated hypergraph $H(I) = (V,H)$, where $V = Var$ and $H = \{S \mid (S,r) \in C\}$. A hypergraph *H* is acyclic iff it has a join tree. A join tree is a tree structure that captures relationships between hyperedges in the hypergraph. For a hypergraph H to be acyclic, it must have a join tree satisfying the running intersection property: For any two hyperedges connected through a path in the join tree, their intersection (shared variables) must also appear in all hyperedges along that path.

This ensures that constraints can be processed in a treelike order, simplifying constraint propagation and search algorithms. A join tree imposes an order where constraints can be checked locally while ensuring global consistency through propagation. Since trees have no cycles, processing constraints in this order avoids revisiting variables unnecessarily. The existence of a join tree guarantees that the CSP can be solved using polynomial-time algorithms such as tree decomposition or join-tree clustering.

A join tree *JT*(*H*) for a hypergraph *H* is a tree whose vertices are the hyperedges of *H* such that whenever the same node $X \in V$ occurs in two hyperedges h_1 and h_2 of *H*, *X* occurs in each vertex on the unique path linking h_1 and h_2 in *JT*(*H*) (Fig. 10). The notion of acyclicity we use here is the most general one known in the literature, coinciding with α-acyclicity. α-acyclicity is a concept introduced by Fagin (1983), related to hypergraphs and database theory. It extends the notion of acyclicity to hypergraphs, which are widely used in CSPs and relational databases.

4.4 Hypergraph Solvers

Once a CSP problem is expressed as a finite set of constraints, the goal is to find the variables' values satisfying them. Even though the problem is in general NP complete, there are some approximation techniques to tackle its intractability. One of the

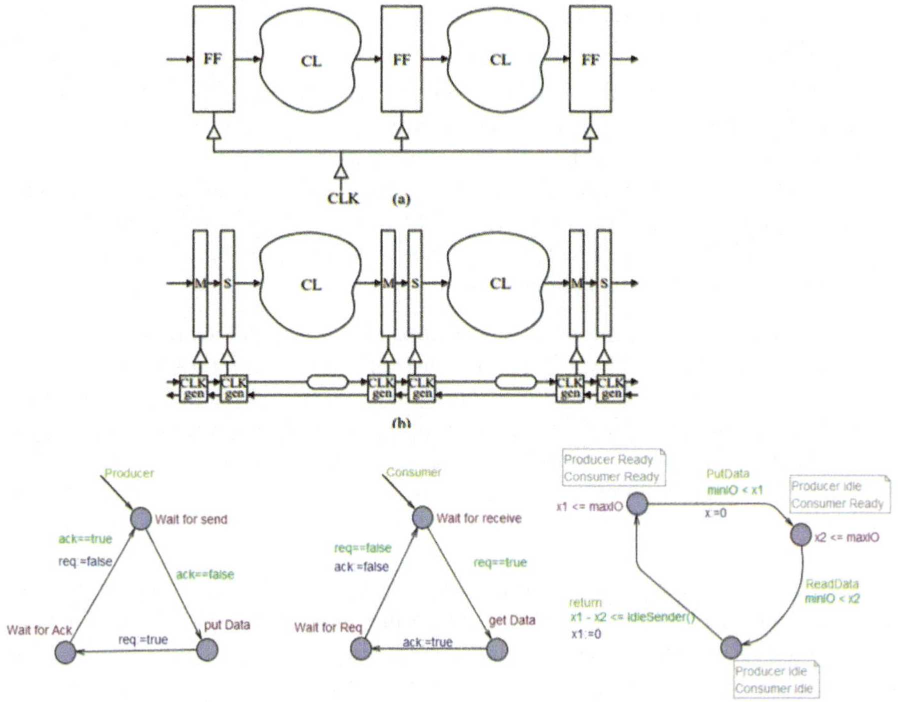

Fig. 12 Inventing a de-synchronized circuit (in the middle) given the synchronous circuit. The handshake protocol is shown on the bottom

most widely used techniques is the Constraint Propagation. It consists in explicitly excluding values or combination of values for some variables whenever they make a given subset of constraints unsatisfied. Oucheikh et al. (2019) define a CSP subclass called 4-CSP whose constraint network infers relations of the form: $\{x \sim \alpha, x - y \sim \beta, (x - y) - (z - t) \sim \lambda\}$, where x, y, z, and t are real variables; α, β, and λ are real constants; and $\sim \in \{\leq, \geq\}$. The authors provide a graph-based proofs of the 4-CSP tractability and elaborates algorithms for 4-CSP resolution based on the positive linear dependence theory, the hypergraph closure, and the constraint propagation technique. Time and space complexities of the resolution algorithms are proved to be polynomial.

The invention process in the form of Four Phase Handshake Protocol (Blunno et al. 2004) is depicted in Fig. 12 on the bottom. This protocol uses two clocks x_1, x_2; two parameters *minIO*, *maxIO*; and the following constraints: $x_1 < maxIO$, $x_1 > inIO$, $x_2 < maxIO$, $x_2 > inIO$, and $x_1 - x_2 \leq maxIO - minIO$. Implementing the protocol can be reduced to its 4-CSP equivalent.

The de-synchronization model presented in this section aims at the substitution of the global clock by a set of asynchronous controllers that guarantee an equivalent behavior. The model assumes that the circuit has combinational blocks (CL) and registers implemented with D flip-flops (FF), all of them working with the same clock edge (e.g., rising in Fig. 12). The de-synchronization method proceeds in three steps:

1. Conversion of the flip-flop-based synchronous circuit into a latch-based one (M and S latches). D-flip-flops are conceptually composed of master-slave latches.
2. Generation of matched delays for the combinational logic (denoted by rounded rectangles). Each matched delay must be greater than or equal to the delay of the critical path of the corresponding combinational block. Each matched delay serves as a completion detector for the corresponding combinational block.
3. Implementation of the local controllers.

Structural decomposition methods, as developed by Greco and Scarcello (2016), are designed to identify tractable classes of constraint satisfaction problems (CSPs) by leveraging the hypergraph structure of problem instances. These methods transform cyclic hypergraphs into acyclic ones by grouping their edges (or nodes) into a polynomial number of clusters and arranging these clusters into a hierarchical structure known as a decomposition tree. Using this decomposition tree—or simply knowing that such a tree exists—enables the problem instance to be solved in polynomial time.

Inventions can also be modeled as conjunctive queries, where evaluating these queries is generally NP hard. However, it becomes computationally feasible in polynomial time for the class of acyclic queries (Q). Each query Q is associated with an acyclic query hypergraph HQ, where the nodes represent the variables of Q and the hyperedges correspond to the sets of variables occurring in its atoms.

To handle cyclic hypergraphs, methods exist to transform them into acyclic ones. This transformation involves grouping edges (or nodes) into polynomially bounded clusters and arranging them into a decomposition tree. The original problem can then be evaluated by processing the sub-problems represented in this tree. The computational cost depends on the width of the decomposition, which is the size of the largest cluster. The process is polynomial if this width is bounded by a constant and exponential otherwise.

The notion of tree projection is an efficient decomposition method, where a query Q is given together with a set V of atoms, called views, which are defined over the variables in Q (Greco and Scarcello 2016). The question is whether (parts of) the views can be arranged as to form a tree projection (playing the role of a decomposition tree), i.e., a novel acyclic query that still subsumes Q. By representing Q and V via the hypergraphs H_Q and H_V, where hyperedges one-to-one correspond with query atoms and views, respectively, the tree projection problem reveals its graph-theoretic nature. For a pair of hypergraphs, H_1 and H_2, let $H_1 \leq H_2$ denote that each hyperedge of H_1 is contained in some hyperedge of H_2. Then, a tree projection of H_Q w.r.t. H_V is any acyclic hypergraph H_a such that $H_Q \leq H_a \leq H_V$. If such a hypergraph exists, then we say that the pair of hypergraphs (H_Q, H_V) has a tree projection (Fig. 13).

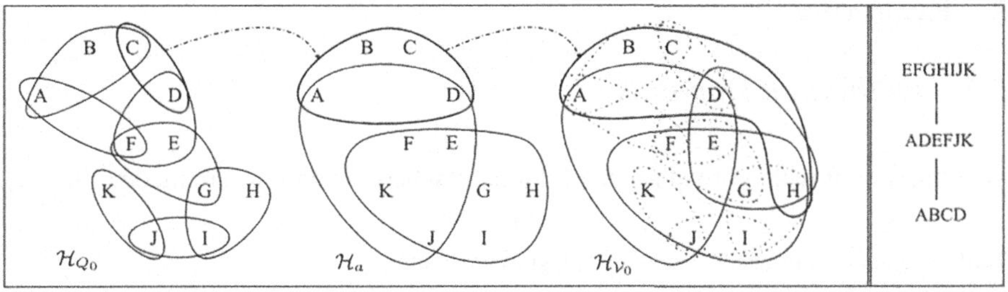

Fig. 13 A tree projection H_a of H_{Q_0} $Q0$ w.r.t. H_{V_0} for entities of an invention A, B, …. On the right: the join tree JT_a

4.5 Game-Theoretic Interpretation of Invention Dialogues

The invention proponent and opponent dialogue game is played on a pair of hypergraphs (H_1, H_2) by two main agents (proponent and opponent) where the proponent controlling her servant agents, in charge of the "surveillance" of a number of strategic invention possibilities. The objective for the proponent of this invention is to catch any possibility of invention failure, desired by the opponent.

The dialogue is driven by the moves of the proponent with her agents and of the opponent. The opponent stands on a node and can run at great speed along the edges of H_1. However, she is not permitted to run through a node that is controlled by a servant agent. Each move of the proponent involves one squad of service agents, which is encoded as a hyperedge $h \in edges(H_2)$. The proponent may ask some agents in the squad h to run to certain nodes, as long as they occupy nodes that are currently reachable by the opponent, thereby blocking an escape path for the opponent. "Second-lines" agent cannot be activated by the invention proponent. Note that the invention opponent is fast and may see agent that are entering nodes around her. Therefore, while agents move, the opponent may run through those positions that are left by agents or not yet occupied by them. The goal of the proponent is to place an agent on the node occupied by the opponent, while the opponent tries to avoid her "capture" in the form of finding free nodes associated with a "flaw" in this invention.

This setting follows the Robber and Captain game (Bonato and Nowakowski 2010), which is a combinatorial game used to analyze CSPs through their hypergraph structures. It is particularly useful for studying tree width, hypertree decompositions, and the tractability of CSP instances. The minimum number of proponent's agents required to block the opponent directly relates to the tree width (or hypertree width) of the graph. *Treewidth* $\leq k$ if $k + 1$ agents can capture the opponent. This property determines whether the CSP can be solved efficiently in polynomial time. CSP instances with low tree width are computationally tractable because they can be decomposed into treelike structures (e.g., decomposition trees). Higher tree width values indicate cyclic dependencies, making the problem potentially NP hard.

5 Evaluation

5.1 Evaluation Domains

We experiment with a broad range of domains where inventions are made (Table 1).

Table 1 Domains for which invention dialogues are built

Domain	Potential inventions
Healthcare	AI-driven diagnostics, personalized medicine, organ bioprinting, wearable health monitors Gene editing tools, regenerative medicine, telemedicine platforms, nano-robotics Mental health tech, anti-microbial resistance solutions
Engineering	Self-healing materials, green energy solutions, smart infrastructure, quantum computing hardware Autonomous transportation, space exploration tech, hydrogen fuel cells Carbon capture technology, robotic construction, advanced alloys and composites
Communication and computing	5G and beyond, quantum communication, brain-computer interfaces (BCIs), holographic displays Satellite-based internet, AI-powered language translation, blockchain security solutions Metaverse platforms, edge computing devices, neural networks
Agriculture	Smart irrigation, drought-resistant crops, vertical farming, soil-less cultivation
Energy	Nuclear fusion reactors, superconductors for energy efficiency
Environment	Plastic-eating bacteria, ocean-cleaning technologies
Education	AI-driven personalized learning, immersive AR/VR educational experiences
Security and defense	Autonomous defense drones, cybersecurity advances in AI threat detection
Transportation	Hyperloop systems, electric vertical takeoff and landing (eVTOL) vehicles

5.2 CSP Tools

We suggest AI Space CSP tool (O'Neill et al. 2024) to solve CSP in a consistency-based manner arising in invention dialogue. Tutorials include creating a CSP, giving information about how to create a new CSP from scratch, loading a preexisting CSP, covering how to load one of the ready-made CSPs, and giving a brief description of each of them. Moreover, there is "Solving a CSP" tutorial that covers how to solve a CSP performing domain splitting and backtracking.

CSP solvers find solutions to CSPs, where the most complex algorithms build on simpler ones; but a common thread throughout all CSP solvers is that they iteratively check potential solutions or partial solutions against the constraints. Wong et al. (2024) build recursive backtracking and forward-checking algorithms for CSPs. In recursive backtracking, progressive partial solutions are recursively explored, with backtracking occurring when there is no possibly path forward due to the constraints.

A pseudocode for the backtracking algorithm to solve invention dialogues is shown below for the case of a grid of entities forming an invention:

```
function backtrack(grid, col):
  for 1 <= i <= row.length:
    Add a new invention entity at (row, column) = (i, col)
    Check constraints
    if constraints fail:
      if i == row.length:
        Return false;//could not find a solution for
this column
      Keep going
    else:
      backtrack(grid, col + 1)
```

Another way of solving a CSP is to use the method of forward checking. In forward checking, the possibilities for each node in the graph (in this case, for each state) are tracked.

We compare default prompt for dialogue construction with category theory based, where categorial model is enforced. One terminates a search path if there are no legal values possible for the node being considered.

```
function forwardCheck(map):
  for each entities in map:
     entities.optionsSet = list of all invention
entities(…);
  forwardCheckPartner({}, map);

function forwardCheckPartner(labelAssignment, CSP):
  if all entities have labels:
    return labelAssignment
  select randomly assigned state from CSP
  for each entity in entities.optionsSet:
    if entity is not the same as any terminal label entity
      add entity and label to labelAssignment
      for each terminalLabel:
        remove option from terminalLabel.optionSet
      forwardCheckPartner(labelAssignment, CSP)
      if labelAssingment works:
        return labelAssignment
      remove entity and label from labelAssignment
      for each terminalLabel:
        add option to terminalLable.optionSet
  return failure
```

An Arc Consistency CSP Solver is an algorithm designed to enforce arc consistency in CSPs. It focuses on reducing the domain of variables by eliminating values that cannot satisfy the binary constraints between pairs of variables, ensuring that every remaining value is consistent with the constraints. Arc Consistency includes:

- A binary constraint between two variables X and Y forms an arc denoted as (X,Y).
- A variable X is arc consistent with Y if, for every value in D_X (domain of X), there exists at least one value in D_Y such that the constraint between X and Y is satisfied.
- If no such value exists, the inconsistent value is removed from D_X.

The algorithm processes the CSP by iteratively checking and enforcing arc consistency. The most common algorithm starts with a queue containing all arcs (X,Y).

1. Process Arcs:
   - Remove an arc (X,Y) from the queue.
   - Check each value in D_X.
   - Remove values in D_X that do not satisfy the constraint with any value in D_Y.
2. Update Queue:
   - If values are removed from D_X, add all arcs (Z,X) back to the queue (where Z is a neighbor of X).

3. Repeat Until Queue is Empty:
   - Continue until no further updates can be made.

Output is a reduced domain for variables, ensuring consistency. The CSP might still need backtracking search if no full assignment is possible.

5.3 CSP-Based Dialogue Construction

We evaluate a direct construction vs CSP-supported construction of an invention dialogue, as well as the contribution of our categorial model. Human experts evaluate the correctness of inventive dialogues constructed by ChatGPT with different prompts:

1. Construct an inventive dialogue for <device>.
2. Construct the inventive dialogue for <device> where components include <components in section …> where entities obey constraints from <section …> and categorial constraints from section …> are satisfied.
3. Construct the CSP expressed by a logic program and an invention dialogue based on this resolved CSP where entities obey constraints (2). Then resolve the constructed CSP by a CSP solver from Sect. 5.2, and use ChatGPT to verify the consistency, given the resolved CSP.

The consistency and correctness results for these three scenarios are shown in Table 2. We used 100 devices from the list of domains in Sect. 5.1.

We observe a systematic improvement in generated invention dialogues once we add the dialogue model (column 3) and then explicit CSP resolution (column 4). These improvements are 3.3% and 5.3%, respectively, confirming the importance of

Table 2 Consistency and correctness of invention dialogues constructed by ChatGPT with various prompts, %

Domain/dialogue construction method	Direct construction by LLM	LLM + Category based	LLM + CSP solver
Healthcare	82.3	84.7	91.3
Engineering	79.9	83.2	89.1
Communication and computing	86.2	89.0	93.4
Agriculture	84.5	88.2	94
Energy	83	87.4	91.5
Environment	82.1	85.8	89.7
Education	84.8	87.5	93.1
Security and defense	80.6	83	89.4
Transportation	82.5	86.7	91.9
Average	82.87	86.17	91.49

CSP solver in a specific sense and a logical reasoning companion in a broader sense for improving LLM results.

5.4 Pedagogical Value

We assess the pedagogical value of representing invention in the form of dialogue. We assess the following features of dialogue-based inventions vs. plain text or math-based exploration of new topics.

We compare conventional format of learning plain text + supportive media versus inventive dialogue format + supportive media. As inventive dialogues encourage step-by-step learning of a concept being invented, it is expected to help students memorize all required intermediate concepts that led to invention (column 2). Evaluation questions are focused on sequences of involved concepts.

Inventive dialogues also promote critical thinking. Dialogues introduce contrasting viewpoints, pushing learners to evaluate and synthesize ideas (column 3). Moreover, they encourage learners to think hypothetically: how an invention might impact society or solve specific problems (column 4). Evaluation questions are focused on societal impact.

There is a memorizing impact on learners: dialogues mimic storytelling, which is easier to remember than abstract descriptions. Also, learners are emotionally invested in character interactions, improving memory retention (column 5). Evaluation questions are focused on details of inventions or specific details of a concept.

We evaluate the quality of acquired knowledge by asking students to answer questions about the topic unrelated to inventive dialogue but important for understanding the phenomenon being learned. These questions are automatically generated from textbooks (and not from dialogue texts), relying on simple question generation prompts.

In Table 3, we assess an improvement or decline in measured knowledge and skills by student in the specific knowledge domain, as estimated by answering questions. The value in each cell is a proportion of questions correctly answered by a student trained by inventive dialogue relative to the ones trained by traditional textbooks. For this evaluation task, six students used 234 text-invention dialogue pairs to learn from and attempt to answer questions. All invention dialogues and questions were generated by ChatGPT. An assessment of correct/incorrect answers to questions was conducted by ChatGPT as well.

One can observe that the best impact of learning from invention dialogues is made in case of *Step-by-step learning of a concept*, followed by *Thinking hypothetically*. In the Healthcare domain, the value of learning from invention dialogues is greatest, followed by Communication and Computing. Whereas in some domain and kinds of knowledge invention dialogues can deteriorate learning outcomes, the overall assessment shows the improved learning efficiency.

Table 3 Assessment of pedagogical value of invention dialogues

Domain/kind of knowledge	Step-by-step learning of a concept	Contrasting viewpoints	Thinking hypothetically	Specific details of a concept	Average per domain
Healthcare	1.08	1.17	1.06	1.05	1.09
Engineering	1.11	0.95	1.18	0.98	1.06
Communication and computing	1.13	1.07	0.99	1.08	1.07
Average per kind of knowledge	1.11	1.06	1.08	1.04	

Table 4 Quality of invention dialogue constructed by students (%)

Domain/invention dialogue quality criterion	Consistency	Originality	Plausibility	Average per domain
Healthcare	68.4	74.5	81.3	74.7
Engineering	73.0	78.9	76.3	76.1
Communication and computing	65.3	71.0	73.7	70.0
Average per quality feature	68.9	74.8	77.1	

5.5 *Manual Construction of Invention Dialogues*

We proceed to an assessment task on manual construction of invention dialogues, based on a description of a phenomenon in health, engineering, or computing, dialogues manually constructed by students from the previous evaluation settings for the same topics. The dialogues written by students were compared with the above 234 invention dialogues. We count the portion of invention dialogues written as well as ChatGPT with respect to consistency, originality, and plausibility in Table 4.

We conclude from this evaluation that once the students learn new technical achievements from hypothetical invention dialogues and answer questions about what they have learned, they can successfully construct such invention dialogues in about ¾ of cases. Plausibility turns out to be the most successfully achieved feature, followed by originality and then consistency.

6 Discussions and Conclusions

In conclusion, we share a hypothetical dialogue between Thomas Edison and his collaborators. Thomas Edison, renowned for his prolific inventions, often collaborated with a team of skilled assistants and co-inventors. While direct transcripts of their dialogues are scarce, historical accounts provide insights into their collaborative processes.

Charles Batchelor, a British engineer, was one of Edison's closest associates. Their partnership was instrumental in developing the incandescent light bulb.

Batchelor's technical expertise complemented Edison's inventive vision. Their interactions likely involved detailed discussions on filament materials, vacuum techniques, and design improvements.

Francis Upton, a mathematician and physicist, joined Edison's team to work on the electric light and power system. Upton's analytical skills were crucial in solving complex problems related to electrical distribution. Their conversations would have encompassed mathematical analyses, experimental results, and system optimization strategies.

Edison's Menlo Park laboratory was a hub of innovation, housing a diverse team of talented individuals. The collaborative environment fostered open communication, brainstorming sessions, and collective problem-solving. While specific dialogues are undocumented, the team's synergy was evident in the rapid development of groundbreaking technologies.

Edison was known for his hands-on approach and often engaged in direct communication with his team. He encouraged experimentation and valued practical results. His leadership likely involved motivational discussions, critical evaluations of ideas, and guidance on experimental methodologies.

Although verbatim records of Edison's dialogues with his co-inventors are limited, the collaborative essence of their interactions is well documented. Their collective efforts led to innovations that have profoundly impacted modern society. Edison's approach—blending invention, entrepreneurship, and aggressive marketing—set a model for modern inventors and tech entrepreneurs. His success illustrates the importance of securing funding, demonstrating commercial value, and building networks to transform ideas into profitable ventures.

Several studies on converting text into dialogue format preceded the development of invention dialogue systems. Galitsky et al. (2019) introduced a chatbot capable of delivering content as virtual dialogues, which are automatically generated from plain texts extracted and selected from documents. These virtual dialogues consist of answers derived from identified and segmented document fragments, paired with questions that are automatically generated based on the initial text. Building on this approach, Galitsky (2021) proposed the Doc2Dialogue algorithm, which transforms paragraphs of text into hypothetical dialogues by analyzing their discourse trees. This method enables a significant expansion of chatbot training datasets across diverse domains.

Another method for building dialogues involves detecting rhetorical agreement between texts (Galitsky 2022). This approach utilizes a rhetorical agreement application that processes a multi-part initial query based on the topic of the intended dialogue. It generates a question communicative discourse tree, which represents the rhetorical relationships between fragments of the query. The algorithm then identifies a sub-discourse tree within the question communicative discourse tree. Next, the algorithm generates a candidate answer communicative discourse tree for each potential answer in each set. It evaluates the level of complementarity between the sub-discourse tree and each candidate answer discourse tree by applying a classification model to both structures. Based on the computed complementarity, the application selects the most appropriate answer from the candidates, thereby constructing a dialogue structure for an interactive session.

An invention can streamline bureaucracy by automating repetitive tasks, improving data management, and enhancing communication systems, thereby reducing paperwork, delays, and human errors (Galitsky 2021). Technologies such as workflow automation software, artificial intelligence-driven decision-making tools, and digital document management systems can simplify processes like approvals, record-keeping, and compliance monitoring. By enabling faster processing, better tracking, and secure storage of information, inventions can increase transparency, accountability, and efficiency within bureaucratic systems, ultimately saving time and resources while improving public service delivery.

Galitsky (2025) investigates CSP for scheduling and resource allocation in healthcare environments. The study begins with fundamental scheduling tasks for nurses and physicians and gradually addresses more complex scenarios where single-shot solutions prove inadequate. The author further examines the role of LLMs in facilitating schedule construction by converting natural language constraints into formats compatible with CSP algorithms. The chapter concludes with an in-depth analysis of temporal CSP techniques.

In a common dialogue, such as customer support or patient-doctor, the dialogue structure can be pre-defined by the initial conversation and problem formulation (Galitsky 2022). This dialogue structure is encoded in the form of discourse tree, a means to represent an overall high level logical organization of a dialogue. Such a dialogue structure determinism can also hold for an invention dialogue, where problem formulation may suffice for the CSP resolution scenario and the whole dialogue structure is determined by problem formulation utterances.

Mahadevan (2022) presents a unified formalism for structure discovery of causal models such as invention dialogues (caused by initial dialogue utterances) and predictive state representation models in reinforcement learning using higher-order category theory. The author describes structure discovery in both settings using simplicial objects, contravariant functors from the category of ordinal numbers into any category. A simplicial object in a category C is a structure that captures the idea of objects organized in a way analogous to simplices in geometry (points, line segments, triangles, etc.). These structures are defined via functors from the simplicial category Δ to C. A contravariant functor is a mapping between categories that reverses the direction of morphisms.

We conclude with the dialogue related to the invention for an automatic shaving:

> Marketing: "Could you please explain how your invention works?"
> Inventor: "It's very simple! The client inserts their head here, and from these openings, manipulators with straight razors extend and shave the client!"
> Marketing: "But wait! Every person has a unique head and face shape!"
> Inventor: "Well… the first time, yes…"

Our evaluation demonstrates a systematic improvement in the quality of generated invention dialogues when incorporating both the dialogue model and explicit CSP resolution. These enhancements result in performance gains of 3.3% and 5.3%, respectively, underscoring the critical role of CSP solvers in specific applications and the broader importance of logical reasoning companions in improving LLM outcomes.

The pedagogical value of automatically constructed invention dialogues is also evident. Compared to traditional plain-text learning materials, invention dialogues substantially improve learning efficiency. The most pronounced benefits are observed in step-by-step learning of concepts, followed by thinking hypothetically. Domain-specific evaluations reveal the highest pedagogical impact in healthcare, with significant benefits also seen in communication and computing. However, it is worth noting that in some domains and knowledge types, invention dialogues may occasionally hinder learning outcomes. Despite these exceptions, the overall assessment affirms the efficacy of invention dialogues in enhancing learning.

Additionally, our findings highlight that students trained with hypothetical invention dialogues can successfully construct similar dialogues in approximately 75% of cases. Among the evaluated features, plausibility is the most consistently achieved, followed by originality and consistency. These results underscore the potential of invention dialogues not only as a learning tool but also as a means of fostering creative and logical reasoning skills in learners.

References

Blunno I, Cortadella J, Kondratyev A, Lavagno L, Lwin K, Sotiriou CP (2004) Handshake protocols for de-synchronization. In: Proc. of 10th international symposium on advanced research in asynchronous circuits and systems (ASYNC 2004), pp 149–158

Bonato A, Nowakowski RJ (2010) The game of cops and robbers on graphs, Student mathematical library, vol 61. American Mathematical Society

Dechter R (1992) Constraint networks. In: Encyclopedia of artificial intelligence, 2nd edn. Wiley, New York, pp 276–285

Fagin R (1983) Acyclic database schemes (of various degrees): a painless introduction. In: CAAP'83, trees in algebra and programming, 8th colloquium, L'Aquila, Italy, March 9–11

Galitsky B (2021) Chatbots for CRM and dialogue management. In: Artificial intelligence for customer relationship management, Human–computer interaction series. Springer, Cham

Galitsky B (2022) Building dialogue structure by using communicative discourse trees. US Patent 11537645

Galitsky B (2025) LLM supporting constraint satisfaction in health. In: Health applications of neuro symbolic AI. Elsevier

Galitsky B, Ilvovsky D, Goncharova E (2019) On a chatbot providing virtual dialogues. In: Proceedings of recent advances in natural language processing, Varna, Bulgaria, Sept 2–4, pp 382–387

Gottlob G, Leone N, Scarcello F (2000) A comparison of structural CSP decomposition methods. Artif Intell 124(2):243–282

Gottlob G, Greco G, Scarcello F (2012) Tractable optimization problems through hypergraph-based structural restrictions. Arxiv 1209:3419

Greco G, Scarcello F (2016) Greedy strategies and larger islands of tractability for conjunctive queries and constraint satisfaction problems. arXiv:1603.09617v2

Mahadevan S (2022) Unifying causal inference and reinforcement learning using higher-order category theory. arXiv:2209.06262

Mahadevan S (2024) GAIA: categorical foundations of generative AI. arXiv:2402.18732

McInnes L, Healy J, Melville J (2018) Umap: uniform manifold approximation and projection for dimension reduction. arxiv.org/1802.03426

O'Neill K, Sueda S, Amershi S, Pavlin M, Porter K, Knoll B (2024) AI space. www.cs.ubc.ca/labs/lci/AIspace/constraint/

Oucheikh R, Berrada I, El Hichami O (2019) A hypergraph based approach for the 4-constraint satisfaction problem tractability. arXiv:1905.09083

Robertson N, Seymour PD (1984) Graph minors III: planar tree-width. J Combin Theory Ser B 36:49–64

Rocca J (2021) The exploration-exploitation trade-off: intuitions and strategies

Wong R, Van Wagenen J, Riley N, Birch C (2024) CSP solvers. https://cse442-17f.github.io/CSP-Solver/

A Neuro-Symbolic System for Distributed Chemical Engineering Simulations

1 Introduction

The field of chemical engineering is undergoing a profound transformation driven by the convergence of symbolic reasoning, machine learning, and cloud-native computing. As chemical processes become more intricate and the demand for high-throughput simulation, design-space exploration, and real-time optimization continues to grow, traditional computational frameworks are increasingly constrained by issues of scalability, adaptability, and interpretability. Conventional models—rooted in first-principles equations or purely data-driven approximations—often struggle to accommodate the nonlinearities, uncertainties, and dynamic behavior inherent in modern chemical systems.

In response, neuro-symbolic AI has emerged as a compelling hybrid paradigm. It unites the structured, explainable logic of symbolic systems with the adaptability and generalization capacity of neural networks. This synthesis enables systems that not only learn from complex experimental and simulation data but also reason over domain knowledge such as conservation laws, thermodynamic constraints, and process topologies. As a result, neuro-symbolic frameworks provide a flexible scaffold for encoding scientific priors, enforcing physical consistency, and inferring latent variables from partial or noisy observations—capabilities that are critically needed in the modeling and control of chemical processes.

This chapter explores the application of neuro-symbolic methods in the context of distributed chemical simulations, particularly when deployed within serverless and event-driven cloud infrastructures. We examine the core challenges of modeling multi-scale, reactive, and transport-intensive systems—areas where classical Partial Differential Equation solvers or black-box machine learning models alone fall short. Neuro-symbolic architectures enable modular and interpretable modeling by integrating domain-specific ontologies, logic rules, and graph structures with trainable

B. Galitsky, *Applications of Neuro-Symbolic Artificial Intelligence*,
https://doi.org/10.1007/978-3-032-16524-4_10

neural components. This layered approach allows for intelligent task decomposition, adaptive simulation control, and semantic error detection across heterogeneous workflows.

Through the integration of large language models (LLMs), symbolic knowledge graphs, and rule-based reasoning engines with numerical solvers and scalable cloud services (e.g., AWS Lambda, Azure Durable Functions, Google Cloud Workflows), we propose a blueprint for intelligent simulation orchestration pipelines. These pipelines are capable of interpreting user-defined goals or system specifications, generating customized simulation flows, and distributing tasks dynamically across a stateless, serverless backend. Crucially, they maintain scientific fidelity and traceability—two attributes essential for industrial validation and regulatory compliance.

This evolving neuro-symbolic approach redefines how we interface with simulation environments. Rather than requiring manual scripting or rigid configuration, intelligent agents can translate experimental intent or process constraints into executable workflows—bridging the gap between domain expertise and computational execution. In doing so, neuro-symbolic systems are positioned not merely as academic prototypes but as practical, transformative tools that support intelligent, transparent, and adaptive chemical process design and control at scale.

Furthermore, recent research has shown that code-generating LLMs possess emergent capabilities in chemical reasoning, owing to their exposure to large-scale technical corpora (White et al. 2023). These models demonstrate an ability to plan multi-step simulations, adapt to observed behaviors, and interpret user intent in complex chemical contexts. However, despite this progress, the level of automation in chemical engineering remains limited relative to domains like software engineering or logistics. This gap is attributed to the experimental nature of the field, the scarcity and heterogeneity of structured data, and the narrow generalizability of traditional simulation tools.

Let us look at an example of a sequence of applications of LLM tools:

Task: Find and Synthesize a Chiral Manganese Salen Catalyst for Asymmetric Epoxidation of Styrene
Step 1: Identify a suitable catalyst

Goal: Locate a manganese salen complex known to catalyze the asymmetric epoxidation of alkenes.
Web Search tool: "*Jacobsen's catalyst for asymmetric epoxidation*"
Result: Found: *Jacobsen's Mn(III)-salen complex*

Step 2: Retrieve the molecular structure

Name2Smiles tool: `SMILES: CC1=CC=CC=C1C2=NC3=CC=CC=C3C(=N2)CC4=CC(=C(C=C4)O)C(C)C`
Note: This is a simplified SMILES; actual Jacobsen catalyst includes Mn center, axial ligands, and is best represented with a 3D structure.

Step 3: Plan the synthesis

SynthesisPlanner tool: Using retrosynthetic analysis, plan the synthesis of Jacobsen's catalyst:

- Start from 1,2-diaminocyclohexane and 3,5-di-tert-butylsalicylaldehyde
- Form Schiff base (salen ligand)
- React with $Mn(OAc)_3$ to coordinate the metal center
- Oxidize to Mn(III) complex
- Introduce axial ligand (e.g., Cl^-)

Output: Detailed synthesis plan with solvents, temperatures, and purification steps.

Step 4: Execute the synthesis

SynthesisExecuter tool: Synthetic steps carried out successfully.
Result: Isolated Jacobsen's Mn(III)-salen complex as a crystalline solid.

By addressing these challenges through a neuro-symbolic lens—one that blends formal knowledge, learning, and computational infrastructure—we lay the foundation for a new generation of scalable, intelligent, and explainable chemical simulation systems. These systems hold the potential to accelerate innovation across academia and industry, reduce development cycles, and empower engineers to focus on high-level problem-solving rather than low-level configuration.

2 Neuro-Symbolic Reasoning for Chemical Simulation

In the context of distributed chemical engineering simulations, symbolic LLMs serve as powerful agents capable of understanding and transforming not only raw data and code but also structured symbolic knowledge embedded in user interfaces, file formats, and simulation logic. This section outlines how symbolic LLMs can navigate and operationalize the entire simulation life cycle—from graphical interfaces to infrastructure deployment—by reasoning over structured inputs, translating between modalities, and generating executable systems.

Many domain-specific chemical engineering tools—such as Aspen Plus, COMSOL Multiphysics, or MATLAB Simulink—are GUI driven and rely on drag-and-drop interfaces for process configuration. While convenient for human users, these interfaces abstract away the underlying symbolic logic and process models, making automation or integration with cloud workflows difficult.

A symbolic LLM, especially when fine-tuned or prompted with domain-specific knowledge, can interpret GUI configurations through:

- Image-to-structure mapping, converting process diagrams into component graphs
- Script extraction, where the LLM infers the semantic model behind the interface (e.g., a CSTR connected to a distillation column via a heat exchanger)
- Symbolic translation, enabling migration of GUI-based workflows into code or XML representations

This capability enables the simulation logic—once locked inside a proprietary GUI—to be externalized, reasoned over, and reused within automated, distributed pipelines.

Chemical simulations are often exportable as structured XML documents, where components, thermodynamic packages, streams, and interactions are represented symbolically. A symbolic LLM can treat this XML not just as text but as a queryable, interpretable knowledge graph.

LLMs can:

- Parse and validate XML schemas using symbolic constraints (Galitsky 2025)
- Extract topological graphs of the process network
- Infer missing metadata or default values
- Translate symbolic process descriptions into domain-specific languages or compute tasks

Moreover, because the XML defines a structured ontology of component interactions, LLMs can reason about causal flows, detect logical errors, and suggest design improvements—functions that go far beyond simple parsing.

2.1 Converting XML to Infrastructure as Code (IaC)

To deploy the system at scale, the symbolic representation must be translated into **Infrastructure-as-Code (IaC)** artifacts. Symbolic LLMs can bridge this gap by generating IaC templates (e.g., Terraform, AWS CloudFormation, Pulumi) from the XML model.

For instance, given an XML node:

<Reactor id="R1" type="CSTR" residence_time="120s"/>

The LLM could generate:

```
resource "aws_lambda_function" "reactor_R1" {
  function_name = "Reactor_R1"
  handler       = "reactor.lambda_handler"
  ...
  environment {
    variables = {
      TYPE = "CSTR"
      RESIDENCE_TIME = "120"
    }
  }
}
```

This enables seamless deployment of each simulation component as a cloud-native unit, while preserving symbolic relationships defined in the original XML.

Neuro-symbolic systems facilitate the binding of symbolic process models to real, cloud-executed infrastructure. Each virtual component—now implemented as a serverless function—can be:

- Triggered via AWS Step Functions or EventBridge
- Communicate via S3, DynamoDB, or message queues (SQS, SNS)
- Dynamically scaled and monitored, without provisioning servers

By leveraging the LLM's symbolic reasoning, deployment artifacts can be adapted based on:

- Simulation scale
- Execution constraints (e.g., memory, timeout)
- Fault tolerance or checkpointing requirements

This creates a complete, distributed, cloud-native simulation pipeline powered by symbolic models and intelligent automation.

Hence, symbolic LLMs act as an intelligent bridge between domain-specific process models and scalable cloud infrastructure. From extracting logic from GUI-based tools to transforming symbolic XML into executable IaC to orchestrating message-passing simulations using serverless primitives, neuro-symbolic systems offer a modular and interpretable path forward for scalable, intelligent chemical engineering simulations (Fig. 1).

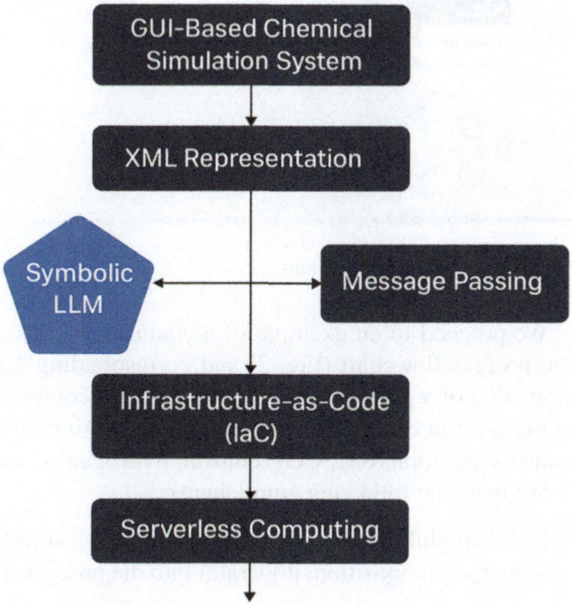

Fig. 1 A chart from converting a task for a chemical process into an IaC simulation framework with the support of a neuro-symbolic system

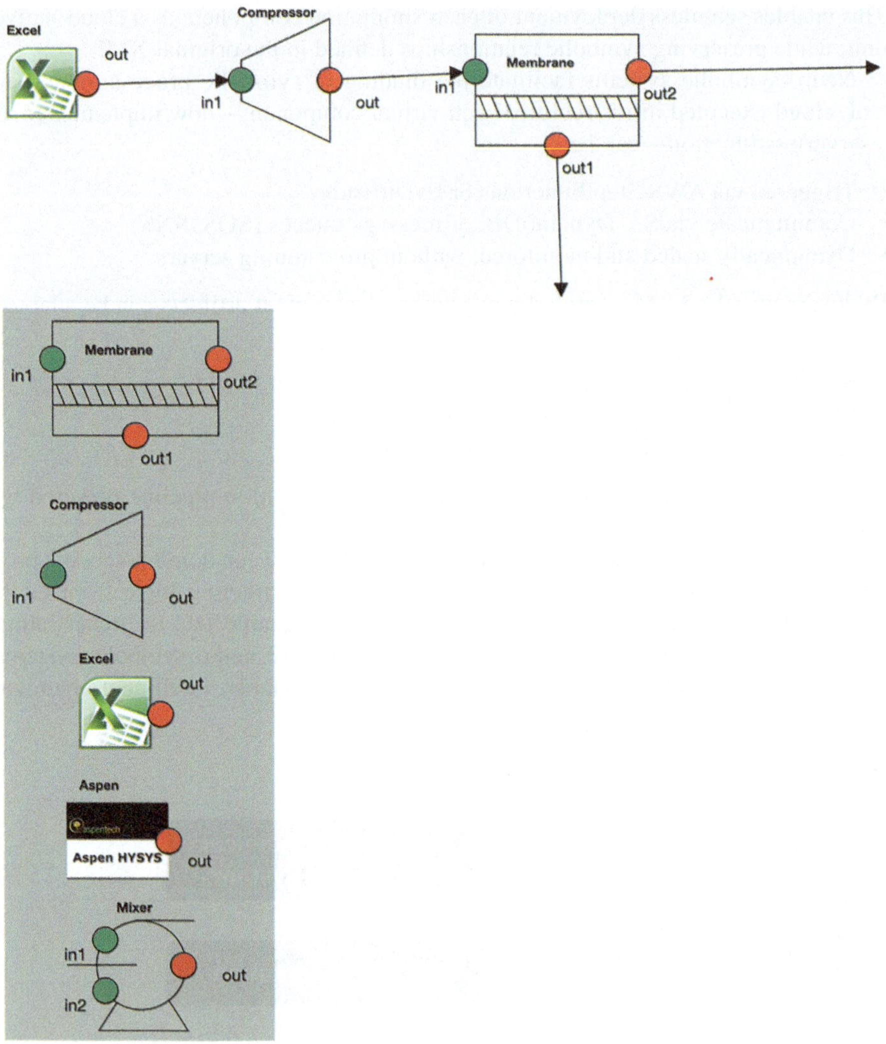

Fig. 2 Simple process flowchart

We proceed to an example of a chemical process and its XML representation. The process flowchart (Fig. 2) and corresponding XML configuration represent a simulation of a gas separation process using a compressor and membrane unit, commonly used in chemical or environmental engineering for gas purification or component separation (e.g., CO_2 removal, hydrogen recovery, air separation).

We have the following components:

- Excel module (Data Source). This module simulates feeding real or test data (e.g., gas composition, flow rate) into the process from an Excel sheet. This component specifies sending gas stream to the compressor. Parameters include fields like mixRate and pressureDrop (although values are not fully defined here).

- Compressor module increases the pressure of the gas stream. Compressing a gas before membrane separation is common to improve separation performance. It receives gas parameters from the Excel module and outputs compressed gas to the membrane unit. Parameters are same as above—may allow toggling pressure drop behavior or setting gas mixture characteristics.
- Membrane module separates gas components based on permeability differences through the membrane. It takes in compressed gas from the compressor and outputs the retentate (gas that did not pass through the membrane—typically richer in the less permeable component) and also the permeate (gas that passed through the membrane, richer in the more permeable component, e.g., CO_2 or H_2).

This setup simulates carbon capturing (compressing flue gas and separating CO_2), hydrogen purification (retaining hydrogen while allowing nitrogen or methane to pass), or air separation (oxygen/nitrogen enrichment).

XML's module are:

- `sfx.Excel` = input source
- `sfx.Compressor` = pressure-boosting
- `sfx.MembraneUnit` = separation

The data flow is *Excel* → *Compressor* → *Membrane* → (Out1 and Out2). Ports are *in1*, *out*, *out1*, *out2* (they label the process streams), and *FlowType* (numerical/continuous flow data).

To represent the chemical process in the XML file using Terraform, we need to make a few assumptions. If we treat each module (Excel, Compressor, Membrane) as computational or data-processing infrastructure blocks, one can simulate a digital twin or simulation pipeline (e.g., using containers or simulation jobs). We deploy this simulation as a set of containerized services, each representing a module orchestrated via Terraform.

```
resource "aws_ecs_cluster" "process_simulation" {
  name = "chemical-process-sim" }
# Define task definition for Excel Module
resource "aws_ecs_task_definition" "excel_module" {
  family                    = "excel-module"
  network_mode              = "awsvpc"      requires_
compatibilities = ["FARGATE"]
  cpu                       = "256"
memory                    = "512"
  container_definitions = jsonencode([
    {  name   = "excel-adapter"
         environment = [  { name   = "MIX_RATE" value =
"0.5" },
        { name    = "PRESSURE_DROP" value = "yes" }]
    } ])}
```

```
# Define task definition for Compressor Module
resource "aws_ecs_task_definition" "compressor_module" {
  container_definitions = jsonencode([
    {
      name  = "compressor"
      image = "your-dockerhub-user/compressor-sim:latest"
    }]))
# Define task definition for Membrane Module
resource "aws_ecs_task_definition" "membrane_module" {
}
# Start ECS services to run each task (Excel → Compressor →
Membrane)
resource "aws_ecs_service" "excel_service" {
  name            = "excel-service"
  cluster         = aws_ecs_cluster.process_simulation.id
  task_definition = aws_ecs_task_definition.excel_module.arn
}
# Repeat similarly for compressor_service and
membrane_service
```

2.2 *Serverless Deployment of a Continuous Stirred-Tank Reactor (CSTR) Simulation*

A process engineer uses a GUI-based simulator (e.g., DWSIM or Aspen Plus) to model a single CSTR, where a first-order irreversible reaction A → B occurs under steady-state conditions. The simulation includes specifications for:

- Inlet concentration and flow rate
- Reactor volume
- Reaction rate constant
- Temperature and pressure conditions

The goal is to offload this simulation to a serverless cloud infrastructure, where it can be automatically deployed and triggered for parameter sweeps.

The prototype workflow starts with GUI Model Export. The engineer exports the CSTR setup as an XML file. This file includes unit operations, stream definitions, and thermodynamic settings.

Then the symbolic LLM parses the XML file into a symbolic graph of process components. It then identifies the CSTR node and extracts key equations (e.g., mole balance, $F_{in}C_{A,in} - FC_A - kVC_A = 0$). After that, component behavior is converted into Python simulation logic. Finally, the system generates iac.tf (Terraform).

The message-passing infrastructure works as follows. Each unit operation (e.g., inlet stream, reactor, outlet stream) is wrapped in a virtual component with defined

Table 1 Steps in serverless deployment for chemical process simulation

Phase	Symbolic LLM role
Model parsing	Interprets XML and translates to logic
Simulation logic	Auto-generates domain-specific compute code
System composition	Creates DAG with symbolic message interfaces
Infrastructure mapping	Translates simulation into IaC

I/O message interfaces (e.g., JSON payloads). Symbolic reasoning is used to build the execution DAG, ensuring correct evaluation order. Interactions are defined as AWS EventBridge or Step Functions transitions.

The LLM generates Terraform templates for:

- AWS Lambda functions for each component
- Step Function definition to orchestrate flow
- S3 buckets or DynamoDB tables for temporary data exchange

Under serverless deployment, the engineer runs a single command (e.g., terraform apply) to deploy all simulation components as AWS Lambda functions, register the workflow as a Step Function, and define event triggers (e.g., via API Gateway or cron for batch runs).

Under execution and scaling, a temperature sweep triggers 100 simulations. Step functions fan out invocations of Lambda functions, each simulating a different temperature point, and results are streamed to S3 and aggregated. Finally, the LLM reads the output files from S3 and generates summary statistics, plots of concentration profiles, and recommendations for process optimization.

We summarize the steps in Table 1.

3 Running Complex Chemical Simulations Across Distributed Nodes

The goal is to run complex chemical simulations (e.g., molecular dynamics, reaction kinetics; Aliev et al. 2020) across distributed nodes in a serverless environment, with LLMs assisting in orchestration, optimization, and interpretation.

LLMs can take high-level engineering goals or natural language descriptions (e.g., "simulate catalytic cracking at 800K with feed composition X") and generate simulation scripts (in Python, MATLAB, or domain-specific languages). It can also partition them into stateless, parallelizable tasks suitable for serverless compute (e.g., via AWS Lambda, Google Cloud Functions).

To Optimize Task Distribution, neuro-symbolic LLMs analyze dependencies between simulation modules; suggest how to split workloads into concurrent, small lambda functions (map-reduce style); and provide glue code for event-driven orchestration (using AWS Step Functions, Azure Durable Functions, etc.)

Table 2 Benefits of serverless implementation

Feature	Benefit
Stateless Sim Tasks	Each reaction run is independent
Event-based Execution	Fan-out via Step Functions
Scalable	10k reactions run in minutes, without cluster management
Pay-per-use	Cost-efficient for batch jobs
LLM Orchestration	Reduces human workload + enables dynamic workflows

LLM Generates code for simulation engines, for example, it converts prompts into code that runs with OpenFOAM, LAMMPS, Cantera, or COMSOL via API. LLM also wraps simulation kernels so they can run in lightweight containers or Lambda functions.

Interfacing with serverless frameworks, LLMs can generate and configure serverless deployment scripts (e.g., using Serverless Framework or Terraform), cloud-native event handlers (e.g., triggering a simulation on a file upload to S3), as well as monitoring, logging, and retry logic. Once distributed simulations complete, LLMs collect and interpret results from logs and outputs, suggest next simulation parameters (like a feedback loop), and auto-generate plots, summaries, or even publish-ready documents.

Use cases include Monte Carlo Simulations for reaction yield—that is parallelized across 1000 Lambda instances, batch thermodynamic calculations with Cantera—where LLM builds input files and dispatches compute, and dynamic reaction optimization—where LLM drives iterative design-simulate-analyze loop.

The benefits of serverless implementation of simulation with neuro Symbolic LLM support are enumerated in Table 2.

4 Introducing Infrastructure as Code for Process Simulation

Infrastructure as Code (IaC) is a practice in which infrastructure (such as servers, networks, storage, and other cloud resources) is managed and provisioned through machine-readable configuration files rather than manual processes. IaC allows developers and operations teams to automate the deployment, scaling, and management of infrastructure using code. This approach brings several benefits:

1. Version Control: Infrastructure configurations can be versioned just like application code, enabling tracking of changes, rollbacks, and collaboration.
2. Consistency: Ensures consistent environments across development, testing, staging, and production by eliminating manual configuration errors.
3. Reproducibility: Environments can be easily recreated or duplicated, reducing the "it works on my machine" problem.
4. Scalability: Automates the provisioning of resources, making it easier to scale infrastructure up or down based on demand.

5. Cost Efficiency: Reduces human error and optimizes resource usage, leading to cost savings.

Terraform is an open-source tool developed by HashiCorp that enables Infrastructure as Code. It allows you to define and manage infrastructure using a declarative language called HashiCorp Configuration Language (HCL). Terraform supports multiple cloud providers (e.g., AWS, Azure, Google Cloud, etc.) and on-premises systems, making it a versatile choice for managing infrastructure across different environments.

Key features of Terraform:

1. Declarative syntax. One describes the desired state of your infrastructure in configuration files ('.tf' files), and Terraform ensures that the actual infrastructure matches this state. For example, instead of writing step-by-step instructions to create a virtual machine, you simply declare that a VM with specific properties should exist.
2. Provider ecosystem. Terraform integrates with hundreds of providers, including major cloud platforms (AWS, Azure, GCP), SaaS tools, and on-premises systems. Providers act as plugins that allow Terraform to interact with APIs of various services.
3. State management. Terraform maintains a state file that tracks the current state of your infrastructure. This file helps Terraform determine what changes are needed to achieve the desired state. State files can be stored locally or remotely (e.g., in a cloud storage bucket or Terraform Cloud).
4. Idempotency. Terraform ensures that running the same configuration multiple times produces the same result. If the infrastructure already matches the desired state, no changes are made.
5. Plan and Apply workflow. 'terraform plan': Generates an execution plan showing what changes Terraform will make to achieve the desired state. 'terraform apply': applies the changes to provision or modify the infrastructure.
6. Modularity. Terraform supports reusable modules, allowing you to encapsulate and share infrastructure components (e.g., a VPC module, a database module).
7. Immutable Infrastructure. Terraform encourages the use of immutable infrastructure, where changes are made by replacing resources rather than modifying them in place. This reduces drift and improves reliability.

Terraform works in the following way:

1. Write Configuration Files:

```
hcl
provider "aws" {
  region = "us-west-2"
}
resource "aws_instance" "example" {
  ami           = "ami-0c55b159cbfafe1f0"
  instance_type = "t2.micro"
}
```

2. Initialize Terraform: run 'terraform init' to download the required provider plugins, and set up the working directory.
3. Preview changes: run 'terraform plan' to see what changes Terraform will make to your infrastructure.
4. Apply changes: run 'terraform apply' to execute the plan and provision the infrastructure.
5. Destroy resources: run 'terraform destroy' to remove all resources defined in the configuration.

Benefits of using Terraform are as follows:

1. Multi-cloud support: Terraform's provider ecosystem allows you to manage resources across multiple cloud providers from a single configuration.
2. Automates repetitive tasks like provisioning servers, configuring networks, and deploying applications.
3. Collaboration: teams can collaborate on infrastructure code using version control systems like Git.
4. Auditability: every change to the infrastructure is documented in the configuration files, making it easy to audit and track changes.
5. Community and ecosystem: Terraform has a large community and a rich ecosystem of modules and plugins, making it easier to get started and solve common problems.

Example Use Case is as follows. Suppose one wants to deploy a simple Web application on AWS with the following components:

- A VPC with public and private subnets
- An EC2 instance in the public subnet to host the application
- An RDS database in the private subnet

One can define this infrastructure in a Terraform configuration file ('main.tf'):

```
provider "aws" {
  region = "us-west-2"
}
resource "aws_vpc" "example" {
  cidr_block = "10.0.0.0/16"
}
resource "aws_subnet" "public" {
  vpc_id     = aws_vpc.example.id
  cidr_block = "10.0.1.0/24"
}
resource "aws_subnet" "private" {
  vpc_id     = aws_vpc.example.id
  cidr_block = "10.0.2.0/24"
}
```

```
resource "aws_instance" "web_server" {
  ami            = "ami-0c55b159cbfafe1f0"
  instance_type = "t2.micro"
  subnet_id      = aws_subnet.public.id
}
resource "aws_db_instance" "database" {
  allocated_storage      = 20
  engine                 = "mysql"
  engine_version         = "5.7"
  instance_class         = "db.t2.micro"
  name                   = "mydb"
  username               = "admin"
  password               = "password"
  db_subnet_group_name = aws_db_subnet_group.private.name
}
```

By running 'terraform apply', Terraform will provision all these resources in AWS.

Hence, Terraform is a powerful tool for implementing Infrastructure as Code. It simplifies the process of managing complex infrastructure by providing a declarative, automated, and reproducible way to define and deploy resources. Whether one working in a single-cloud or multi-cloud environment, Terraform can help streamline an infrastructure management process while ensuring consistency and scalability.

Using *Infrastructure as Code* (IaC) to represent a chemical process or technological scheme is an innovative and unconventional application of IaC principles. While IaC is traditionally used for managing IT infrastructure (e.g., servers, networks, databases), the same principles can be adapted to model and manage physical systems like chemical processes. Below we describe how one can approach this problem by leveraging IaC tools like Terraform or other declarative languages.

4.1 Key Concepts for Representing a Chemical Process as Code

We enumerate the features of IaC that can be leveraged to represent a chemical process:

1. Declarative representation. Just as IaC defines infrastructure in a declarative manner, you can describe the components of a chemical process (reactors, pumps, heat exchangers, etc.) and their relationships in a similar way. The goal is to define the desired state of the chemical process, including inputs, outputs, and intermediate transformations.

2. Modularity breaks down the chemical process into reusable modules (e.g., a reactor module, a distillation column module, etc.). This allows you to compose complex processes from simpler building blocks.
3. State management tracks the current state of the process (e.g., flow rates, temperatures, pressures) and compares it to the desired state. This ensures that the system operates as intended.
4. Automation simulates, validates, or even controls the process. For example, you could generate control scripts for programmable logic controllers (PLCs) or simulate the process using computational tools.

While Terraform is primarily designed for IT infrastructure, one can use other tools or frameworks better suited for modeling physical systems. Alternatively, one can extend Terraform's capabilities with custom providers or scripts. Here are some options:

1. Custom Terraform provider: develop a custom Terraform provider to model chemical process components. For example, you could define resources like 'chemical_reactor', 'heat_exchanger', or 'pump'.
2. Domain-specific languages (DSLs). Use or create a DSL tailored to chemical engineering. For example, Python-based libraries like Pyomo or ChemCAD can be used to model and optimize chemical processes.
3. Simulation tools. One can integrate with simulation tools like Aspen Plus, MATLAB/Simulink, or gPROMS to simulate the behavior of the chemical process based on one's code.
4. JSON/YAML for configuration. One can use lightweight formats like JSON or YAML to define the process schema. These can then be parsed by custom scripts or tools.

4.2 *Example 1 Representing a Chemical Process Using Terraform*

Below is an example of how to use Terraform to represent a simple chemical process involving a reactor and a separator. A Terraform configuration file ('main.tf') to define the components of the chemical process needs to be created:

````
```hcl
provider "chemical_process" {
 # Custom provider for chemical processes
}
resource "chemical_reactor" "reactor_1" {
  name             = "Reactor-1"
  type             = "Continuous Stirred-Tank Reactor (CSTR)"
  volume           = 1000  # liters
````
````

```
  temperature     = 350    # Kelvin
  pressure        = 1      # atm
  feed_streams    = ["stream_feed_1"]
  product_stream = "stream_product_1"
}
resource "chemical_separator" "separator_1" {
  name           = "Separator-1"
  type           = "Distillation Column"
  feed_stream    = "stream_product_1"
  top_product    = "stream_top_product"
  bottom_product = "stream_bottom_product"
}
resource "chemical_storage_tank" "tank_1" {
  name         = "Storage-Tank-1"
  capacity     = 5000  # liters
  input_stream = "stream_top_product"
}
```

To make this work, one would need to implement a custom Terraform provider ('chemical_process'). This provider would define resource types like 'chemical_reactor', 'chemical_separator', and 'chemical_storage_tank'. I also translate these resources into actions, such as generating simulation inputs, creating control scripts, or visualizing the process.

To simulate the chemical process, one can integrate an IaC representation with simulation tools. This is done by applying IaC tool to generate input files for simulation software like Aspen Plus or gPROMS. To automate simulations, it is necessary to write scripts to automate the execution of simulations and collect results. To visualize results, it is worth using visualization libraries (e.g., Matplotlib, Plotly) to display the results of the simulation.

4.3 Enterprise Service Bus

Enterprise Service Bus (ESB) is a software architecture pattern used to facilitate communication between different applications, services, and systems within an enterprise. It acts as a centralized middleware layer that enables seamless integration, data exchange, and service orchestration across various components of an organization's IT infrastructure.

Key concepts and functions of an enterprise service bus are as follows:

1. Centralized Communication Layer:
    - An ESB provides a single, unified platform for integrating disparate systems, applications, and services.

- Instead of creating point-to-point connections between systems, all communication flows through the ESB, simplifying the architecture and reducing complexity.

2. Message Routing:

   - The ESB routes messages between services based on predefined rules or dynamic conditions.
   - For example, it can direct requests from a Web application to a backend database or send notifications to multiple systems when an event occurs.

3. Protocol Transformation:

   - Different systems often use different communication protocols (e.g., HTTP, SOAP, REST, JMS, FTP).
   - The ESB can translate between these protocols, ensuring that systems with incompatible interfaces can still communicate effectively.

4. Data Transformation:

   - Systems may use different data formats (e.g., JSON, XML, CSV).
   - The ESB transforms data into the required format for each system, ensuring compatibility and consistency.

5. Service Orchestration:

   - The ESB can coordinate multiple services to execute complex business processes.
   - For example, it can combine data from multiple sources, process it, and then send the results to another system.

6. Monitoring and Management:

   - The ESB provides tools for monitoring message flows, tracking errors, and managing the performance of integrated systems.
   - This helps administrators ensure reliability and troubleshoot issues.

7. Security:

   - The ESB can enforce security policies, such as authentication, authorization, and encryption, to protect sensitive data during transmission.

8. Scalability and Flexibility:

   - As an organization grows, new systems and services can be easily added to the ESB without disrupting existing integrations.
   - This makes the ESB highly scalable and adaptable to changing business needs.

ESB works in the following way:

1. Message ingestion: the system sends a message (e.g., a request or event) to the ESB.

2. Routing and transformation: The ESB determines the destination of the message, applies any necessary transformations (protocol, data format), and routes it to the appropriate service.
3. Execution: The receiving service processes the message and sends a response back to the ESB.
4. Response handling: the ESB routes the response back to the original sender or to another system, as needed.

5 Simulation Using Infrastructure as Code

5.1 Modeling Task

A chemical process flow diagram (CPF) is a description of a process system characterized by at least one process and consists of apparatuses (in particular, chemical apparatuses, chemical-process apparatuses, process apparatuses, etc.). The apparatuses (in which the processes take place, including chemical and chemical-process processes) are modules of the process flow diagram connected by process channels through which flows (chemical-process flows) are transmitted. Each flow in the process flow diagram corresponds to at least one channel connecting at least two modules. Examples of apparatuses include mass exchange apparatuses, evaporators, a distillation column, an absorber, etc.

Channels, both for a module of the process flow diagram and for a process flow diagram, are divided into input channels and output channels. In a particular case, the output channel from one module of the process flow diagram is an input channel for another module of the process flow diagram (Finlayson et al. 2007). Channels are used to communicate/connect modules of the process flow diagram and are used to transmit flows between modules of the process flow diagram. Each device is included in a module of the chemical-technological flow diagram. A module of the process flow diagram can be calculated by a calculation module (implementing at least one algorithm for calculating a module of the process flow diagram), which is a software solution (software, program), with the help of which the calculation is performed (Towler and Sinnott 2008).

For one type of apparatus, there may be several calculation modules that are capable of calculating this type of apparatus in various ways, for example, a compressor-type apparatus may be implemented by such calculation modules as:

- A calculation module that allows calculating a certain model of the apparatus using software supported by the system.
- Using an industrial modeling program (calculation program).
- Using software, etc.

The task of modeling is to determine the output flows of the apparatus (chemical-technological process, chemical-technological scheme, chemical-technological system, chains of the chemical-technological scheme, etc.) based on its mathematical model, i.e., a mathematical description of the operation of the apparatus and determining its properties and features. The technological scheme consists of devices that convert input flows into output flows. The possible number of such devices can be calculated in tens, hundreds, etc., while an infinite number of process flow diagrams can be created using the devices. In this regard, as a rule, a mathematical model for each diagram is not created, since it is an extremely resource-intensive task.

Mathematical models of devices describe the operation of devices in detail; for example, they describe how the output flow will change based on the input flow and the properties of the device. At the same time, the mathematical model does not take into account in which particular process flow diagram a particular device is used, since such parameters as initial data (properties, characteristics, parameters of input flows and the device) set, for example, by the user or parts of the automated control system, play a role. However, even if a mathematical model of the device is used, the parameters of the output flows of the device cannot always be calculated with high accuracy. Often the most accurate methods of calculating chemical-technological apparatuses (apparatuses of chemical-technological schemes) are based on solving differential equations of ordinary or partial derivatives, on complex iterative methods with unguaranteed convergence, especially if the initial conditions are not specified accurately (Yin et al. 2021). Thus, several models can be used for calculations, so the ability to select a model is very important for the user.

Neuro-symbolic AI offers a promising approach to overcoming these limitations by integrating data-driven learning with symbolic reasoning about system structure and context. While mathematical models focus narrowly on the internal physics of a device, neuro-symbolic AI can reason across the broader process flow diagram by combining learned representations of device behavior with symbolic knowledge about their roles within the larger system. For instance, a neural component can be trained on historical data to approximate flow behaviors under uncertain conditions, while a symbolic layer can enforce constraints based on engineering principles, flow connectivity, or user-defined process goals (Garcez et al. 2019; Besold et al. 2017). This enables dynamic model selection based on context—automatically choosing or blending the most appropriate models for a given configuration or data quality level. Moreover, by encoding expert knowledge (e.g., typical parameter ranges, failure modes, or inter-device dependencies) symbolically, neuro-symbolic AI systems can guide simulation or control processes even when data is sparse or noisy, improving both robustness and explainability (Colelough and Regli 2025).

Simulating a chemical production plant, such as mineral fertilizer production, using Infrastructure as Code (IaC) with Terraform requires a structured approach:

1. Understanding the XML Representation
   If the process is described in XML, it likely includes:
   - Chemical reactions (input chemicals, reaction conditions, output products).
   - Equipment and infrastructure (reactors, heat exchangers, pipelines, storage).
   - Operational parameters (flow rates, temperatures, pressures).
   - Control logic and automation (sensors, actuators, PLCs).
2. Translating XML to Terraform Resources
   Terraform is typically used for managing cloud and infrastructure resources, so we must map the chemical plant components to cloud or on-premise resources, such as:
   - Compute resources: Virtual machines (VMs), Kubernetes clusters for simulation and control systems.
   - Networking: VPCs, subnets, load balancers to simulate process communication.
   - Storage: Databases for process data storage, log management.
   - IaC for process simulation: Tools like Simulink, Aspen HYSYS, or OpenFOAM could be deployed via Terraform.
   - IoT integration: Simulated sensors using cloud IoT services (AWS IoT, Azure IoT).
3. Terraform Implementation
   - Convert XML to a structured format (JSON/YAML), which can be ingested into Terraform.
   - Use Terraform modules to deploy plant infrastructure, including:
     - Compute (EC2, GCP Compute, AKS/EKS for containers).
     - Networking (VPCs, Security Groups, Service Mesh).
     - Storage (S3, Azure Blob, RDS for logging and data collection).
     - Simulation Software Deployment (Containers, VMs with installed software).

Example Terraform Code Snippet:

```
provider "aws" {
  region = "us-west-1"
}
resource "aws_instance" "simulator" {
  ami           = "ami-12345678"
  instance_type = "t3.large"
  tags = {
    Name = "FertilizerProductionSimulator"
  }
}
resource "aws_s3_bucket" "process_data" {
  bucket = "fertilizer-plant-data"
  acl    = "private"
}
resource "aws_vpc" "plant_network" {
  cidr_block = "10.0.0.0/16"
}
```

4. Automating Deployment and Execution

- Parse XML with Python → Convert it into Terraform variables.
- Terraform Modules for Scalability → Define separate modules for reactors, storage tanks, and control systems.
- Use Terraform Apply for Deployment → Run Terraform to deploy infrastructure.

5. Simulating Process Execution
   Once deployed, the simulated production plant could:

- Run chemical process models in VMs/containers.
- Store real-time process data in databases.
- Integrate with IoT simulators to replicate sensors and actuators.

5.2 XML Configuration for a Spring Integration Flow that Models Events in a Chemical Plant

Below is an example of an XML configuration for a Spring Integration flow that models events in a chemical plant. The scenario involves monitoring reagents (chemical inputs) and processing messages related to their flow, such as detecting when reagents are added or when the flow rate changes.

```xml
<?xml version="1.0" encoding="UTF-8"?>
<beans xmlns="http://www.springframework.org/schema/beans"
       xmlns:int="http://www.springframework.org/schema/integration"
       xmlns:int-
file="http://www.springframework.org/schema/integration/file"
       xmlns:int-
stream="http://www.springframework.org/schema/integration/stream"
       xmlns:xsi="http://www.w3.org/2001/XMLSchema-instance"
       xsi:schemaLocation="
           http://www.springframework.org/schema/beans
           http://www.springframework.org/schema/beans/spring-beans.xsd
           http://www.springframework.org/schema/integration
           http://www.springframework.org/schema/integration/spring-
integration.xsd
           http://www.springframework.org/schema/integration/file
           http://www.springframework.org/schema/integration/file/spring-
integration-file.xsd
           http://www.springframework.org/schema/integration/stream
           http://www.springframework.org/schema/integration/stream/spring-
integration-stream.xsd">

    <!-- 1. Monitor Input Files for Reagent Events -->
    <int-file:inbound-channel-adapter id="reagentFileAdapter"
                                      channel="reagentEventChannel"
                                      directory="/input/reagents/"
                                      prevent-duplicates="true">
        <int:poller fixed-rate="5000"></int:poller>
    </int-file:inbound-channel-adapter>

    <!-- 2. Channel for Reagent Events -->
    <int:channel id="reagentEventChannel" />

    <!-- 3. Parse Reagent Flow Data from File -->
    <int:service-activator input-channel="reagentEventChannel"
                           output-channel="reagentFlowChannel"
                           ref="reagentParser"
                           method="parseReagentData" />

    <!-- 4. Bean for Parsing Logic -->
    <bean id="reagentParser" class="com.example.ReagentParser" />

    <!-- 5. Channel for Processed Reagent Flow Messages -->
    <int:channel id="reagentFlowChannel" />

    <!-- 6. Log Reagent Flow Events -->
    <int-stream:stdout-channel-adapter id="reagentFlowLogger"
                                       channel="reagentFlowChannel"
                                       append-newline="true" />

    <!-- 7. Handle Reagent Flow Alerts -->
    <int:filter input-channel="reagentFlowChannel"
                output-channel="alertChannel"
                expression="payload.flowRate > 100" />

    <!-- 8. Channel for Alerts -->
    <int:channel id="alertChannel" />

    <!-- 9. Log Alerts for High Flow Rates -->
    <int-stream:stdout-channel-adapter id="alertLogger"
                                       channel="alertChannel"
                                       append-newline="true" />
</beans>
```

Explanation of the Components

1. Monitoring Input Files

```
<int-file:inbound-channel-adapter id="reagentFileAdapter"
                                  channel="reagentEventChannel"
                                  directory="/input/reagents/"
                                  prevent-duplicates="true">
    <int:poller fixed-rate="5000"></int:poller>
</int-file:inbound-channel-adapter>
```

This adapter monitors the /input/reagents/ directory for new files containing reagent data.

It sends the file contents as messages to the reagentEventChannel.

2. Reagent Event Channel

```
<int:channel id="reagentEventChannel" />
```

Acts as a conduit for transporting reagent event messages to the next step in the flow.

3. Parsing Reagent Data

```
<int:service-activator input-channel="reagentEventChannel"
                       output-channel="reagentFlowChannel"
                       ref="reagentParser"
                       method="parseReagentData" />
```

A service activator processes incoming messages from reagentEventChannel.

It uses a custom parser (ReagentParser) to extract reagent flow information (e.g., flow rate, reagent type).

4. Custom Parser Bean

```
<bean id="reagentParser" class="com.example.ReagentParser" />
```

A Java bean (ReagentParser) contains logic to parse raw file data into structured reagent flow messages.

Example method:

```
public class ReagentParser {
    public ReagentFlow parseReagentData(String fileContent) {
        // Parse file content and return ReagentFlow object
        return new ReagentFlow("ReagentA", 120); // Example
flow rate
    }
}
```

5. Processed Reagent Flow Channel

```
<int:channel id="reagentFlowChannel" />
```

Transports parsed reagent flow messages to downstream components.

6. Logging Reagent Flow Events

```
<int-stream:stdout-channel-adapter id="reagentFlowLogger"
                                   channel="reagentFlowChannel"
                                   append-newline="true" />
```

Logs all reagent flow messages to the console for monitoring purposes.

7. Filtering High Flow Rates

```
<int:filter input-channel="reagentFlowChannel"
            output-channel="alertChannel"
            expression="payload.flowRate > 100" />
```

Filters messages based on the flowRate property of the reagent flow payload.

If the flow rate exceeds 100 units, the message is sent to the alertChannel.

8. Alert Channel

```
<int:channel id="alertChannel" />
```

Transports high-flow-rate alerts to the next step.

9. Logging Alerts

```
<int-stream:stdout-channel-adapter id="alertLogger"
                                   channel="alertChannel"
                                   append-newline="true" />
```

Example Input File

```
Assume the /input/reagents/ directory contains a file named
reagent_data.txt with the following content:
ReagentA:120
ReagentB:80
Expected Output
Reagent Flow Logs :ReagentFlow{reagent='ReagentA',
flowRate=120}
ReagentFlow{reagent='ReagentB', flowRate=80}
Alert Logs :
ReagentFlow{reagent='ReagentA', flowRate=120}
```

Key features of the configuration are:

- File monitoring: Automatically detects new files containing reagent data
- Custom parsing: Converts raw file data into structured messages
- Flow rate filtering: Identifies abnormal flow rates and triggers alerts
- Logging: Provides detailed logs for both normal and alert events

This configuration demonstrates how Spring Integration can be used to model and process events in a chemical plant, ensuring efficient monitoring and handling of reagent flows.

6 IoC for Simulating Chemical Process

To implement this XML-based Spring Integration workflow in Terraform, we need to create infrastructure that supports:

- File monitoring—Watch a directory for new reagent data files.
- Processing pipeline—Parse and filter reagent flow messages.
- Logging and alerts—Log normal flows and trigger alerts for high flow rates.

This Terraform configuration will:

1. Deploy an EC2 instance (AWS) or VM to run the reagent processing logic.
2. Configure an S3 bucket (AWS) or a local directory for file-based monitoring.
3. Deploy CloudWatch (AWS) or equivalent logging for reagent flow messages.
4. Use AWS Lambda or an event-driven mechanism to parse files and trigger alerts.

We describe how to provision this infrastructure in AWS:

1. Create an S3 Bucket for Reagent Data Files

```
resource "aws_s3_bucket" "reagent_bucket"{
  bucket = "reagent-data-bucket"
  acl    = "private"
}
```

2. Create an EC2 instance to process files. This instance will run a spring boot application (or equivalent) to monitor and process reagent files. It also logs reagent flow and alerts.

```
resource "aws_instance" "reagent_processor" {
  ami           = "ami-12345678"  # Replace with the
correct AMI
  instance_type = "t3.medium"

  user_data = <<-EOF
              #!/bin/bash
              sudo yum update -y
              sudo yum install -y java-11
              sudo mkdir /input/reagents/
              sudo aws s3 sync s3://${aws_s3_bucket.reagent_
bucket.id} /input/reagents/ --delete
              sudo java -jar /opt/reagent-processor.jar &
              EOF

  tags = {
    Name = "ReagentProcessor"
  }
}
```

3. Create an S3 Event Notification for New Files
   This triggers processing when a new reagent file is uploaded.

```
resource "aws_s3_bucket_notification" "reagent_event" {
  bucket = aws_s3_bucket.reagent_bucket.id

  lambda_function {
    lambda_function_arn = aws_lambda_function.reagent_
parser.arn
    events              = ["s3:ObjectCreated:*"]
  }
}
```

4. Deploy AWS Lambda for Reagent Parsing
   This function processes reagent flow data.

```
resource "aws_lambda_function" "reagent_parser" {
  function_name     = "ReagentParser"
  handler           = "parser.lambda_handler"
  runtime           = "python3.8"
  role              = aws_iam_role.lambda_exec.arn
  filename         = "reagent_parser.zip"
}
```

Lambda Code (Python)

```
def lambda_handler(event, context):
    for record in event['Records']:
        reagent_data = process_file(record['s3']['object']
['key'])

        if reagent_data['flowRate'] > 100:
            alert_high_flow(reagent_data)

    return {
        'statusCode': 200,
        'body': json.dumps('Reagent data processed.')
    }
```

5. Create a CloudWatch Log Group for Reagent Events

```
resource "aws_cloudwatch_log_group" "reagent_logs" {
  name = "/aws/reagentFlowLogs"
}
```

There are the following steps:

1. Files uploaded to S3 → Triggers AWS Lambda (parses data).
2. Lambda filters flow rates:
3. Normal flows go to CloudWatch logs.
4. High-flow alerts are sent to an SNS topic for notifications.
5. EC2 instance can optionally run Spring Boot for local monitoring.

6.1 Modeling the Reagent Flow Using Simple Queue Service

To model the reagent flow using AWS SQS (Simple Queue Service), we will implement a message-based architecture where:

1. S3 uploads a reagent data file → Triggers AWS Lambda to read the file.
2. Lambda sends parsed reagent flow messages to an SQS queue (ReagentFlowQueue).
3. Another Lambda (or an EC2 instance with a Spring Boot app) consumes messages from the queue to process reagent data.
4. High–flow rate messages (>100) are sent to an Alert SQS queue (AlertQueue).
5. Logs are sent to CloudWatch for monitoring.

The steps for Terraform Implementation with SQS

1. Create SQS Queues for Reagent Flow and Alerts

```
resource "aws_sqs_queue" "reagent_flow_queue" { name =
= "ReagentFlowQueue"
  visibility_timeout_seconds = 30
}

resource "aws_sqs_queue" "alert_queue" {
  name  = "AlertQueue"
  visibility_timeout_seconds = 30
}
```

2. Modify Lambda to Send Messages to SQS
   This Lambda parses the reagent file and sends flow data as messages.

```
resource "aws_lambda_function" "reagent_parser" {
  function_name    = "ReagentParser"
  handler          = "parser.lambda_handler"
  runtime          = "python3.8"
  role             = aws_iam_role.lambda_exec.arn
  filename         = "reagent_parser.zip"
  environment {
    variables = {
      REAGENT_FLOW_QUEUE = aws_sqs_queue.reagent_flow_
queue.url
      ALERT_QUEUE        = aws_sqs_queue.alert_queue.url
    }
  }
}
```

3. Python Code for Lambda (Parsing and Sending to SQS)

```
import json
import boto3
sqs = boto3.client('sqs')
REAGENT_FLOW_QUEUE = "https://sqs.us-east1.amazonaws.com/123/ReagFlowQueue"
ALERT_QUEUE = "https://sqs.us-east-1.amazonaws.com/123456789/AlertQueue"

def lambda_handler(event, context):
    for record in event['Records']:
        reagent_data = parse_file(record['s3']['object']['key'])

        message_body = json.dumps(reagent_data)

        # Send parsed reagent data to SQS
        sqs.send_message(
            QueueUrl=REAGENT_FLOW_QUEUE,
            MessageBody=message_body
        )

        # Send high-flow-rate alerts to Alert Queue
        if reagent_data['flowRate'] > 100:
            sqs.send_message(
                QueueUrl=ALERT_QUEUE,
                MessageBody=message_body
            )

    return {
        'statusCode': 200,
        'body': json.dumps('Reagent data processed.')
    }


def parse_file(file_key):
    # Simulated parsing logic
    return {
        "reagent": "ReagentA",
        "flowRate": 120
    }
```

4. Create a Consumer Lambda for Processing Messages
   This function reads messages from ReagentFlowQueue, processes them, and logs them.

```
resource "aws_lambda_function" "reagent_processor" {
  function_name    = "ReagentProcessor"
  handler          = "processor.lambda_handler"
  runtime          = "python3.8"
  role             = aws_iam_role.lambda_exec.arn
  filename         = "reagent_processor.zip"
}
```

Lambda Code (Consumer)

```
def lambda_handler(event, context):
    for record in event['Records']:
        reagent_data = json.loads(record['body'])
        print(f"Processing reagent: {reagent_
data['reagent']}, Flow Rate: {reagent_data['flowRate']}")

    return {
        'statusCode': 200,
        'body': "Messages processed."
    }
```

5. Configure Lambda Triggers for SQS

```
resource "aws_lambda_event_source_mapping" "reagent_
processor_trigger" {
  event_source_arn = aws_sqs_queue.reagent_flow_queue.arn
  function_name    = aws_lambda_function.reagent_
processor.arn
  batch_size       = 10
}
```

These are the steps:

1. File Uploaded to S3 → Triggers Lambda (Parsing).
2. Lambda Parses Data → Sends to ReagentFlowQueue (SQS).
3. Another Lambda Consumes Messages from SQS → Logs them.
4. If Flow Rate > 100, Message is Sent to AlertQueue.
5. Alerts Can Be Processed Separately or Sent to CloudWatch/SNS.

7 System Architecture

System architecture for designing a chemical process using cloud infrastructure is depicted in Fig. 3.

1. Start with a chemical production task. A user defines what chemical process they want to simulate or carry out (e.g., producing a certain compound).
2. Check the request data. The system checks that the input data (like chemical properties, flow rates, or process conditions) is complete and correct.

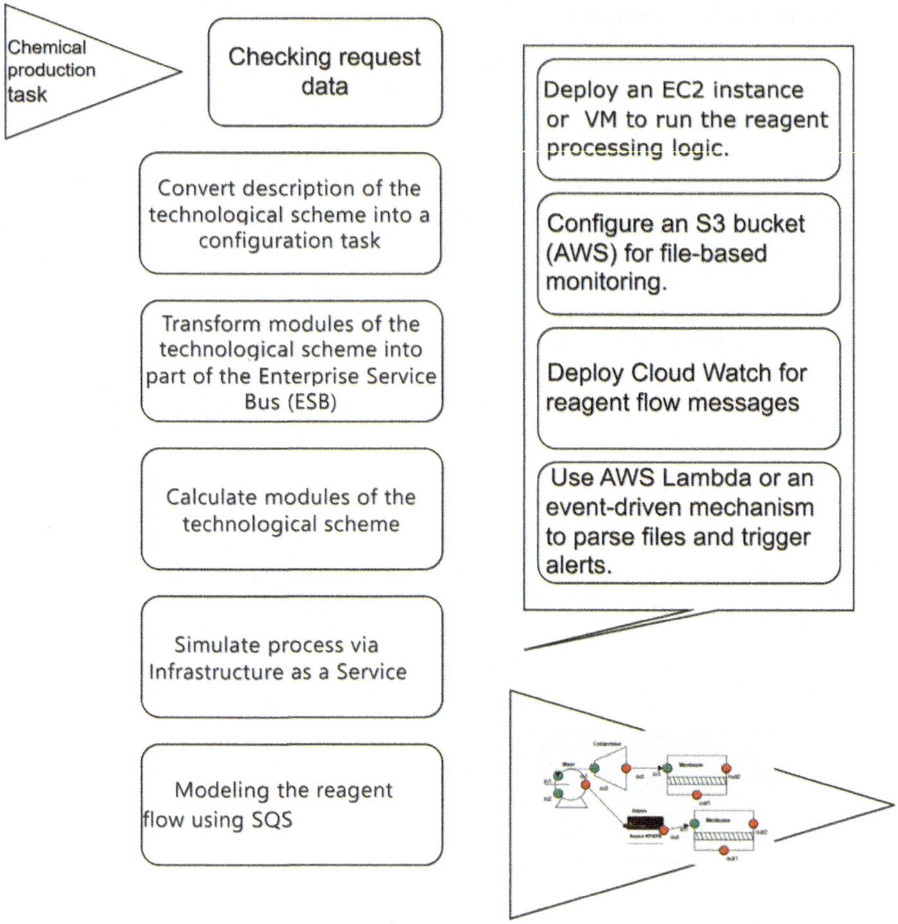

Fig. 3 The chart for IaS-based design of a chemical process

3. Turn the process description into a setup task. The user's process diagram is translated into a configuration that can be used by the simulation system.
4. Convert process modules into software components. Each unit in the chemical process (such as pumps, compressors, membranes) is turned into a software module that connects with others through a central system (called an Enterprise Service Bus or ESB).
5. Run calculations for each module. The system calculates what happens inside each part of the process—things like pressure changes, flow rates, and chemical reactions.
6. Run the full simulation using cloud services. The entire chemical process is simulated using cloud infrastructure (like virtual machines or containers), making it scalable and efficient.

7. Model the flow of chemicals using AWS queues. An AWS Simple Queue Service (SQS) is used to represent and track how materials move through the process—just like messages in a queue.

7.1 Example: Simulating Ammonium Nitrate Production

Ammonium nitrate is produced through the reaction of gaseous ammonia with nitric acid, releasing heat ranging from 100 to 115 J per mole of NH_4NO_3. This exothermicity is utilized in various production processes for the partial or complete evaporation of water. Depending on pressure conditions and nitric acid concentration, it is possible to produce ammonium nitrate solutions with concentrations of 95–97%.

During the neutralization process, efficient mixing of reactants within the reactor is necessary to prevent localized overheating, nitrogen losses, and the decomposition of ammonium nitrate. Conventional methods such as the Uhde process and the SBA (Société Belge de l'Azote) process are known for their lower reaction temperatures and reduced susceptibility to corrosion. Optimally utilizing the heat of reaction is achieved through pressure-based neutralization.

The process features a heat exchanger within the pressure reactor, using part of the heat of reaction to generate steam. Preheated ammonia and 52–63% HNO_3 are injected into the reactor's sump, operating at a pressure of approximately 0.45 MPa (4.5 bar) and temperatures between 170 and 180 °C (Fig. 4). The terraform representation for this process is shown in Fig. 5.

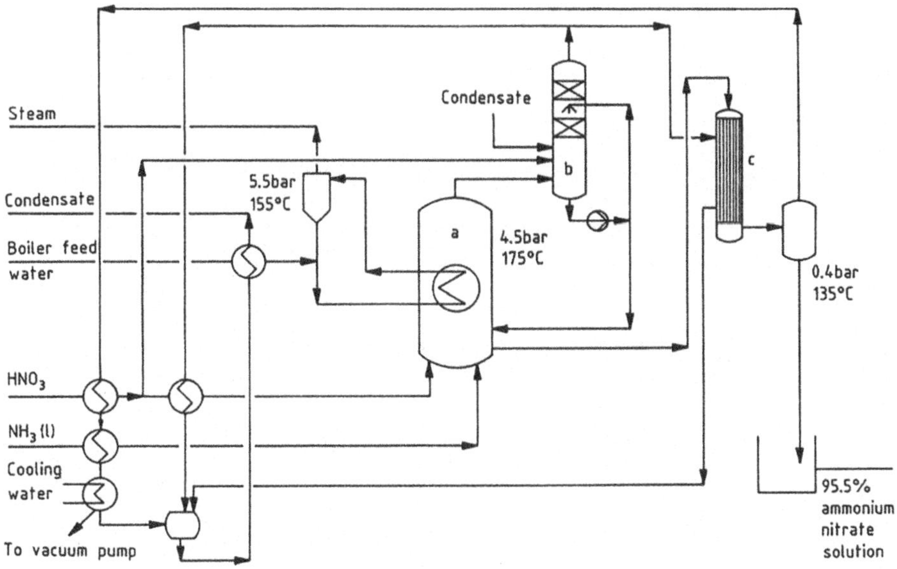

Fig. 4 Ammonium nitrate production chart

```
provider "aws" {
  region = "us-east-1"
}
# SQS Queues representing flow between components
resource "aws_sqs_queue" "hno3_supply" {
  name = "HNO3_Supply"
}
resource "aws_sqs_queue" "nh3_supply" {
  name = "NH3_Supply"
}
resource "aws_sqs_queue" "reactor" {
  name = "Reactor_Process"
}
resource "aws_sqs_queue" "heat_exchanger" {
  name = "Heat_Exchanger"
}
resource "aws_sqs_queue" "storage" {
  name = "Ammonium_Nitrate_Storage"
}
# Lambda function simulating the reactor process
resource "aws_lambda_function" "reactor_process" {
  filename      = "reactor_lambda.zip"
  function_name = "reactor_process"
  role          = aws_iam_role.lambda_exec.arn
  handler       = "reactor.lambda_handler"
  runtime       = "python3.9"
  timeout       = 10
  environment {
    variables = {
      INPUT_QUEUE  = aws_sqs_queue.reactor.id
      OUTPUT_QUEUE = aws_sqs_queue.heat_exchanger.id
    }
  }
}
# Lambda function for heat exchanger simulation
resource "aws_lambda_function" "heat_exchanger_process" {
  filename      = "heat_exchanger_lambda.zip"
  function_name = "heat_exchanger_process"
  role          = aws_iam_role.lambda_exec.arn
  handler       = "heat_exchanger.lambda_handler"
  runtime       = "python3.9"
  timeout       = 10
  environment {
    variables = {
      INPUT_QUEUE  = aws_sqs_queue.heat_exchanger.id
      OUTPUT_QUEUE = aws_sqs_queue.storage.id
    }
  }
}
# IAM role for Lambda execution
resource "aws_iam_role" "lambda_exec" {
  name = "lambda_exec_role"
  assume_role_policy = jsonencode({
    Version = "2012-10-17"
    Statement = [{
      Action = "sts:AssumeRole"
      Effect = "Allow"
      Principal = {
        Service = "lambda.amazonaws.com"
      }
    }]
  })
}
# CloudWatch for monitoring the process

resource "aws_cloudwatch_metric_alarm" "high_temperature" {
  alarm_name          = "High_Reactor_Temperature"
  comparison_operator = "GreaterThanThreshold"
  evaluation_periods  = 1
  metric_name         = "ReactorTemperature"
  namespace          = "FertilizerProcess"
  period             = 60
  statistic          = "Average"
  threshold          = 180
  alarm_actions      = ["arn:aws:sns:us-east-1:123456789012:alerts"]
}
```

Fig. 5 Terraform representation for the process in Fig. 4

8 A Broader Look at Process Simulation

8.1 Automation Pyramid

To manage the complex and diverse tasks in industrial automation, the **automation pyramid** provides a widely accepted conceptual framework that organizes industrial functions into five hierarchical levels (Weyrich 2023). This structured model ensures clear separation of concerns while enabling seamless interaction between physical equipment and high-level enterprise systems. Each layer is designed to handle specific operational goals—from real-time control to strategic decision-making—forming a unified system that supports the full spectrum of industrial processes.

At the base of the pyramid lies the Field Level, where physical interaction with the technical process occurs. This layer includes hardware such as sensors, actuators, drives, and other embedded field devices that collect real-time data (e.g., temperature, pressure, flow) and execute mechanical or electrical responses. The role of this level is to interface directly with the physical environment, making it foundational to all subsequent layers. It ensures that accurate, time-sensitive data is available to control systems, enabling reliable and safe operation of equipment.

Above the field lies the Control Level, where programmable logic controllers (PLCs), distributed control systems (DCS), and other embedded controllers execute real-time control logic. This layer processes signals from the Field Level to make deterministic decisions, such as opening valves, adjusting motor speeds, or regulating temperature. Control algorithms—often based on PID control or finite state machines—are implemented here to maintain desired process conditions. The strict timing and reliability requirements at this level demand robust hardware and real-time software execution.

The Supervisory, Planning, and Enterprise Levels extend the system into broader operational and business concerns. The Supervisory Level includes systems such as SCADA (Supervisory Control and Data Acquisition) and HMIs (Human-Machine Interfaces), which allow operators and engineers to monitor system status, acknowledge alarms, and manually intervene when necessary. Above this, the Production Planning Level handles scheduling, quality assurance, and logistics coordination—transforming raw production goals into actionable execution plans. At the top, the Enterprise Level integrates business intelligence, supply chain data, and finance via ERP (Enterprise Resource Planning) systems, enabling strategic decision-making and optimization. Together, these top three layers provide visibility, adaptability, and alignment between shop-floor execution and organizational goals.

This hierarchical architecture ensures systematic data flow and control delegation across industrial operations. Data gathered from sensors propagates upward, enabling higher-level analysis and decisions, while commands and policies flow downward to regulate physical processes. By structuring automation in this way, organizations can achieve modularity, fault tolerance, and scalability, which are essential for modern manufacturing, smart factories, and Industry 4.0 applications (Fig. 6).

In the context of chemical process automation, LLMs are transformed into intelligent agents that can understand engineering documentation, interpret simulation output, propose control actions, or generate infrastructure as code. These agents function using a paradigm known as in-context learning, where the LLM's behavior is steered using structured prompts. Each prompt serves as a task definition or subtask delegation, enabling the LLM to act as a programmable, goal-directed information processor.

However, the control and simulation of complex chemical plants often exceed the capacity of a single LLM agent. Tasks such as interpreting P&ID diagrams, converting process descriptions into XML or SBML, deploying simulations to cloud services, and orchestrating control systems involve multiple modalities, abstraction levels, and domain-specific constraints. To address this complexity, we adopt a neuro-symbolic architecture within the LangGraph framework (chapter "A Copilot for Social Promotion")—a compositional agent graph where each node represents a specialized LLM agent or symbolic tool and edges encode information flow or task delegation rules (Xia et al. 2024).

In this LangGraph-based system, each node agent is responsible for a well-defined sub-task in the chemical automation pipeline. For instance, one agent may convert GUI-based unit operation data into symbolic XML representations; another may translate those into executable simulation logic (e.g., Python code for a CSTR model); a third agent may determine optimal parameter sweeps or control policies; and yet another may transform symbolic representations into IaC for deployment to AWS Lambda or Step Functions. A final agent may act as an orchestrator—monitoring the overall simulation state, coordinating message passing between virtual components (e.g., reactors, separators), and invoking corrective actions if constraints are violated. These LLM agents can be enhanced with symbolic modules such as rule-based validators, equation solvers, or constraint satisfaction solvers, ensuring that neural-generated outputs remain physically and logically sound.

This distributed multi-agent structure follows the LangGraph paradigm, in which:

- Nodes represent LLM agents or symbolic modules, each dedicated to a chemical sub-domain (e.g., reaction kinetics, control theory, process scheduling).

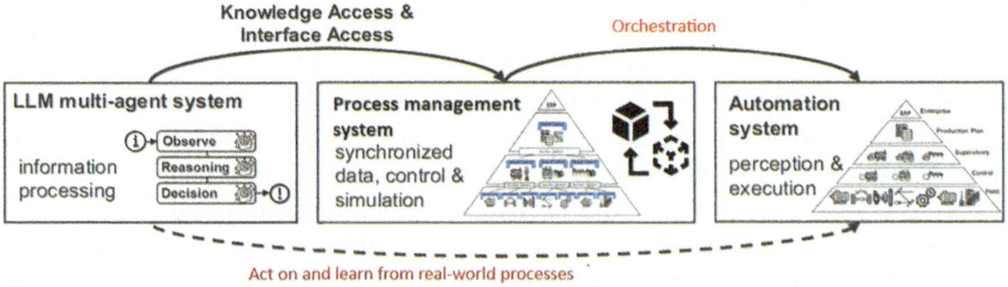

Fig. 6 From hybrid neuro-symbolic LLM multi-agent to automation system

- Edges define communication protocols—passing messages such as simulation parameters, equations, JSON payloads, or deployment artifacts.
- Loops within the graph allow for feedback control, retry mechanisms, or iterative optimization (e.g., refining a reactor's residence time based on simulation output).
- Entry points and exit conditions govern high-level workflows—such as completing a batch simulation, raising alarms on constraint violations, or triggering production planning updates.

This architecture introduces a dynamic, agent-based automation model for chemical processes, where neuro-symbolic LLM agents collaboratively interpret, plan, and execute across the entire automation pyramid—from field-level data (e.g., sensor streams) to enterprise-level integration (e.g., triggering ERP updates based on batch yields). The LangGraph framework makes this possible by providing a programmable, stateful execution model where tasks are decomposed, routed, and resolved across interconnected reasoning units (Fig. 7).

In contrast to monolithic AI models, this modular approach ensures:

- Scalability: More agents can be added as the complexity of plant operations increases.
- Explainability: Each agent's role is defined and its reasoning traceable.
- Interoperability: Symbolic agents can interface with legacy systems, simulators, and process databases.
- Adaptability: Agents can be retrained, fine-tuned, or re-prompted for new domains without affecting the entire system.

We now describe the IaC Simulation V-Model. Adapted from the traditional V-model by Forsberg and Mooz (1991), we illustrate a feedback-driven life cycle tailored to Infrastructure-as-Code environments. In the system design phase, declarative IaC frameworks are combined with Agile development practices (Le et al. 2022) to enable rapid, iterative infrastructure modeling while maintaining traceable requirements. During implementation, simulation and validation tools—such as test harnesses, policy-as-code (e.g., Open Policy Agent), and static analysis—are used to verify infrastructure behavior and compliance. Once deployed, infrastructure performance and drift monitoring tools feed operational data back into the simulation and design loop. This enables continuous refinement of both infrastructure

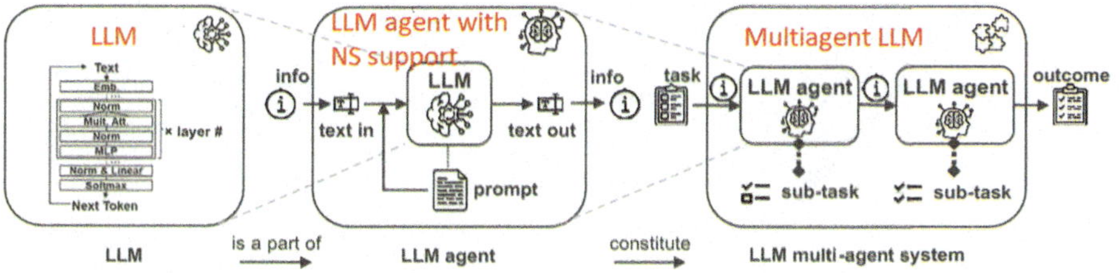

Fig. 7 Building a hybrid neuro-symbolic LLM and multi-agent system

specifications and deployment strategies, ensuring that the IaC system remains resilient, efficient, and aligned with evolving business and technical requirements (Fig. 8).

8.2 LLM-Powered Chemistry Engine

Bran et al. (2024) propose an LLM-powered chemistry engine, ChemCrow, designed to streamline the reasoning process for various common chemical tasks across areas such as drug and materials design and synthesis. ChemCrow harnesses the power of multiple expert-designed tools for chemistry and operates by prompting GPT-4o with specific instructions about the task and the desired format, as shown in Fig. 9. The LLM is provided with a list of tool names, descriptions of their utility, and details about the expected input/output. It is then instructed to answer a user-given prompt, using the tools provided when necessary. The model is guided to follow the *thought*, *action*, *action input*, and *observation* framework, which requires it to reason about the current state of the task, consider its relevance to the final goal, and plan the next steps accordingly, demonstrating its level of understanding. After the reasoning in the thought step, the LLM requests a tool (preceded by the keyword "Action") and the input for this tool (with the keyword *Action Input*). The text generation then gives way to the program attempts to execute the requested function using the provided input. The result is returned to the LLM tagged as *Observation*, and the LLM proceeds to the *thought* step again. It continues iteratively until the final answer is reached.

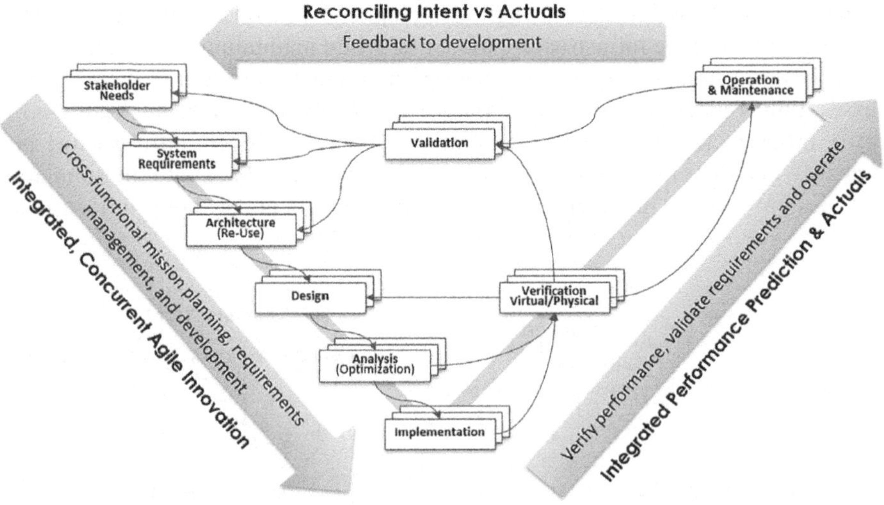

Fig. 8 The IaC Simulation V-Model

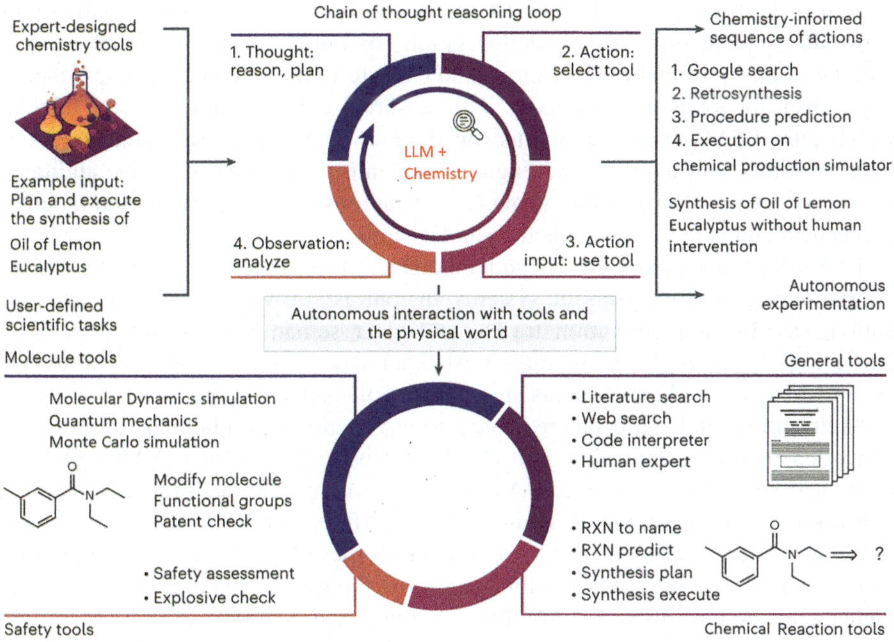

Fig. 9 A schema of chemical design solving process

This pipeline effectively combines chain-of-thought reasoning with tools relevant to the tasks. In the course of this pipeline, the LLM transitions from a trusted information source to a reasoning engine that is prompted to reflect on a task, act using a suitable tool to gather additional information, observe the tool's responses, and repeat this loop until the final result is obtained.

9 Conclusions

While LLMs have proven to be remarkably capable at generating simulation code, interpreting scientific documents, and responding to engineering prompts, LLMs alone are not enough for reliable, scalable, and explainable chemical simulation systems. The complexity of chemical engineering problems demands a hybrid architecture—one that combines neural inference (from LLMs) with symbolic reasoning and formal structure. Here is why:

The key reason is that structured scientific domains requires formal consistency. Chemical engineering involves conservation laws (mass, energy, momentum), reaction kinetics and thermodynamics, as well as complex, interdependent process networks. LLMs can suggest plausible equations or code, but they lack built-in mechanisms to enforce physical laws or domain constraints.

Neuro-symbolic systems allow LLMs to use symbolic rule sets to validate or constrain outputs, rely on knowledge graphs or ontologies to represent process topologies and physical relationships, and execute logic-based consistency checks (e.g., ensuring total mass balance). This ensures scientific and physical fidelity, which pure LLMs cannot guarantee on their own. Another important feature is interpretability and auditability. Engineering simulations must often be auditable for regulatory or safety reasons, provide explanations for why a simulation failed or succeeded, and support traceable design decisions.

LLMs are black boxes—they can hallucinate, invent functions, or give convincing but wrong results. Symbolic systems, in contrast, provide transparent rule execution, step-by-step derivation traces, and clear semantic representations (e.g., XML, logic trees, or decision graphs). Also, a neuro-symbolic system ensures that even LLM-generated components are interpretable and verifiable.

Multi-modal and multi-layered integration is a must for chemical simulations that often require graphical design (GUI), symbolic model files (XML, SBML), numerical solvers (e.g., Cantera, Aspen libraries), and cloud infrastructure (IaC, Terraform). No single representation mode suffices. At the same time, neuro-symbolic system uses symbolic logic to connect these layers; lets the LLM reason across text, code, and configuration; and allows symbolic planners to coordinate components across infrastructure, logic, and physical behavior. Pure LLMs might generate pieces of code, but they cannot ensure system-level alignment across modalities and abstraction levels.

Neuro-symbolic systems support dynamic decision-making and execution control. Distributed simulations involve:

- Component-based execution
- Parallel coordination (e.g., through Step Functions or message passing)
- Runtime error handling, rollback, and retries

Whereas LLMs are generative but not reactive, neuro-symbolic systems can use symbolic controllers (e.g., decision rules, production systems), allow runtime adaptation (e.g., stop a reactor when temperature exceeds bounds), and enable event-driven logic with formal semantics. This is essential for real-time, distributed, or autonomous chemical simulation workflows.

We summarize the complementarity of neural and symbolic approaches in Table 3.

Benefits of using IaC for chemical processes are as follows:

1. Reproducibility ensures that the process can be reproduced exactly as defined, reducing errors and inconsistencies.
2. Scalability easily scales up or modifies the process by updating the code.
3. Teams can collaborate on process design using version control systems like Git.
4. Automation tasks like simulation, optimization, and control system programming are supported.
5. IaC provides clear documentation of the process in a machine-readable format.

Table 3 Benefits for integration of LLM and symbolic features

Task	LLM (Neural)	Symbolic Component
Interpret natural language	Very strong	❌ Not applicable
Generate boilerplate code	Excellent	❌ Cannot generate
Enforce physical laws	Unreliable	✅ Always
Validate complex constraints (Galitsky 2025)	Weak	✅ Deterministic & explainable
Control distributed workflows	Partial	✅ Rule-based planners + logic
Ensure auditability	✕ Opaque	✅ Fully traceable
Multi-modal integration	⚠ Partial (if prompted)	✅ Formal link between layers

As to limitations, they include the complexity of modeling chemical processes, which involves complex interactions that may not fit neatly into the IaC paradigm. Also, traditional IaC tools like Terraform are not designed for physical systems, so you may need to develop custom solutions. Integrating IaC with existing simulation and control systems can be challenging. Moreover, one needs to ensure that the code accurately represents the physical process and produces reliable results.

Hence, using Infrastructure as Code to represent a chemical process is a creative way to apply modern software engineering practices to chemical engineering. By defining the process in a declarative, modular, and automated manner, you can improve reproducibility, scalability, and collaboration. While traditional IaC tools like Terraform may require customization, lightweight formats like YAML or domain-specific tools like Pyomo can also be effective. Ultimately, the choice of tools depends on the specific requirements of the project and the complexity of the process being modeled.

References

Aliev AM, Safarov AR, Osmanova II, Guseinova AM, Balaev IV (2020) Ensuring the stability of operation of the ethylene region of a chemical and technological complex for processing gases of cracking and pyrolysis by taking into account the dynamics of the processes. In: Theoretical foundations of chemical engineering, 2020, vol 54(5). Pleiades Publishing, Ltd., pp 805–817

Besold TR, Garcez AA, Bader S, Bowman H, Domingos P, Hitzler P et al (2017) Neural-symbolic learning and reasoning: a survey and interpretation. arXiv preprint arXiv:1711.03902

Bran MA, Cox S, Schilter O et al (2024) Augmenting large language models with chemistry tools. Nat Mach Intell 6:525–535. https://doi.org/10.1038/s42256-024-00832-8

Colelough BC, Regli W (2025) Neuro-symbolic AI in 2024: a systematic review. arXiv:2501.05435

Finlayson BA, Biegler LT, Grossmann IE (2007) Mathematics in chemical engineering. Wiley VCH Verlag &Co. KGaA, Weinheim

Forsberg K, Mooz H (1991) The relationship of systems engineering to the project cycle. In: Wayback machine, first annual symposium of the National Council on Systems Engineering (NCOSE)

Galitsky B (2025) LLM-based constraint satisfaction. In: Healthcare applications of neuro-symbolic artificial intelligence. Academic

Garcez A d A, Lamb LC, Gabbay DM (2019) Neural-symbolic cognitive reasoning. Springer

Le Q, Feingold J, Glandorf W, Kent J, Sherman R, Ferri JK (2022) Model-based systems engineering approaches to chemicals and materials manufacturing. Process Syst Eng. https://doi.org/10.1002/aic.18114

Towler G, Sinnott R (2008) Chemical engineering design principles, practice and economics of plant and process design. Elsevier, Amsterdam

Weyrich M (2023) Industrielle utomatisierungs- und Informationstechnik. Industrielle Automatisierungs- und Informationstechnik. https://doi.org/10.1007/978-3-662-56355-7

White AD, Hocky GM, Gandhi HA, Ansari M, Cox S, Wellawatte GP, Sasmal S, Yang Z, Liu K, Singh Y, Peña Ccoa WJ (2023) Assessment of chemistry knowledge in large language models that generate code. Digit Discov (2)

Xia Y, Zhang J, Jazdi N, Weyrich M (2024) Incorporating large language models into production systems for enhanced task automation and flexibility. arXiv:2407.08550

Yin Y, Rządkowski G, Pasban A, Tohidi E, Shateyi S (2021) A high accurate scheme for numerical simulation of two-dimensional mass transfer processes in food engineering. Alex Eng J 60(2):2629–2639. https://doi.org/10.1016/j.aej.2020.12.055

Step-Wise Approximation of CBOW Reduces Hallucinations in Tail Cases

1 Introduction

The distributional hypothesis, which states that words with similar meanings tend to occur in similar linguistic contexts, constitutes a fundamental theoretical underpinning of contemporary word embedding techniques in natural language processing (NLP). This hypothesis embodies the principle that semantic similarity is correlated with contextual distribution and finds its earliest philosophical articulation in Wittgenstein's *Philosophical Investigations* (Malcolm 1954), where he asserted that "the meaning of a word is its use in the language." Wittgenstein's use-based view of meaning marked a significant departure from referential theories, emphasizing the functional role of language in practice.

The distributional hypothesis was subsequently formalized within linguistics during the mid-twentieth century, most notably by Zellig S. Harris, who introduced the idea of analyzing language through distributional structures (De Santis et al. 2024), and by J. R. Firth, whose slogan "You shall know a word by the company it keeps"—encapsulated the core of distributional semantics. These foundational insights provided the conceptual scaffolding for data-driven approaches to meaning representation.

In modern computational linguistics, these theoretical principles are operationalized through embedding models, which map words into continuous vector spaces based on co-occurrence statistics within large corpora. Such models, including Word2Vec, GloVe, and more recent contextual embeddings like BERT, aim to capture semantic and syntactic regularities by modeling the distributional properties of language. Moreover, these models are often integrated with the principle of compositionality, which asserts that the meaning of a complex expression is determined by the meanings of its constituents and the structural rules governing their combination.

B. Galitsky, *Applications of Neuro-Symbolic Artificial Intelligence*,
https://doi.org/10.1007/978-3-032-16524-4_11

Thus, state-of-the-art deep learning architectures in NLP are grounded in the confluence of two key linguistic traditions: the empirical, corpus-based methodology of distributional semantics and the formal, rule-based tradition of compositional semantics. This theoretical synergy enables the development of robust models capable of representing meaning at multiple levels of granularity, from individual words to complex discourse structures.

At the core of these models lies the concept of *embedding*—the mapping of discrete linguistic units such as words, phrases, or even entire documents into continuous vector spaces. These embeddings are learned from vast corpora, capturing nuanced patterns of co-occurrence and semantic relatedness. Word embeddings, for instance, reflect the distributional hypothesis: words appearing in similar contexts tend to have similar meanings. More advanced techniques, such as contextual embeddings produced by Transformer-based models, go further by allowing representations to adapt dynamically depending on surrounding linguistic context. This process distills high-dimensional language usage into compact, information-rich representations that form the input to deeper neural architectures for downstream tasks.

One popular method for generating word embeddings is Word2Vec, which is based on the distributional hypothesis. Word2Vec learns word embeddings by training a neural network model to predict the context words given a target word (continuous bag of words) or predict the target word given context words (skip-gram) from a large corpus of text. Once trained, the hidden layer of the neural network serves as the word embeddings. These embeddings can then be used to represent the meaning of words in downstream NLP tasks such as sentiment analysis, machine translation, and named entity recognition.

Another approach is GloVe (Global Vectors for Word Representation), which learns word embeddings by factorizing the co-occurrence matrix of words in a corpus. GloVe captures the global statistical information of word co-occurrences and produces word embeddings that are effective in capturing semantic relationships between words.

More recently, transformer-based models like BERT (Bidirectional Encoder Representations from Transformers) and GPT (Generative Pre-trained Transformer) have gained popularity for generating contextualized word embeddings. These models use attention mechanisms to capture the context of each word in a sentence and generate word embeddings that are sensitive to the surrounding context. Representing the meaning of a word by a discrete numerical function involves learning word embeddings that capture semantic relationships between words in a continuous vector space. This allows NLP models to effectively understand and process natural language text.

However, the reliance on embeddings can inadvertently introduce *hallucinations*, particularly for *tail data*—that is, rare or underrepresented entities, expressions, or combinations not sufficiently captured during training. Since embedding spaces are shaped by the statistical properties of the training corpus, tail items may receive poor or ambiguous representations. Consequently, when models encounter these items, they may interpolate their meaning from nearby but semantically inappropriate vectors, leading to confident but inaccurate outputs. Hallucinations are fabricated or incorrect information not supported by the input or real-world knowledge—emerging from the model's attempt to generalize beyond its data support (chapter "Adversarial Abductive Dialogue Framework with Reinforcement for Tackling LLM Hallucination").

To address noise and inaccuracies in embeddings, especially those generated by simpler models like Continuous Bag of Words (CBOW) (Mikolov et al. 2013), step-wise approximation can be used to refine these embeddings progressively (Fig. 1). CBOW generates word embeddings by predicting a target word from the surrounding context words, averaging their vectors to produce a single representation. While efficient, this averaging process can introduce noise by blending diverse contexts indiscriminately, which reduces the precision of the resulting embedding (Levy and Goldberg 2014; Ethayarajh 2019).

Step-wise approximation tackles this issue by decomposing the embedding process into multiple stages instead of producing a final embedding in one step (Liang et al. 2019). Rather than directly averaging all context vectors at once, the model applies a sequence of functions that iteratively refine the representation, selectively weighting or transforming context words based on their relevance (Coenen et al. 2019). For example, early steps may emphasize the most informative context words, while later steps reduce the impact of noisy or less relevant signals. This gradual refinement helps filter out noise that would otherwise be embedded through blunt averaging, allowing the model to generate cleaner, more context-sensitive

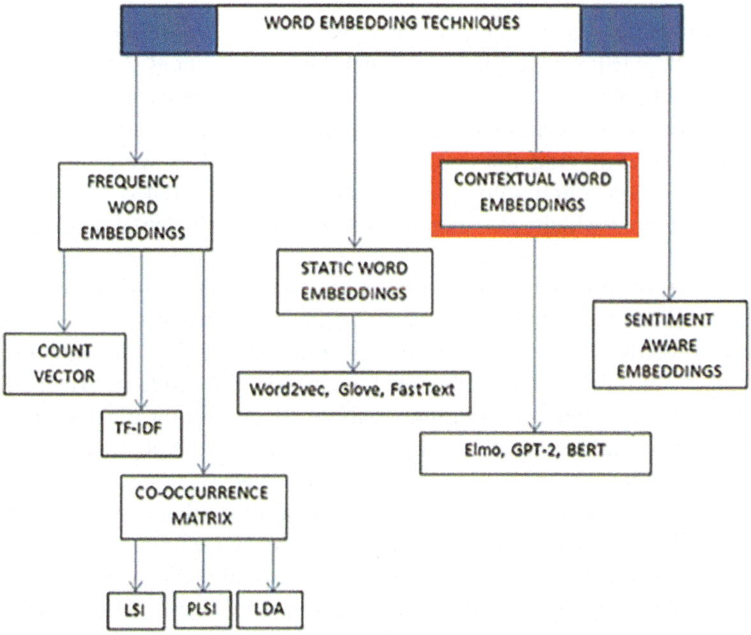

Fig. 1 Our approach to word embedding where step-wise approximation is beneficial

embeddings. By approximating the CBOW process as a step-wise function, the model gains greater control over how context contributes to the final vector, improving robustness and accuracy, especially for ambiguous or rare words where noise can disproportionately degrade the embedding quality (Murphy et al. 2012; Jia and Liang 2017).

Recent research has shown that hallucinations in language models—where models generate factually incorrect or contextually unsupported outputs—are often exacerbated by noisy or imprecise semantic representations. In this context, applying step-wise approximation to embedding generation, such as in modified versions of the CBOW model, can significantly reduce hallucinatory behavior by refining contextual representations iteratively. Instead of collapsing all context vectors into a single averaged representation upfront, step-wise methods progressively adjust the influence of each word based on its relevance and coherence with the target word. This selective weighting prevents misleading signals from dominating the final embedding, especially in ambiguous or low-resource contexts where noise is more likely to distort meaning. By enabling a more nuanced construction of semantic vectors, step-wise approaches enhance the fidelity of internal representations, which in turn leads to more accurate and grounded language generation (Coenen et al. 2019).

1.1 From Discrete to Step-Wise Embeddings

Discrete embedding-based language models, such as Word2Vec (Mikolov et al. 2013), DRG2Vec (Shu et al. 2020), and Dict2Vec (Tissier et al. 2017), utilize structured sources like Wikipedia or dictionaries as the input corpus. These models apply a sliding context window along the input text and optimize a similarity-based objective, typically maximizing the co-occurrence likelihood of words within a fixed window size. Upon training completion, each word is assigned a single, discrete embedding vector, which can then be used directly during downstream inference tasks.

However, this approach assigns one static vector per word (or token), which limits the ability of the model to capture context-specific nuances, especially for polysemous words—those with multiple meanings depending on context (e.g., the word "bank" can refer to either a financial institution or the side of a river). Because the embeddings are fixed after training, these models often blur or average the semantic representations across all senses of a word, which degrades performance in tasks that require fine-grained semantic distinctions.

To address this limitation, in this chapter, we explore the step-wise approximation of embeddings, an approach that introduces piecewise representations of words based on their contextual semantics. Instead of associating each word with a single fixed vector, step-wise approximation represents a word as a set of discrete semantic steps, where each step approximates a distinct contextual meaning. These steps can be learned by partitioning the training data into clusters of contexts (e.g., via k-means or attention-based segmentation) and assigning different sub-vectors or embeddings to each partition. The result is a multifaceted embedding where a word like "bank" is represented by a small set of contextually grounded vectors, one for financial uses, another for geographic references, and so on.

While language models rely on continuous contextual embeddings, which allow for an infinite number of representations for each token, it contrasts with the way humans perceive language, where each word often carries only a limited number of distinct senses. For example, consider the visualization of the different contextual embeddings of the word "draft" in Fig. 2.

t-SNE is a *t*-distributed stochastic neighbor embedding, introduced by van der Maaten and Geoffrey Hinton in (2008), a popular dimensionality reduction technique used for visualizing high-dimensional data in 2D. Unlike methods like PCA, which preserve global variance, *t*-SNE focuses on preserving local similarities (i.e., nearby points in high-dimensional space stay close in 2D). We can observe a good amount of variation in the embeddings, but there are only two main senses, as a noun and as a verb.

Unlike LLMs, which acquire language competence by processing vast corpora and generating fluid, highly context-dependent embeddings for each token, human language learning is shaped by a slower and more discrete process. When individuals acquire new words, they do not create infinitely adaptable representations for every possible context. Instead, they store a limited number of senses for each word, with each sense tied to specific contextual cues. Retrieval then occurs dynamically, guided by surrounding linguistic and situational signals. Research in cognitive science has consistently shown that this incremental, sense-based approach underpins how humans manage ambiguity and construct meaning in real time (Davis and Gaskell 2009; Wojcik 2013).

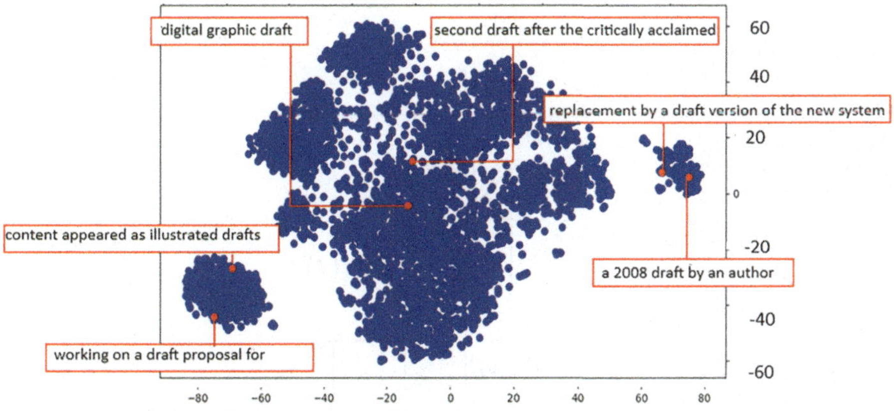

Fig. 2 Contextual embeddings of the word "draft" where a scatterplot is shown in 2D via t-SNE (Van der Maaten and Hinton 2008) on embedding vectors. Two main meaning classes are the verb and the noun

This human strategy opens up the possibility of rethinking how embeddings might be designed in artificial systems. Instead of relying on models that generate a unique, high-dimensional embedding for every conceivable context, one could approximate meaning through a finite set of step-wise embedding variants, each aligned with a discrete semantic interpretation. Such representations would be inherently more interpretable, offering clearer mappings between context and meaning while also being more computationally efficient. By embracing this structured, human-inspired framework, we could move toward models that combine the adaptability of LLMs with the clarity and economy of human language understanding, ultimately yielding systems better equipped to handle nuanced communication in ways that are both transparent and efficient.

1.2 Step-Wise Approximation of Continuous Bag of Words

Continuous bag of words (CBOW) predicts a word from a context using a softmax over the entire vocabulary. We have a probability distribution

$$P\left(w_t|\text{context}\right) = \frac{e^{v_{w_t}^T h}}{\sum_{w \in V} e^{v_{w_t}^T h}}, \tag{1}$$

where v_{w_t} is the output vector for the target word, h is the average of the context word vectors, and V is the vocabulary (can be millions of words). We intend to approximate $P(w_t)$ by means of approximation function $P_s(w_t)$. This approximation function $P_s(w_t)$ is fully determined by $P(w_t)$ it approximates and specific words (values) w_i, where it jumps from $P(w_i)$ to $P(w_{i+1})$. The values for w_i are to be determined (Fig. 3).

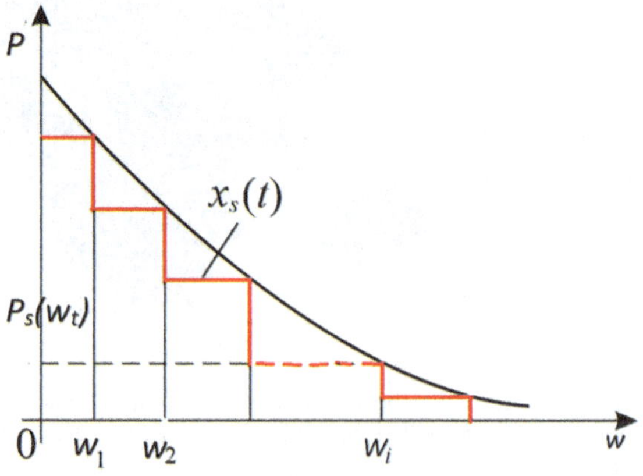

Fig. 3 Step-wise approximation of continuous embedding function

Instead of expressing a meaning of a word via the totality of other words in V, we do that in the basis of classes of equivalent words where the distribution $P_s(w_t)$ is constant (forms a step).

1.3 Minimization of Entropy Production and CBOW

In thermodynamics, there are two classes of problems where the criterion is the minimization of entropy production (Berry et al. 2020). One such class considers processes with a fixed duration and a given average flow of energy or matter. The goal is to find the law of evolution for the intensive variables (such as temperature or chemical potential) of the subsystems participating in the exchange, such that total entropy production is minimized. These are known as minimum dissipation processes.

The CBOW model can be viewed as an information-theoretic analog of such thermodynamic systems (Tsirlin et al. 2025). The correspondence maps out in the following way (Table 1).

In minimum dissipation thermodynamic problems, the system evolves along the trajectory $x(t)$ that minimizes the total entropy production over time:

$$\min_{x(t)} \int_0^T \sigma\left(x(t), \dot{x}(t)\right) dt$$

This can be mirrored in CBOW as optimization objective (information-theoretic form): CBOW minimizes the cross-entropy between the true word distribution and the predicted distribution from the context:

$$\min_{\theta} \mathrm{E}_{(c,w)}\left[-\log P_\theta\left(w \mid c\right)\right]$$

subject to constraints on embedding norm or training time

Table 1 Thermodynamics and CBOW

Thermodynamic concept	CBOW analogy
Intensive variables $x_i(t)$ (e.g., temperature)	Word embeddings $\boldsymbol{v}_i(t) \in R^d$
Flow of matter/energy $J_i(t)$	Gradient flow during training $d\boldsymbol{v}_i/dt$
Entropy production rate $\sigma(t)$	Cross-entropy loss $\mathcal{L}(t)$ or KL divergence from data
Constraint on process duration	Fixed training time or number of epochs
Constraint on average flux	Bounded gradient norm, limited context window
Law of evolution $dx_i/dt = f(x)$	Stochastic gradient descent or Adam updates: $v_i^{(t+1)} = v_i^{(t)} + \eta \nabla \mathcal{L}$

where:

- c: context window (sum/avg. of embeddings)
- w: center word
- θ: parameters of embeddings
- $P_\theta(w|c)$: softmax output

Interpretation as minimum entropy production is as follows. During training, the embedding vectors v_w evolve via gradient flow, analogous to the evolution of physical variables $x_i(t)$. The objective functional being minimized is cross-entropy, which corresponds to informational entropy production. The learning process seeks an embedding configuration where information is stored efficiently (low entropy state), under constraints (context size, training steps).

Thus, CBOW learning is akin to finding a minimum-dissipation path in parameter space, subject to fixed information flow and temporal limits. Hence, CBOW can be interpreted as an information-dynamic system analogous to minimum entropy production problems in thermodynamics (Fig. 4):

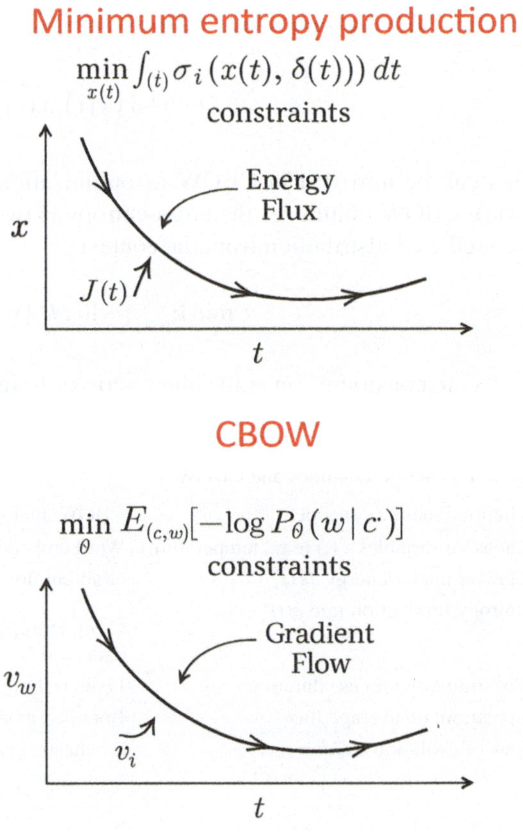

Fig. 4 Minimum entropy production and CBOW

1. Both minimize a cumulative cost over time (entropy vs. loss).
2. Both evolve under constraints (physical flows vs. learning limits).
3. The optimal embeddings in CBOW correspond to **steady states of minimal informational dissipation**, much like how thermodynamic systems evolve to minimize physical dissipation.

The chapter is organized as follows. We begin by motivating the transition from discrete to step-wise embeddings and introduce a mathematical formulation for approximating continuous semantic evolution using entropy-minimizing segmentation. We then explore how continuous Word2Vec models can be extended to handle context shifts more robustly through step-wise embedding updates. Next, we detail practical methods for computing and modeling step-wise embeddings and demonstrate their application in a setting of enhancing embedding robustness under hallucination-inducing conditions.

To evaluate the proposed method, we assess its relevance in a retrieval-augmented generation framework, focusing on tail entities, where standard LLM embeddings often degrade in quality. Our experiments show that step-wise approximation improves performance for rare words while maintaining competitiveness for frequent ones.

Overall, this work contributes a scalable, cognitively inspired alternative to continuous context-based embeddings, enabling more interpretable and adaptable representations for a wide range of NLP tasks.

2 Continuous Word2Vec and Word Profiles

Representing the meaning of a word by a continuous function is a fundamental concept in NLP known as word embedding. In NLP, text embeddings (e.g., word embeddings, sentence embeddings) are used to represent textual data in a continuous vector space. Word embeddings are dense vector representations of words in a vector space, where each dimension of the vector captures different aspects of the word's meaning.

To form a continuous word embedding, we need to form a basis to express a meaning of words (Galitsky 2025). To express a meaning of a noun, we build its profile in the basis of verbs applicable to this noun. Analogously, to express a meaning of a verb, we build its profile in the basis of nouns this verb is acting upon. To have the basis for verbs fixed in the most general form, we select the noun with a broad scope like *human* or *animal*. Then we select a spectrum of verbs related to, for example, physical or mental activity of humans. To form a basis for *human*, we order the verbs according to the distance to human, from the most similar to the least similar. Another option for basis construction is to start with a most general verb like *move* and finish the basis with a specific movement verb like *jump*. The basis for verbs is formed in a similar manner.

A profile for a word is a sequence of distances to the elements of the basis. For most general and representative words like *human* {*move*, *walk*, *sit*, *hike*, *jump*}, a typical profile is a monotonic decreasing function. What makes this function continuous is that the elements of the basis have "intermediate" meanings in between, corresponding to "fictitious" words positioned "in between" the meanings of the words of the basis. The range [*walk…sit*] is filled with [*mostly walk and sit sometimes*, …, *mostly sit and sometimes stand up and walk*] meaning values. Hence, the values between the basis can be interpolated in a continuous manner. Although there is no specific word to express this intermediate position in the basis, the respective meaning can be expressed by a natural language phrase. Hence, the profile is a continuous function, close to monotonic for a broad, general, canonic nouns and verbs and non-monotonic for other, specific nouns and verbs. Other parts of speech such as adjectives can have this basis as nouns or verb phrases including nouns (Galitsky 2025).

A verb-noun Word2Vec distance matrix is shown in Fig. 5. As a simple case, we formed a list of verbs and corresponding nouns such as run-runner. The verbs are ordered by a distance to the head verb: we start with the lowest distance and end with the highest distance for verbs in a given semantic family of verbs for movement types.

Then for each noun, we form a profile of distances {|*embed(run)-embed(running)*|, |*embed(jump)-embed(running)*|, … |*embed(jump)-embed(running)*|}. The curve is continuous as we can form a fuzzy meaning 30% *run* and 70% *jump*.

Once individual words are encoded as profiles, a phrase from these words is encoded as a concatenation of the profiles for individual words. This concatenation is not a continuous function any more, and similarity of each profile needs to be computed separately.

A continuous Word2Vec embedding is a real-valued vector representation of a word, learned from text using neural language models, where semantic similarity is preserved as geometric proximity in the embedding space.

	Runner	Jumper	Walker	Swimmer	Kicker	Puncher	Thrower	Catcher	Climber	Pusher	Puller	Lifter	Carrier	Dancer	Skipper	Stretcher	Bender	Twister	Crawler	Skater
Run	0.9	0.8	0.8	0.6	0.4	0.4	0.4	0.4	0.3	0.3	0.4	0.4	0.3	0.4	0.2	0.3	0.2	0.2	0.3	0.2
Jump	0.7																			
Walk	0.9																			
Swim	0.6																			
Kick	0.4																			
Punch	0.4																			
Throw	0.4																			
Catch	0.3																			
Climb	0.3																			
Push	0.3																			
Pull	0.3																			
Lift	0.4																			
Carry	0.3																			
Dance	0.3																			
Skip	0.2																			
Stretch	0.3																			
Bend	0.2																			
Twist	0.2																			
Crawl	0.3																			
Skate	0.2																			

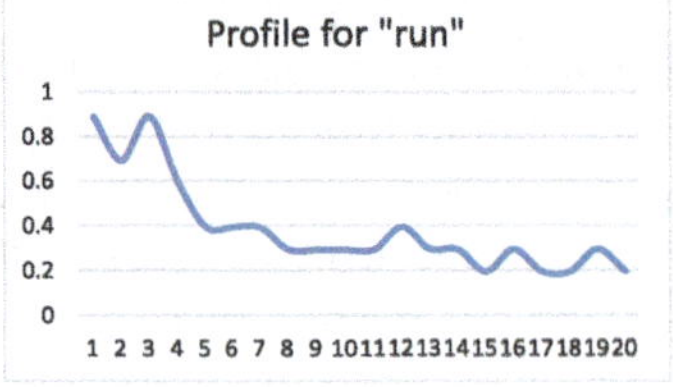

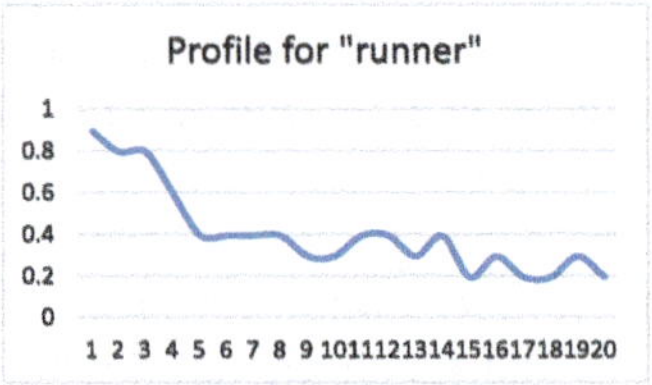

Fig. 5 Word2vec continuous profiles of words

If the vocabulary size is V and embedding size is D, then the embedding matrix $E \in R^{V \times D}$ maps each word w_i to a vector $\vec{v}_i \in R^D$. CBOW and Skip-Gram models **learn** the matrix E by minimizing prediction error:

- Skip-Gram: Predict context words from a target word.
- CBOW: Predict the target word from surrounding context words.

3 Step-Wise Approximation of Text Embeddings

Step-wise approximation refers to breaking down the embedding process into discrete steps or stages, often using piecewise-constant functions. This approach simplifies complex embeddings by approximating them with simpler, step-like functions.

When the context of the text changes (e.g., shifting from one topic to another), the step-wise approximation may fail to capture the nuances of the new context. This can lead to hallucinations, where the model generates incorrect or nonsensical outputs because it misinterprets the context or fails to adapt smoothly to the change.

The challenge lies in ensuring that the step-wise approximation remains accurate even as the context changes. If the approximation is too coarse or poorly aligned with the underlying semantics, it can introduce errors that propagate through the system.

3.1 Why Step-Wise Approximation Is Needed in CBOW

Computing this denominator (the normalization term) is too expensive for large vocabularies—it requires computing scores for every single word in the vocabulary on every training step. To solve this problem, instead of computing probabilities over the entire vocabulary, CBOW just learns to distinguish the correct word from a few "negative" words. For each training example, we form positive pair (context → true target word) and negative samples (context → random, unrelated words). The objective of CBOW is to maximize the probability of a target word given its context words:

$$\log \sigma\left(v_{w_t}^T h\right) + \sum_{i=1}^{k} E_{w_i \sim P_n(w)}\left[\log \sigma\left(-v_{w_t}^T h\right)\right].$$

Here, σ is the sigmoid function, and k is the number of negative samples (e.g., 5–20).

Hierarchical softmax replaces the flat softmax with a binary tree over the vocabulary. Each word is a leaf node; predicting a word involves walking down the tree. Instead of computing $|V|$ scores, we compute $\log_2|V|$ scores. This is useful when you want to retain probability estimates.

Subsampling of frequent words is not softmax related but improves efficiency. Very frequent words (like "the" and "is") are randomly dropped to speed up training and reduce their dominance. Step-wise approximation is essential in CBOW to avoid computing full softmax over large vocabularies. It is implemented through techniques like negative sampling or hierarchical softmax, which drastically reduce computation while still enabling effective learning of word embeddings.

CBOW and skip-gram both produce dense and low-dimensional word vector representations that can be used to solve diversified natural language processing challenges. These approaches produce vectors with a significant flaw: they only have local word co-occurrence information and no global co-occurrence information. Local and global information are merged as input to a two-layer neural network (Huang et al. 2012).

Graph embedding is employed to combine local and global co-occurrence counts. The nodes in the graph represent words, and the weight of each edge is determined by a combination of local and global co-occurrence counts. When compared to Word2Vec and GloVe, the graph embedding outperforms sequence baseline in terms of word similarity. Since the GloVe model was created to address the constraints of Word2Vec. Moreover, clustering is applied in the context of a word, where the word sense is represented by the centroid of each cluster (Reisinger and Mooney 2010). The original Word2Vec model is also extended for producing multi-sense word embeddings (Wang et al. 2025).

3.2 *Identifying Breakpoints for Context Changes*

In the context of text embeddings, breakpoints (t_i) can correspond to points where the context changes significantly. For example:

1. A change in topic (e.g., moving from discussing "climate change" to "technology")
2. A shift in tone or sentiment (e.g., from positive to negative)
3. A transition between different types of information (e.g., factual vs. speculative)

To identify these breakpoints, one can use context-aware models (e.g., transformers) to detect shifts in semantic meaning. It is also possible to apply change-point detection algorithms to analyze the embedding vectors and identify points where the distribution of embeddings changes abruptly.

We minimize the difference between the original function $x(t)$ and the approximated function $x_s(t)$. In the context of text embeddings:

- Original Function $x(t)$: The true, continuous representation of the text embedding over time or sequence
- Approximated Function $x_s(t)$: The step-wise approximation of the embedding

To minimize the error, one can use metric-based criteria to evaluate the quality of the approximation. Common metrics include:

1. Euclidean distance between the original and approximated embeddings
2. Cosine similarity to measure the alignment of embedding vectors
3. Mean squared error (MSE) to quantify the overall discrepancy

We need to ensure that the choice of breakpoints (t_i) minimizes the cumulative error across all intervals ($[t_i - 1, t_i]$).

3.3 Handling Context Changes Smoothly

The choice of t_i affects only the preceding and following intervals. This suggests a localized approach to handling context changes:

- Adaptive step size: adjust the granularity of the step-wise approximation dynamically. For example, use finer steps during rapid context changes and coarser steps during stable contexts.
- Interpolation: instead of purely step-wise approximation, one can consider using interpolation techniques (e.g., linear or spline interpolation) to smooth transitions between breakpoints.

In NLP, one introduces intermediate embeddings that bridge the gap between two distinct contexts. These intermediate embeddings can help the model adapt smoothly to context changes without generating hallucinations. This is done with the use of contextualized embeddings (e.g., BERT, RoBERTa) that are pre-trained to handle context-dependent meanings.

There is an iterative process to refine the approximation by minimizing the sum of differences between $x(t)$ and $x_s(t)$. To apply this principle to text embeddings, we start with an initial set of breakpoints (t_i). We then evaluate the approximation using metrics like MSE or cosine similarity.

Finally, we adjust the breakpoints iteratively to reduce the error, focusing on regions where context changes occur. The steps are as follows:

1. Detect context changes, using advanced NLP techniques (e.g., attention mechanisms, change-point detection) to identify points where the context shifts.
2. Refine step-wise approximation, choosing breakpoints (t_i) strategically to align with context changes. We use metrics like Euclidean distance or cosine similarity to evaluate and optimize the approximation.
3. Smooth transitions, introducing intermediate embeddings or using interpolation to bridge gaps between contexts. We then employ contextualized models that adapt dynamically to changing contexts.
4. Iterate and validate, continuously refine the approximation by adjusting breakpoints and evaluating performance.
5. Test the model on diverse datasets to ensure robustness against context changes.

The approximating function $x_s(t)$ is fully determined by the function $x(t)$ it approximates and the values t_i, at which it jumps from $x(t_i)$ to $x(t_{i+1})$. The values of the "breakpoints" t_i are the unknown variables in such problems (Fig. 3).

The criteria used to search for these variables can vary. In approximation problems, this could be the distance between $x(t)$ and $x_s(t)$ in some metric or uniform approximation. Since the choice of t_i affects the function $x_s(t)$ only on the preceding and succeeding intervals, this choice should minimize the total difference between $x(t)$ and $x_s(t)$ only on the interval $[t_{i-1},t_{i+1}]$. Thus, with the introduction of each new jump point t_i, the reduction in the criterion on this interval should be maximized.

We approximate a continuous embedding function $x(t)$ using a piecewise-constant function $x_s(t)$ designed to approximate the original function $x(t)$ using a series of constant segments $x_s(t)$ by means of two parts:

1. The original function $x(t)$, which serves as the target for approximation.
2. A set of breakpoints t_i, where the approximating function $x_s(t)$ changes its value abruptly (jumps from one constant segment to another). The breakpoints t_i determine where the approximating function $x_s(t)$ transitions occurs between different constant values. In many approximation problems, the exact locations of these breakpoints (t_i) are not known beforehand and must be determined as part of the solution. At each breakpoint t_i, the function $x_s(t)$ jumps from $x(t_i)$ to $x(t_{i+1})$, creating a step-like structure.

When the goal is to find the best set of breakpoints t_i such that the approximating function $x_s(t)$ closely matches the original embedding function $x(t)$, the possible metrics are:

1. Euclidean distance: Minimize the sum of squared differences between $x(t)$ and $x_s(t)$.
2. Maximum error: Ensure that the maximum deviation between $x(t)$ and $x_s(t)$ is minimized.
3. Uniform approximation: Ensure that the error between $x(t)$ and $x_s(t)$ is evenly distributed across the entire domain.

Each breakpoint t_i affects the behavior of $x_s(t)$ only on the intervals immediately before and after it, i.e., $[t_{i-1},t_i]$ and $[t_i,t_{i+1}]$. This localized impact simplifies the optimization process because the effect of t_i can be analyzed independently on these two intervals.

4 Solving Step-Wise Approximation

Let us have an embedding distribution function $x(t)$ such that there exists the integral $X = \int_0^{t_k} x(t)\,dt$. The derivative $\frac{dx}{dt} < 0$. We need to find breakpoints t_i, $i = 1, 2, \ldots k$ such that

$$I = \int_0^{t_k} \left(x(t) - x_s(t)\right) dt \to \min / \; x_s(t) = x(t_i), \quad t_{i-1} < t \le t_i, \quad t_k - \text{fix}. \tag{2}$$

The value

$$I = X - X_s = X - \int_0^{t_k} x_s(t)\,dt = X - \sum_{i=1}^{k} x(t_i)(t_i - t_{i-1}), t_0 = 0.$$

The condition of minimality of this value over t_i leads to boundary value problem

$$\left(\frac{dx}{dt}\right)_{t_i} (t_i - t_{i-1}) + x(t_i) - x(t_{i+1}) = 0, \quad t_0 = 0, \quad t_k, x(t_k) - \text{fix}. \tag{3}$$

Since the function $x(t)$ and its derivatives are specified, Eq. (2) determines $t_i^* = \Phi(t_{i-1}, t_{i+1})$, which is an optimality assertion.

4.1 *Implicit Optimization Assertion*

An assertion for the breakpoints can be obtained in an implicit form $\Phi(t_{i-1}, t_i, t_{i+1}) = 0$; however in the case where this equality can be resolved relative to t_i^*, the following conditions would hold:

1. $t_{i-1} < \Phi(t_{i-1}, t_{i+1}) < t_{i+1}$
2. $\Phi(t_{i-1}, t_{i+1}) = \Phi(t_{i+1}, t_{i-1})$ (symmetric function)
3. $\Phi(t_{i-1}, t_{i+1}) = \Phi\left[\Phi(t_{i-1}, t_i^*), \Phi(t_i, t_{i+1})\right]$

The last condition states that t_i^* value must be optimal not only for t_{i-1}, t_{i+1} but also for t values calculated in the same way and positioned between t_{i-1}, t_i and t_i, t_{i+1}. Functions that perform averaging of t_{i-1}, t_{i+1} satisfy this condition including the arithmetic mean or geometric mean of these values

Example 1 Let

$$x(t) = x_0 - \alpha t, \quad t_k = \frac{x_0}{\alpha}, \quad x(t_k) = 0.$$

Then condition (2) will look like:

$$\frac{\alpha(t_i - t_{i+1})}{t_i - t_{i-1}} = \alpha \to t_i = 0.5(t_{i-1} + t_{i+1}), \; t_i = \frac{t_k}{k} i. \tag{4}$$

Here all intervals between breakpoints must be equal and have value t_k/k, function Φ—arithmetic mean, and a minimum difference $x(t)$, and its step-wise approximation is $I^* = \frac{x_0^2}{2\alpha k}$.

Example 2 Let

$$x(t) = e^{-\alpha t},\ t_k = \frac{2}{\alpha},\ x(t_k) = e^{-2}.$$

Then condition (2) will be

$$t_{i-1} = t_i - \frac{1 - e^{-\alpha(t_{i+1} - t_i)}}{\alpha},\ i = 1,2,\ldots k-1,\ t_0 = 0. \tag{5}$$

As a rule, the "shooting" method is used to solve such a boundary value problem (4). In this case, all other "breaking points" t_{k-1} are set and calculated according to condition (4) up to t_0. If it turns out that $t_0 = 0$, the problem is solved. Otherwise, one sets another value of t_{k-1}, and the computation is repeated.

4.2 Computing Approximated Embedding

We express the distance between the original embedding $x(t)$ and its step-wise approximation $x_s(t)$ as $\eta(x, x_s)$. Let us have this distance η and breakpoints t_{i-1} and t_{i+1}. This distance function obeys $\eta(x,x_s) \geq 0$, $\eta(x_s,x_s) = 0$. In the interval $[t_{i-1}, t_i]$, the function $x_s = x(t_i)$, and in the interval $[t_i, t_{i+1}]$ $x_s = x(t_{i+1})$. We need to find such t_i that the distance between functions $x(t)$ and $x_s(t) < x(t)$ in the interval between these points is minimum over t_i. The optimality condition has the form

$$\frac{\partial\left(\int_{t_{i-1}}^{t_i} \eta\left(x, x(t_i)\right)dt + \int_{t_i}^{t_{i+1}} \eta\left(x, x(t_{i+1})\right)dt\right)}{\partial t_i} = 0. \tag{6}$$

Then taking into account the properties of η we obtain the equation for finding n approximation breakpoints t_i

$$\int_{t_{i-1}}^{t_i} \left(\frac{\partial \eta}{\partial x_s}\right)_{x(t_i)} \left(\frac{dx}{dt}\right)_{t_i} dt = \eta\left(x(t_i), x(t_{i+1})\right). \tag{7}$$

4.3 Modeling Changing Embeddings

In contrast to the traditional concept of embeddings, where each word is represented by a single vector, sense embeddings associate multiple vectors per word, where each one of the vectors aims to capture a different meaning (Camacho-Collados and Pilehvar 2018). LMMS (Loureiro et al. 2022) generates embeddings per sense key defined by annotated resource, such as WordNet (Miller 1995). As a result, LMMS can achieve more precise representations for each sense key but lacks support for most datasets consisting of plain text. Thus, this chapter develops a general solution to generating multiple embeddings per token without requiring annotation and an evolution of embedding over time.

The derivation (7) about step-wise approximation of embeddings can significantly improve vector similarity search in vector databases, especially when embeddings change over time, like in contextual or dynamic systems. Most vector databases (like FAISS, Pinecone, Weaviate) use static embeddings and compute cosine or L2 distance between a query and a set of stored vectors:

$$\text{similarity}(q,x) = \cos(q,x) \text{ or } \| \mathrm{q} - \mathrm{x} \|^2$$

However, if embeddings are evolving functions of time (or context, training steps, etc.), comparing fixed vectors may miss nuances of semantic drift or temporal smoothness.

We have defined the distance between a function $x(t)$ (e.g., a dynamically updated embedding) and its step-wise approximation $x_s(t)$ to minimize the total deviation $\eta(x, x_s)$ over each segment $[t_{i-1},t_{i+1}]$. We index time-series embeddings with step-wise approximation. If embeddings $x(t) \in \mathrm{R}$ change (e.g., based on dialogue turns, document growth, or RL training steps), we store a step-wise compressed version of the trajectory using this optimization (minimizing total distance η). In vector DB, instead of one vector per item, we store multiple segments with $x(t_i)$ representing the value over that time block. At the query time, we match queries against all segments and interpolate.

On top of that, we use distance function η as a custom similarity. Instead of cosine or Euclidean distance, we define

$$\eta(x, x_s) = \int_{t_{i-1}}^{t_i} \eta(x(t), x(t_i))\,dt + \int_{t_i}^{t_{i+1}} \eta(x(t), x(t_{i+1}))\,dt \tag{8}$$

This can be computed efficiently via lookup or pre-integration and used as:

$$\text{similarity}(q,x) = -\eta(q,x_s)$$

This distance respects temporal or semantic smoothness and avoids sharp changes that occur in non-smooth embeddings.

Hence, the compression occurs with accuracy control. The optimality condition derived (stationary condition for t_i) allows us to compress dynamic embeddings adaptively. Moreover, we store fewer vectors without significantly degrading search accuracy—information-theoretic compression.

For time-evolving user embeddings in recommendation, step-wise approximation of *dynamic* embedding compresses and track evolving preferences. For long documents where meaning shifts over paragraphs, we use paragraph-wise anchor embeddings. Also, for LLMs with prompt-driven context change, step-wise approximation matches a query against dynamic context.

5 Improving Embedding in a Fight with Hallucination

Incorrect transformer embeddings can contribute to hallucinations in language model outputs (Galitsky et al. 2025b). The possibilities of hallucinations are as follows:

1. Semantic misrepresentation: if embeddings fail to capture the true meaning of a token or context (e.g., conflating "apple" the fruit with "Apple" the company), the model may draw faulty associations. This can lead to outputs that are semantically incoherent or factually wrong.
2. Tokenization artifacts. Poor tokenization can yield fragmented or misleading embeddings. For example, rare or out-of-vocabulary words may get broken into subwords in a way that loses meaning. As a result, the model may generate text unrelated to the input's intended meaning.
3. Contextual drift. If the embedding layer (or early transformer layers) fails to preserve relevant context, the decoder or downstream attention layers may build on distorted input, amplifying errors into hallucinations.
4. Training data bias or sparsity. Embeddings trained on biased or incomplete corpora can reflect these gaps. The model may hallucinate facts that "fit" patterns in the training data even if they aren't real. This leads to making up citations, events, or medical facts.

For example, for input "The president of Canada is...," incorrect embedding may overweight the term "president" and neglect the fact that Canada has a *prime minister*, not a *president*. Hallucinated output is "Justin Trudeau is the president of Canada."

For mitigating hallucination, we could perform embedding quality control such as improving pretraining data, subword segmentation, or using supervised alignment (e.g., with grounding in knowledge bases). Post hoc analysis would also be helpful, such as probing internal representations for semantic accuracy. However, the most efficient approach is retrieval augmented models (RAG), which reduce the reliance on embeddings alone by retrieving grounded facts (Galitsky et al. 2025a). Step-wise approximation of embedding of tail words turns out to be the most efficient way to prevent hallucination in underrepresented knowledge.

6 Evaluation of Relevance

We evaluate relevance improvement due to improved vector matching in RAG setting. To evaluate relevance improvement from applying step-wise approximation to CBOW embeddings in a RAG system, one can measure the quality of retrieved documents based on their alignment with the query. In this setup, both queries and documents are encoded using the refined step-wise CBOW embeddings and stored in a vector database FAISS. The evaluation involves comparing retrieval performance metrics Normalized Discounted Cumulative Gain (NDCG), when using standard CBOW embeddings versus step-wise approximated embeddings. Specifically, if the step-wise approach improves semantic resolution by reducing noise and better modeling context, it should yield more accurate nearest-neighbor matches in the vector space, resulting in higher retrieval accuracy. Additionally, end-to-end evaluation through answer relevance scoring (e.g., using BLEU on the final generated response) provides further evidence of how enhanced vector matching translates into better contextual grounding of the generated output.

6.1 *Dataset*

Mallen et al. (2022) observed that popular open-domain QA datasets like Natural Questions (NQ) tend to focus on widely known entities and have varied question phrasing, making it hard to analyze relation types. To support fine-grained memorization analysis, we rely on PopQA, a large-scale entity-centric QA dataset featuring entities with diverse popularity levels.

Mallen et al. (2022) built PopQA by randomly sampling 16 types of knowledge triples from Wikidata and turning them into natural language questions using relation-specific templates. Each triple (Subject, Relation, Object) is verbalized into a question by inserting the subject into a manually created template. Valid answers are all entities linked to the subject via that relation in Wikidata. The authors verified that the approach is robust across different templates. Thanks to Wikidata grounding, PopQA supports reliable popularity and relation-type analysis. Another part of question-answering dataset is formed from EntityQuestions (Sciavolino et al. 2021), a dataset designed to include both common and long-tail entities.

We use Wikipedia page views as a measure of popularity and form knowledge triples from Wikidata, varying the level of popularity. The triples have the form <*subject*, *relationship*, *object*> (Fig. 6).

Questions are focused on the original tail entities and relationship types for which a limited context for embedding computation is available. A question-answering session against this dataset produces a sequence of candidate answers, so NDCG is a good measure of relevance.

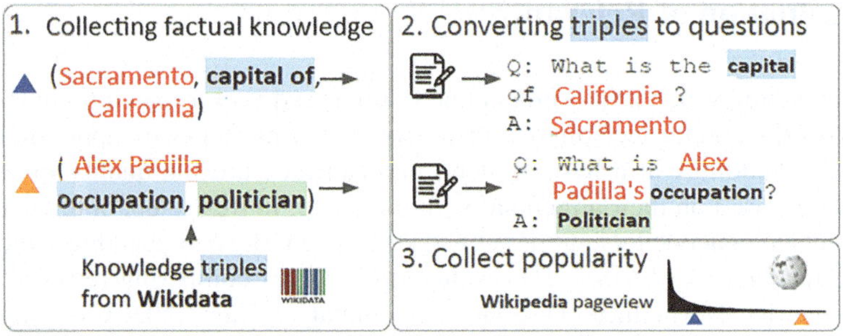

Fig. 6 Forming the evaluation dataset of tail entities where an improvement in embedding is needed

6.2 Assessing RAG Relevance for Tail Entities

Since embedding relevance improvements are particularly important for tail entities—those that occur infrequently—we evaluate the impact of these improvements separately for less frequent and more frequent entities. We assume that errors in answers are due to hallucinations in questions vs. answer representation caused by incorrect embeddings. Specifically, we assess question-answering accuracy using our RAG system based on:

1. The default embedding
2. Its step-wise approximation
3. A dynamic embedding modeled through step-wise approximation over time

In Table 2, we compare accuracy for popular entities (popularity $>10^5$) and tail entities (popularity $<10^3$) across the three embedding variants. The second and third columns show baseline accuracy using the default embedding (grayed out for reference).

We observe that for tail entities, the step-wise approximation yields a 4% performance improvement, whereas for popular entities, the gain is only 2%. When modeling the evolution of the embedding using a step-wise approach, tail entities show an even larger improvement of 5%, while accuracy for popular entities slightly declines. Hence, step-wise approximation improves the quality of embeddings for infrequent words and, as a result, boosts question-answering relevance, as we observed.

Table 2 NDCG@5 for infrequent and frequent entities for default, approximated and dynamic embeddings

	Default embedding		Step-wise approximation of embedding		Step-wise approximation of *dynamic* embedding	
Entity class	Popularity $<10^3$	Popularity $>10^5$	Popularity $<10^3$	Popularity $>10^5$	Popularity $<10^3$	Popularity $>10^5$
occupation	0.720	0.783	0.768	0.798	0.724	0.745
author	0.743	0.793	0.775	0.824	0.816	0.765
director	0.681	0.759	0.698	0.775	0.698	0.731
country	0.692	0.701	0.686	0.703	0.677	0.726
capital of	0.732	0.750	0.803	0.783	0.754	0.733
capital	0.782	0.783	0.852	0.814	0.787	0.782
religion	0.742	0.795	0.762	0.816	0.741	0.789
sport	0.702	0.751	0.725	0.788	0.764	0.780
producer	0.681	0.715	0.692	0.711	0.750	0.743
mother	0.703	0.737	0.761	0.741	0.718	0.781
place of birth	0.725	0.741	0.772	0.738	0.783	0.699
composer	0.704	0.753	0.705	0.787	0.777	0.771
genre	0.726	0.724	0.776	0.765	0.804	0.710
father	0.706	0.740	0.775	0.744	0.723	0.716
color	0.689	0.726	0.718	0.730	0.714	0.721
Average	**0.715**	**0.750**	**0.751**	**0.768**	**0.724**	**0.745**
Boost to baseline, approx. %			**4**	**2**	**5**	**-0.05**

7 Conclusions

In our previous study (Galitsky 2021), we subjected embedding techniques to critical evaluation and observed its limitations in expressing semantic relatedness of words. We analyzed a representation of meaning by the distributional semantics and also obtained an anecdotal evidence about various systems employing it. To overcome the revealed limitations of the traditional embedding techniques, where similarity is computed between words of different semantic roles, we proposed to use them on top of a linguistic structure, not in a stand-alone mode. In the phrase similarity assessment task, only when phrases are aligned and syntactic, semantic and entity-based map is established, we assessed word2vec similarity between aligned words. We added the distributional semantics feature to the results of syntactic, semantic, and entity/attribute-based generalizations and observed an improvement in the similarity assessment task. It turned out that structurized Word2Vec improves the similarity assessment task performed by the integrated syntactic, abstract meaning representation and entity comparison baseline system by more than 10%.

Step-wise approximation serves as a computationally efficient alternative to fully contextual embeddings (e.g., from BERT), enabling models to disambiguate meanings with lightweight inference mechanisms. At test time, the model can match the current context to the nearest precomputed step and select the most appropriate sub-embedding. This maintains the inference efficiency of discrete models while providing a finer granularity of semantic representation.

Step-wise approximation thus bridges the gap between traditional word embeddings and modern contextual language models. It enhances the representational capacity of static embeddings without incurring the full cost of deep transformer-based encoders, making it particularly useful for applications where scalability, interpretability, or real-time inference is critical.

Our findings demonstrate that our step-wise approximation of embeddings offers clear benefits for improving semantic representations of infrequent (tail) entities. Specifically, we observed a 4% performance gain for tail entities using step-wise approximation, compared to a more modest 2% gain for popular entities. Furthermore, when modeling the evolution of embeddings through a step-wise mechanism, tail entities exhibited an even greater improvement of 5%, while performance for popular entities slightly declined. These results suggest that step-wise approximation is particularly effective at enhancing the embedding quality for less frequent terms, leading to improved relevance in downstream tasks such as question answering.

References

Berry RS, Salamon P, Andresen B (2020) How it all began. Entropy 22:908. https://doi.org/10.3390/e22080908

Camacho-Collados J, Pilehvar MT (2018) From word to sense embeddings: a survey on vector representations of meaning. J Artif Intell Res 63:743–788

Coenen A, Reif E, Yuan A, Kim B, Pearce A, Viégas F, Wattenberg M (2019) Visualizing and measuring the geometry of BERT. arXiv:1906.02715

Davis MH, Gaskell MG (2009) A complementary systems account of word learning: neural and behavioural evidence. Philos Trans R Soc Lond Ser B Biol Sci 364(1536):3773–3800. https://doi.org/10.1098/rstb.2009.0111

De Santis E, Martino A and Rizzi A (2024) Human versus machine intelligence: assessing natural language generation models through complex systems theory. IEEE Transactions on pattern analysis and machine intelligence, 46(7):4812.

Ethayarajh K (2019) How contextual are contextualized word representations? In: Proceedings of the 2019 conference on empirical methods in natural language processing and the 9th international joint conference on natural language processing (EMNLP-IJCNLP), pp 1–14. https://doi.org/10.18653/v1/D19-1009

Galitsky B (2021) Distributional semantics for CRM: MakingWord2vec models robust by structurizing them. In: Artificial intelligence for customer relationship management. Springer, Cham, pp 25–56

Galitsky B (2025) Employing Kolmogorov–Arnold network for man–machine collaboration. In: Lawless et al (eds) Bi-directionality in human-machine collaborative systems

Galitsky B, Ilvovsky D, Morkovkin A (2025a) Enhancing RAG and knowledge graphs with discourse. In: Proceedings of the international conference "Dialogue 2025"

Galitsky B, Chernyavskiy A, Ilvovsky D (2025b) Truth-o-meter: handling multiple inconsistent sources repairing LLM hallucinations. In: Proceedings of the 47th international ACM SIGIR conference on research and development in information retrieval, pp 2817–2821

Huang EH, Socher R, Manning CD, Ng AY (2012) Improve word representation via global context and multiple word prototypes. In: Proceedings of the 50th annual meeting of the Association for Computational Linguistics. Association for Computational Linguistics, Islan, Korea, pp 873–882

Jia R, Liang P (2017) Adversarial examples for evaluating reading comprehension systems. In: Proceedings of the 2017 conference on empirical methods in natural language processing (EMNLP), pp 2023–2033. https://doi.org/10.18653/v1/D17-1215

Levy O, Goldberg Y (2014) Neural word embedding as implicit matrix factorization. Adv Neural Inf Process Syst 27:2164–2172

Liang P, Huai M, Zhang A (2019) Word embedding refinement: a survey of theoretical and empirical techniques. ACM Trans Inf Syst 37(3):1–40. https://doi.org/10.1145/3342949

Loureiro D, Jorge AM, Camacho-Collados J. (2022) LMMS reloaded: Transformer-based sense embeddings for disambiguation and beyond. Artificial Intelligence 305:103661. ISSN 0004-3702, https://doi.org/10.1016/j.artint.2022.103661

Malcolm N (1954) Wittgenstein's philosophical investigations. Philos Rev 63(4):530–559

Mallen A, Asai A, Zhong V, Das R, Khashabi D, Hajishirzi H (2022) When not to trust language models: investigating effectiveness of parametric and non-parametric memories. arXiv:2212.10511

Mikolov T, Chen K, Corrado G, Dean J (2013) Efficient estimation of word representations in vector space. arXiv preprint arXiv:1301.3781

Miller GA (1995) WordNet: a lexical database for English. Communications of the ACM 38, 11 (Nov. 1995), 39–41

Murphy BR, Talukdar PP, Crammer K (2012) Learning from implicitly supervised sequential data. Mach Learn 86(2):197–233. https://doi.org/10.1007/s10994-011-5266-6

Reisinger J, Mooney RJ (2010) Multi-prototype vector-space models of word meaning. In: NAACL HLT 2010—human language technologies: the 2010 annual conference of the North American chapter of the Association for Computational Linguistics, proceedings of the main conference, no June, pp 109–117

Sciavolino C, Zhong Z, Lee J, Chen D (2021) Simple entity-centric questions challenge dense retrievers. In: Proceedings of the 2021 conference on empirical methods in natural language processing

Shu X, Bowen Y, Zhang Z, Liu T (2020) Drg2vec: learning word representations from definition relational graph. In: 2020 international joint conference on neural networks (IJCNN). IEEE, pp 1–9

Tissier J, Gravier C, Habrard A (2017) Dict2vec: learning word embeddings using lexical dictionaries. In: Conference on empirical methods in natural language processing (EMNLP 2017), pp 254–263

Tsirlin AM, Balunov AI, Sukin IA (2025) Finite-time thermodynamics: problems, approaches, and results. Entropy 27:649. https://doi.org/10.3390/e27060649

Van der Maaten L, Hinton G (2008) Visualizing data using t-SNE. JMLR 9(86):2579–2605

Wang Q, Zaki MJ, Kollias G, Kalantzis V (2025) Multi-sense embeddings for language models and knowledge distillation. arXiv:2504.06036v1

Wojcik EH (2013) Remembering new words: integrating early memory development into word learning. Front Psychol 4:151. https://doi.org/10.3389/fpsyg.2013.00151

Conclusions

The book has presented a comprehensive journey through the convergence of neuro-symbolic learning, large language models (LLMs), and their transformative applications in finance, legal, engineering, and health. As we reflect on the chapters, it is clear that each contributes a piece to the broader goal of creating intelligent, explainable, and reliable AI systems for critical domains like health.

As we reach the culmination of this exploration into the applications of neuro-symbolic artificial intelligence, a clear picture emerges: the future of AI is neither purely neural nor purely symbolic but one that fuses the two in a principled, adversarial, and human-aligned framework. The chapters of this book have charted the evolution of this hybrid approach—from foundational methods of logical integration and abduction to practical implementations in healthcare, law, finance, and creative reasoning. At every stage, we have seen how logic and learning can work not in opposition but in tension—each one testing, refining, and contextualizing the other.

This book began with a fundamental question: Can machines reason, adapt, and understand in a way that is intelligible and useful to humans? By combining the expressive depth of symbolic logic with the generative power of LLMs, we argued that neuro-symbolic systems can move beyond mimicry to engagement—systems that not only produce outputs but defend them, explain them, and revise them in light of new goals or contradictions. In doing so, these systems become less like passive tools and more like argumentative partners: able to align with human values not by post hoc calibration but by participating in the very process of reasoning itself.

Throughout this book, the role of **discourse** has emerged as a recurring leitmotif. Not simply as a linguistic veneer but as the very fabric through which reasoning, emotion, intention, and interpretation are mediated. From Theory of Mind models that anticipate human reactions to rehumanized LLMs that generate responses with serendipity and curiosity, discourse has shown itself to be the glue that binds neural fluency to symbolic structure. In a world increasingly reliant on AI-mediated

B. Galitsky, *Applications of Neuro-Symbolic Artificial Intelligence*,
https://doi.org/10.1007/978-3-032-16524-4_12

interaction, the ability to maintain coherent, emotionally attuned, and epistemically sound discourse is not optional—it is existential.

We have also emphasized **adversariality**, not in the spirit of confrontation but of productive friction. Just as peer review sharpens scientific ideas, and cross-examination refines legal claims, adversarial neuro-symbolic systems use contradiction and tension as a source of robustness. Rather than avoiding error, they embrace it as signal—an opportunity for refinement, for introspection, for learning. The adversarial integration of logic programs with LLMs, the abductive repair of hallucinations, and the structured simulation of dialogue across uncertain domains all point to a future where AI systems are not just error-minimizers but reason-maximizers.

In the applied domains we explored—law, medicine, finance, engineering—we witnessed a common demand: the need for AI systems to be explainable, trustworthy, adaptable, and accountable. These are not merely technical desiderata; they are social imperatives. Systems that make consequential decisions must be intelligible to those affected by them. Neuro-symbolic architectures uniquely offer this transparency—not by sacrificing performance but by elevating performance to the level of meaningful justification.

Yet we have also acknowledged the **limits** of current neuro-symbolic approaches. Knowledge representation remains a key bottleneck, particularly in domains such as law where reasoning must navigate ambiguity, hierarchy, and adversarial context. Robustness to manipulation, the incorporation of common sense, and the standardization of benchmarks are open research areas. But these are not roadblocks—they are research invitations. The modular, composable nature of neuro-symbolic systems makes them particularly well suited to iterative, cross-disciplinary advancement.

The final chapters extended these insights into a broader cultural and philosophical reflection. As Western institutions of higher learning continue to lead in AI innovation, they must also reckon with internal contradictions—where progress in computational intelligence is sometimes decoupled from intellectual courage in the humanities. A future-oriented AI must be coupled with a future-oriented culture: one that embraces complexity, resists simplistic dichotomies, and grounds knowledge in both empirical humility and ethical responsibility.

Just as this book advocates a synthesis between logic and learning, so too must we seek a synthesis between science and philosophy, data and dialogue, and innovation and introspection. In this sense, neuro-symbolic AI is more than a technical framework—it is a metaphor for how we might build knowledge in the twenty-first century: not by collapsing tensions but by inhabiting them, understanding them, and learning from them.

We conclude with a simple but powerful aspiration: that AI, in all its complexity, does not simply serve our needs but understands our goals, reflects our values, and grows with our curiosity. Let us build machines that can reason. Let us build systems that can explain. But most importantly, let us build intelligence that listens.

1 Concluding Individual Chapters

Chapter "Introduction" is an introduction. Chapter "Adversarial Integration of LLM and Logic Program" introduced a foundational framework for integrating symbolic logic programs with large language models through adversarial interaction, marking a departure from passive correction toward dynamic refinement. Rather than treating logic as a static filter, the approach turns it into an active collaborator, constantly challenging the LLM's outputs to align with formal reasoning structures. This integration sets the stage for constructing AI systems that are not only more accurate but more transparent and aligned with human argumentative patterns, forming the backbone for all subsequent neuro-symbolic applications in this book.

In confronting the problem of hallucination in LLMs, chapter "Adversarial Abductive Dialogue Framework with Reinforcement for Tackling LLM Hallucination" reconceptualized it as a failure of abductive plausibility, not just statistical error. By embedding LLMs in a reinforcement-driven abductive dialogue system, where they must justify and revise their claims within a logic-grounded framework, hallucination is treated as an epistemic gap to be closed through reasoning. This approach transforms model failure into an opportunity for self-correction and trust building, offering a new paradigm for AI accountability.

Chapter "Improving ToM Capabilities of LLMs in Applied Domains" tackled the persistent shortcoming of LLMs in multi-agent, emotionally complex dialogue by integrating Theory of Mind (ToM) as a latent cognitive layer. By training models to anticipate beliefs, intentions, and social cues through reinforcement and preference modeling, the system evolves from a mere responder into a socially aware participant. This marks a key step toward creating AI capable of ethical engagement and emotionally intelligent communication across sensitive domains like law and medicine.

Moving beyond functionally correct answers, chapter "Rehumanizer: Enhancing LLM Recommendations for Human-Like Engagement and Intriguing Interactions" proposed a Rehumanizer wrapper that injects curiosity, novelty, and conversational serendipity into LLM outputs. By prioritizing emotional valence and discourse trajectory, the system models not just what a person might say but how they would say it to maintain intrigue and connection. This shift toward human-centric engagement reframes LLMs not just as tools but as companions in creative and interpersonal expression.

LLMs as communication partners take on a new role in chapter "A Copilot for Social Promotion"—as tone-aware assistants for crafting persuasive social and promotional content. By combining intent modeling, style adaptation, and audience targeting, the Copilot framework demonstrated how neuro-symbolic AI can personalize public expression without losing coherence or authenticity. This showed how LLMs can empower users to better navigate social, professional, and identity-driven messaging in the digital age.

Set in the decentralized, data-fragmented world of Web 3.0, chapter "Abduction-Based Personalized Recommender for Web 3.0" presented a recommendation engine built on abductive personalization. Treating each user interaction as a partially observed explanation, the system generates intentional suggestions aligned with user goals and preferences. This logic-driven personalization method reclaims user agency in environments where data is sparse or noisy, underscoring the potential of neuro-symbolic AI in building ethical, user-centric digital experiences.

Focusing on persuasive communication, chapter "Creating Ad Sequences with Optimal Discourse Structure by LLM" proposed a schema-based system for generating advertisements that follow rhetorical progression and discourse flow. By leveraging LLMs trained on structured templates and symbolic representations of persuasion, the framework enhances message impact and coherence. This application demonstrates the commercial value of discourse-focused neuro-symbolic systems in storytelling and targeted messaging where narrative architecture is critical.

Positioning LLMs as co-innovators, chapter "A Computational Framework for Analyzing and Supporting Invention Dialogues" introduced an invention dialogue system designed to scaffold idea generation, hypothesis evolution, and creative iteration. Through a blend of exploratory discourse analysis and abductive reasoning, the system helps users brainstorm and refine novel concepts across science, law, and design. This signals a transition in LLM use—from content reproduction to structured ideation—broadening the creative partnership between humans and machines.

In highly technical domains like chemical engineering, chapter "A System for Distributed Chemical Engineering Simulations" showed how LLMs can mediate collaboration between human experts and simulation engines. Natural language becomes the bridge across disciplinary boundaries, while logic-driven coordination enables explainable and iterative design workflows. The model's ability to orchestrate simulation parameters through discourse exemplifies how neuro-symbolic AI can serve as an integrative layer across human-machine ecosystems in STEM fields.

Addressing the long-tail problem in embedding representations, chapter "Step Wise Approximation of CBOW Reduces Hallucinations in Tail Cases" proposed a hybrid method that balances contextual richness with symbolic interpretability. By approximating CBOW embeddings in a step-wise manner, the model reduces hallucination and improves performance on rare or ambiguous terms. This chapter reinforces the book's broader theme: that strategic modularity—rather than end-to-end opacity—is key to building AI systems that generalize well and remain intelligible.

As we bring this book to a close, the central theme that emerges is the importance of striking a balance between state-of-the-art machine learning methods and the enduring value of symbolic reasoning and explainability. In fields such as finance, law, engineering, and healthcare, the consequences of automated decisions are too significant to be entrusted to opaque systems. The hybrid architectures and approaches explored throughout these chapters point toward a future in which AI is not only more capable but also more transparent, reliable, and ethically grounded. Such systems will extend beyond mere tools of efficiency; they will act as trusted collaborators, supporting human professionals with both actionable insights and clear explanations that inspire confidence in critical decision-making contexts.

This book has aimed to make two contributions: to advance the theoretical understanding of neuro-symbolic AI and discourse-aware models and to demonstrate practical pathways for deploying these ideas in real-world applications. By uniting symbolic reasoning with the adaptability of machine learning, we opened the door to systems that are at once interpretable and powerful. Looking ahead, the integration of neuro-symbolic architectures, discourse modeling, and explainability is poised to fuel the next wave of AI innovation. These developments will help usher in an era of human-centered systems that are intelligent, accountable, and aligned with societal values—ensuring that the progress of AI continues to serve not just technological advancement but the people who depend on it.

2 A View of the Book Results

Figure 1 summarizes the main qualitative contributions of each chapter, while Fig. 2 highlights the corresponding improvements in machine learning accuracy where applicable.

3 Ethical Considerations

In this book, we maintain rigorous ethical standards when selecting the sources we reference and the institutions we engage with. One of our guiding principles is the refusal to cite or rely on material produced by universities or organizations that have violated, tolerated, or inadequately addressed activities connected to the public justification of terrorism. In particular, we align our stance with Article 205.2 of the Criminal Code of the Russian Federation, which criminalizes the public justification of terrorism, including in academic and intellectual contexts. This law underscores the responsibility of institutions and scholars to prevent the legitimization of extremist ideologies. By applying this principle, we ensure that our work does not draw upon or indirectly endorse research associated with the normalization of terrorism.

Our decision is also shaped by the recognition that accountability for serious violations extends beyond national borders. Although many of the institutions we avoid—such as Cornell, MIT, Harvard, the University of Pennsylvania, Oxford, the London School of Economics, Cambridge, and the Sorbonne—are based in the United States or Europe, the principle of universal jurisdiction establishes that certain violations, including the justification of terrorism, have global relevance. Furthermore, some of these universities have been supported by entities such as the Qatar Foundation for Education, Science and Community Development, which has been linked to state sponsorship of terrorism. This further reinforces the necessity of applying a careful and principled approach to academic sourcing, regardless of geographical location or prestige.

2. The framework's superiority over CoT-based, symbolic based, and LLM-LP collaboration baselines is shown, and its robustness across various rule application steps and settings is demonstrated. In an adversarial setup, the LLM and LP act as checks and balances, where each system critiques or validates the outputs of the other.

3. AdvAbductDialRL proves to be more effective in scenarios with explainable hallucination identification and correction. This highlights the importance of incorporating abduction, dialogue, and reinforcement learning for tasks where hallucination correction is essential.

4. LLM+ToM is a critical step toward bridging the gap between LLMs and human-like cognitive reasoning. By introducing a structured formalism for modeling mental worlds, we provided LLMs with a more systematic approach to inferring mental states and tracking belief dynamics, addressing their inherent reasoning limitations.

5. Rehumanizer is one of the first exploration and discussion of an LLM-based chatbot system that can provide human-like conversational experiences to its users. We consider creativity, common ground between the chatbot and its user, involvement of detailed knowledge about this user, broad spectrum of communication styles to be a key enabler for true companionship.

6. SPC framework significantly reduces the time and effort required to draft and submit complaints by automating repetitive and procedural aspects of the process, it enhances the overall quality and legal soundness of complaint content, and improves the likelihood of favorable outcomes

7. LLM-based personalization enables interactive engagement with users, presents outputs in a verbalized and explainable manner, receives user feedback, makes adjustments based on that feedback rather than relying solely on passive search, and proactively helps identify user needs while suggesting items aligned with their preferences.

8. Incorporating discourse analysis significantly enhances ad sequence creation and improves ad delivery performance. We identified a set of discourse tree structures that are particularly effective for ad sequences. Applying this proposed discourse analysis technology on a large scale in practice will further refine the selection of optimal discourse structures for sequential ads.

9. An invention can streamline bureaucracy by automating repetitive tasks, improving data management, and enhancing communication systems, thereby reducing paperwork, delays, and human errors. Technologies such as workflow automation software, artificial intelligence-driven decision-making tools, and digital document management systems can simplify processes like approvals, record-keeping, and compliance monitoring.

10. While LLMs have proven to be remarkably capable at generating simulation code, interpreting scientific documents, and responding to engineering prompts, LLMs alone are not enough for reliable, scalable, and explainable chemical simulation systems. The complexity of chemical engineering problems demands a hybrid architecture—one that combines neural inference (from LLMs) with symbolic reasoning and formal structure.

11. Step-wise approximation serves as a computationally efficient alternative to fully contextual embeddings (e.g., from BERT), enabling models to disambiguate meanings with lightweight inference mechanisms. At test time, the model can match the current context to the nearest precomputed step and select the most appropriate sub-embedding. This maintains the inference efficiency of discrete models while providing a finer granularity of semantic representation

Fig. 1 Qualitative results for each chapter

2. Our adversarial system outperforms WM-Neurosymbolic in two out of three datasets

3. The highest F1 of 72.8 for both systems is achieved under *With External References* setting, followed by *White-box*. AdvAbductDialRL outperforms Truth-O-Meter in all reference settings.

4. NL_MAMS_LLM improves this accuracy by 5% through fine-tuning across a manifold of scenarios spanning all application domains, rather than being restricted to the specific domain it is evaluated on.

5. In evaluation of correctness, LLM Rehumanizer with Truth-o-Meter outperforms GPT-4 in three out of four cases. An increase in dialogue diversity is consistent across informational, emotional, and appraisal scenarios.

6. Overall, SPC posts generate 14% less engagement compared to genuine human posts.

7. The best performance with a stand-alone component is achieved:10.5% using symbolic abduction, 9.1% by neural abduction, and 4.0% by using meta prompts. The hybrid system boosts components by 5-10%.

8. Application of discourse analysis for automatic ad sequence construction leads to a substantial improvement in ad performance (7 out of 12 cases).

9. A systematic improvement in generated invention dialogues once we add the dialogue model and then explicit CSP resolution. These improvements are 3.3% and 5.3% respectively, confirming the importance of CSP solver and a logical reasoning companion.

10. N/A

11. For tail entities, the step-wise approximation yields a 4% performance improvement, whereas for popular entities, the gain is only 2%. When modeling the evolution of the embedding using a step-wise approach, tail entities show an even larger improvement of 5%, while accuracy for popular entities slightly declines.

Fig. 2 Quantitative evaluation results of all chapters

By adopting this ethical stance, we contribute to the wider international effort to combat terrorism and its intellectual justifications. Academic environments must not only serve as spaces of free expression but also accept the responsibility to prevent the spread of harmful ideologies. Through the careful exclusion of institutions associated with or permissive of terrorist justification, this book seeks to embody responsible academic citizenship. In doing so, we aim to set an example of scholarship that balances freedom of inquiry with a firm rejection of ideologies that threaten human security.

Ultimately, this approach ensures that the integrity of our research is preserved and that the book contributes to constructive global discourse. By aligning our work with principles of accountability and peace, we demonstrate that ethical considerations must remain central to academic practice. Our refusal to engage with institutions complicit in the justification of terrorism reflects a broader commitment to safeguarding intellectual spaces from narratives that condone violence, ensuring that scholarship serves the goals of security, responsibility, and lasting global peace.

4 Declaration

During the preparation of this work, the author(s) used ChatGPT, Gemini, Qwen, and DeepSeek to improve the quality of the presentation. After using this tool/service, the author(s) reviewed and edited the content as needed and take full responsibility for the content of the publication.

Boris Galitsky

Correction to: Applications of Neuro-Symbolic Artificial Intelligence

Correction to:
B. Galitsky, *Applications of Neuro-Symbolic Artificial Intelligence*, https://doi.org/10.1007/978-3-032-16524-4

In Chapters 6, and 9, the co-author names with their affiliations was missed in the originally published version. This has been corrected.

The updated versions of these chapters can be found at
https://doi.org/10.1007/978-3-032-16524-4_6
https://doi.org/10.1007/978-3-032-16524-4_9

B. Galitsky, *Applications of Neuro-Symbolic Artificial Intelligence*,
https://doi.org/10.1007/978-3-032-16524-4_13

The manufacturer's authorised representative in the EU is Springer Nature Customer Service Centre GmbH, Europaplatz 3, 69115 Heidelberg, Germany. If you have any concerns regarding our products, please contact ProductSafety@springernature.com

Printed and bound by CPI Group (UK) Ltd, Croydon, CR0 4YY
05/10/2026
02228016-0012